FOURTH EDITION

Lecture Notebook for MOLECULAR CELL BIOLOGY

Harvey Lodish

Arnold Berk

S. Lawrence Zipursky

Paul Matsudaira

David Baltimore

James Darnell

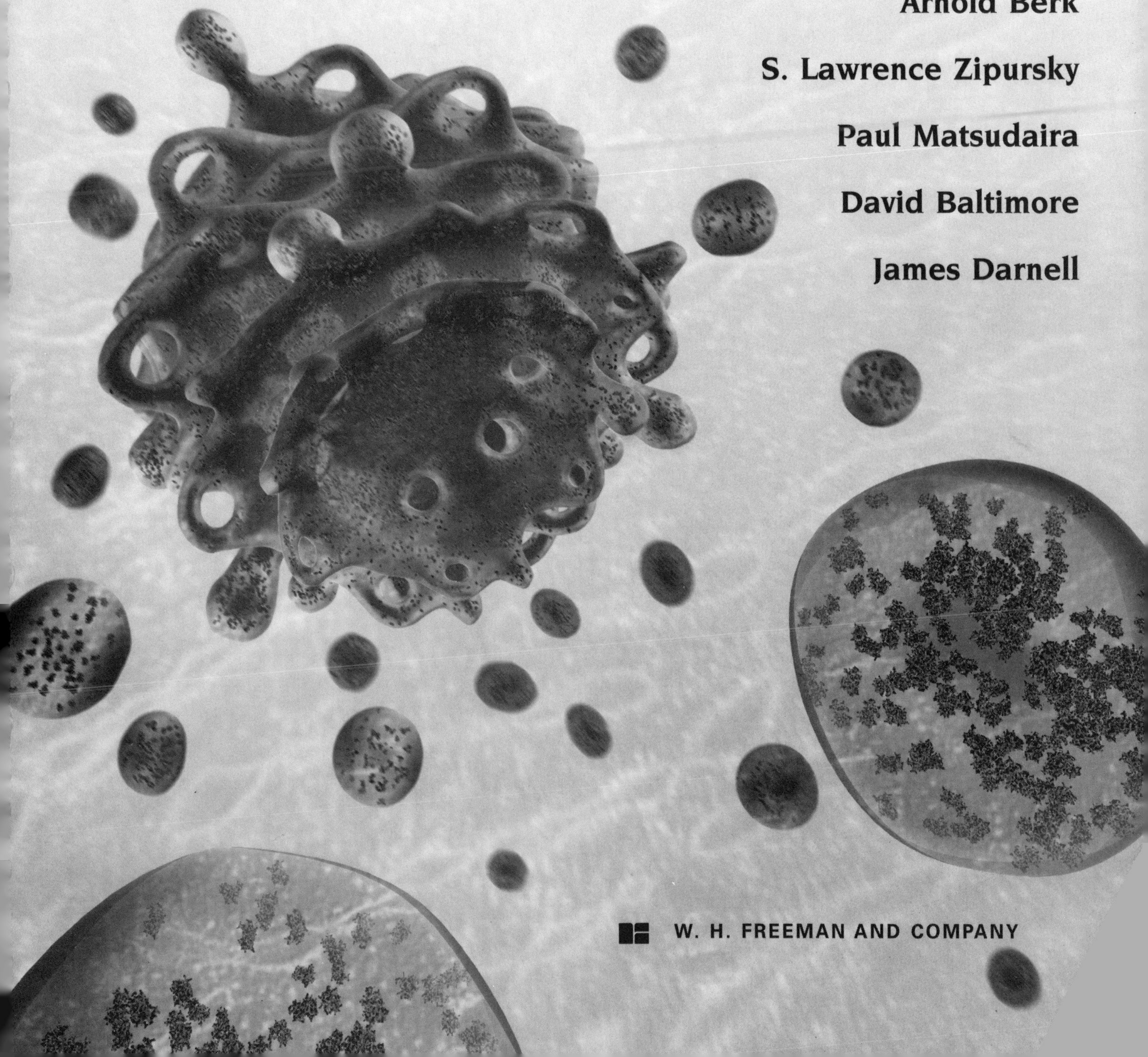

W. H. FREEMAN AND COMPANY

ISBN 0-7167-4078-8

Printed in the United States of America

First printing, 2000

Contents

Acknowledgments

W. H. Freeman would like to thank the following professors for their help in selecting the figures for this notebook.

Mary A. Farwell, East Carolina University
Edward T. Kipreos, University of Georgia
Bruce D. Parker, Utah Valley State College
Trina A. Schroer, Johns Hopkins University
Harriet E. Smith-Somerville, University of Alabama

The Dynamic Cell

1

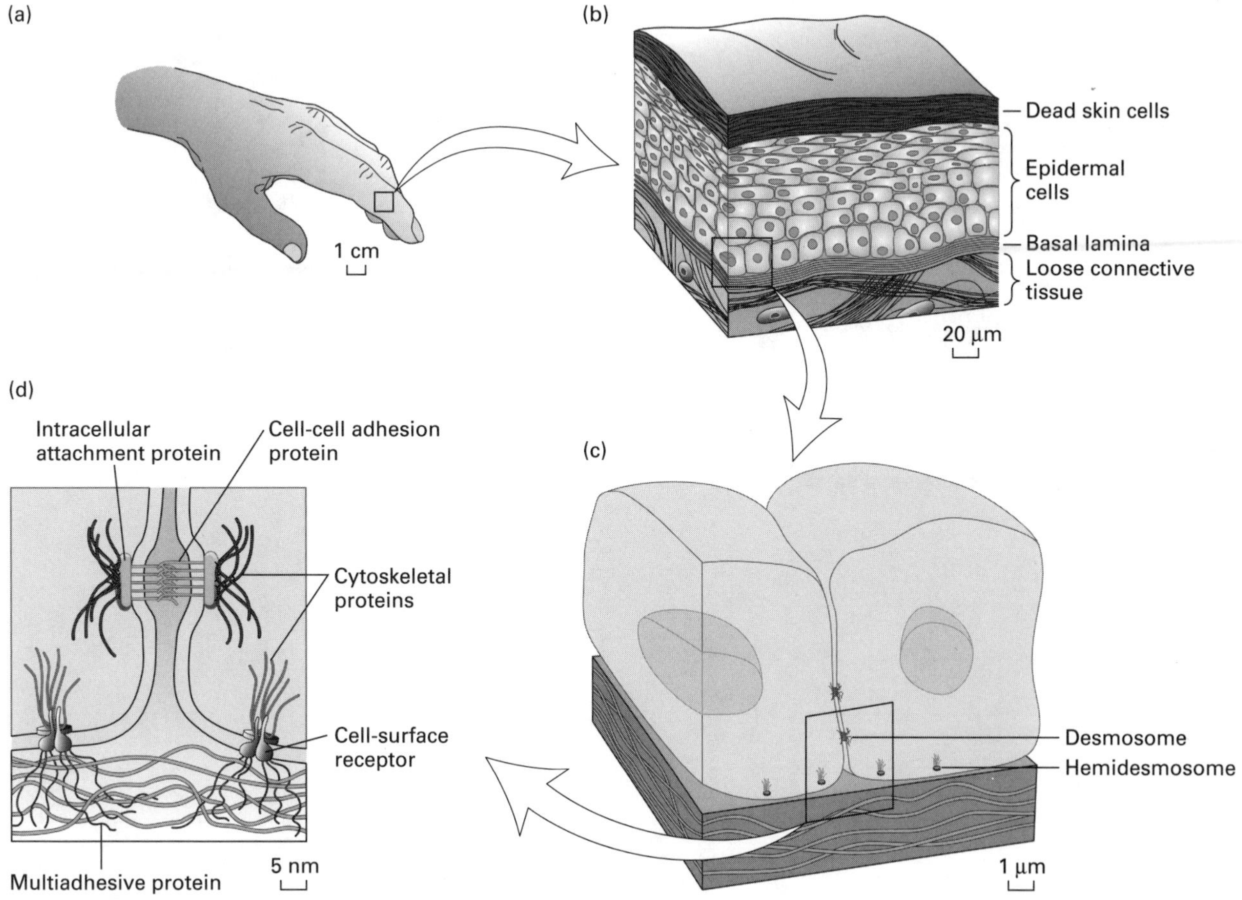

FIGURE 1-1, page 2
Living systems such as the human body consist of closely interrelated elements.

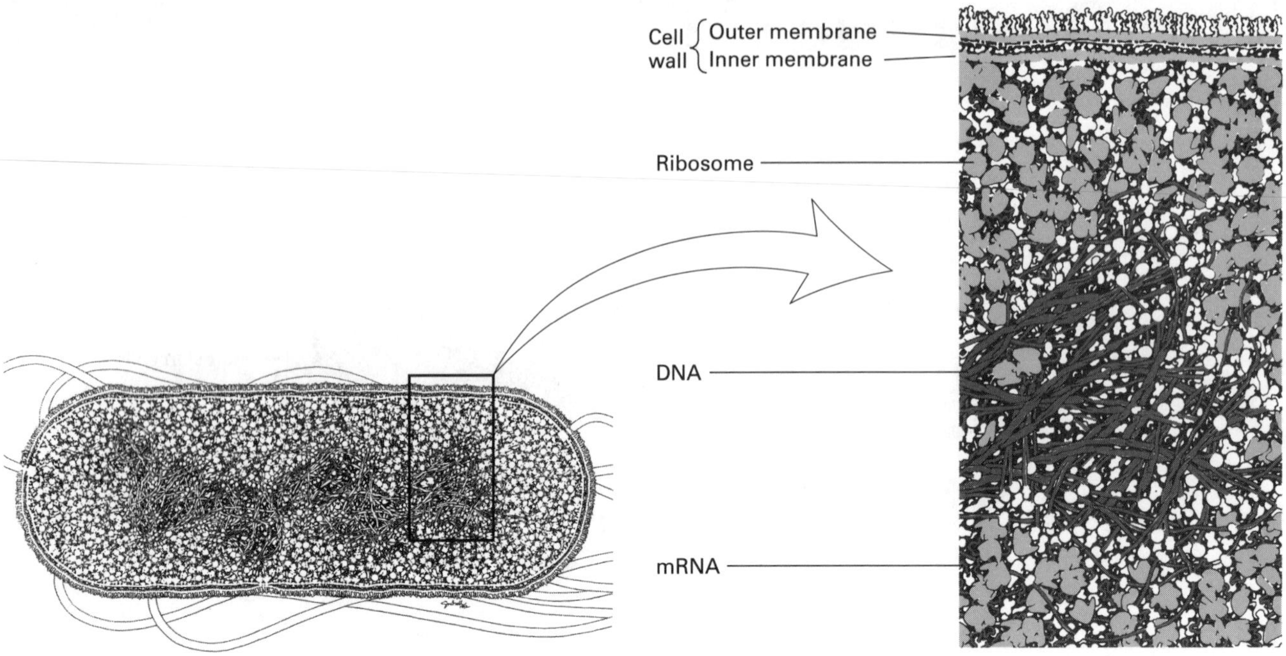

FIGURE 1-2, page 3
Cells are filled with molecules large and small.

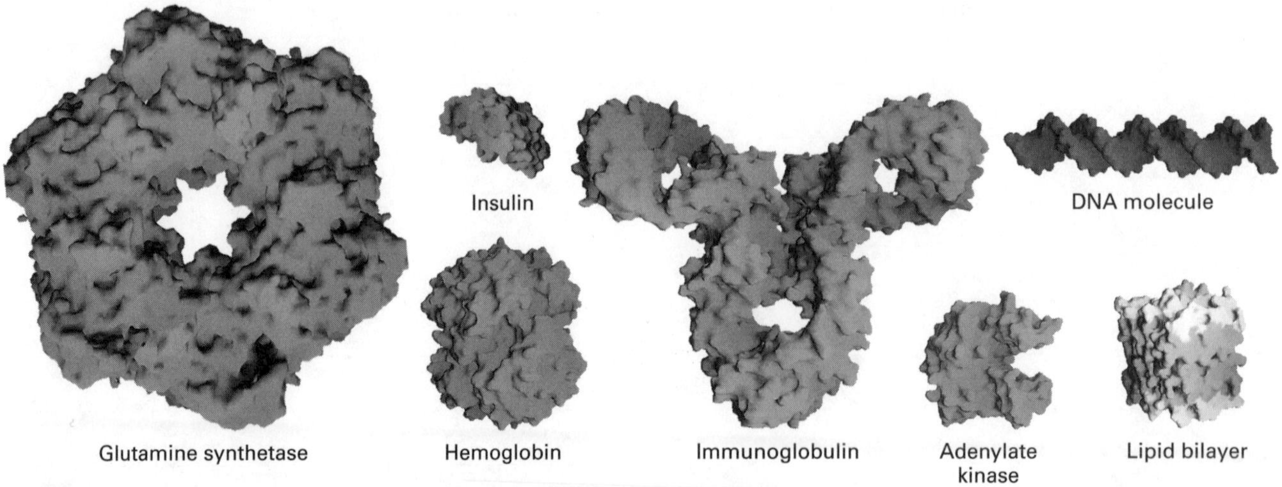

FIGURE 1-3, page 4
Models of some representative proteins drawn to a common scale and compared with a small portion of a lipid bilayer sheet and a DNA molecule.

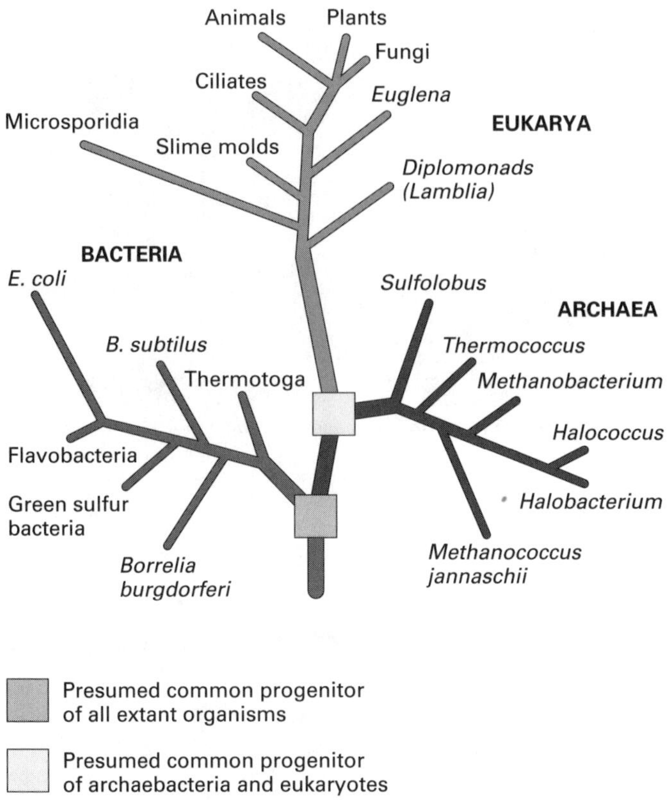

FIGURE 1-5, page 6
The three kingdoms of organisms are related through common sequences of their ribosomal RNAs.

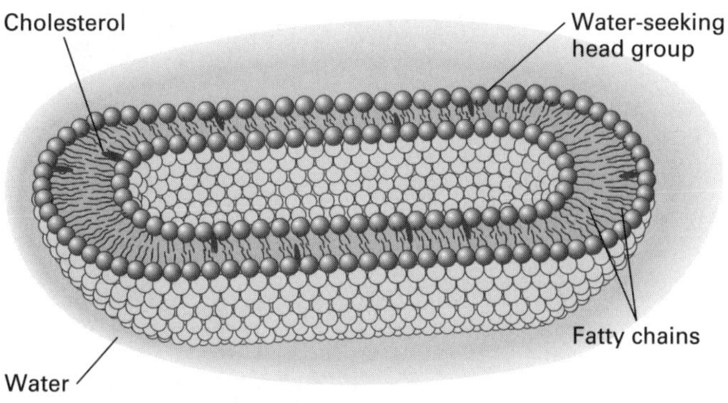

FIGURE 1-6, page 6
The watery interior of cells is surrounded by the plasma membrane, a two-layered shell of phospholipids.

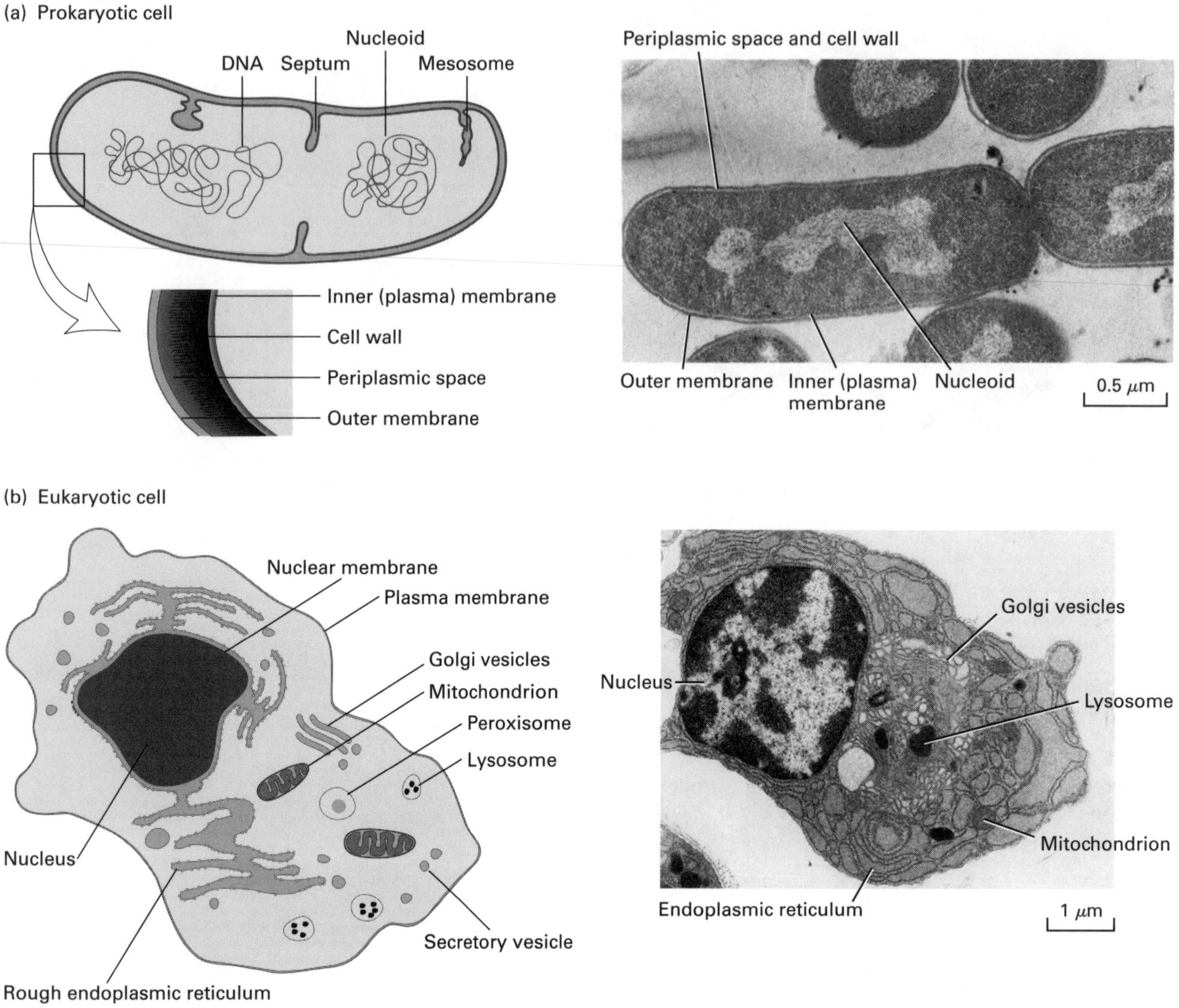

FIGURE 1-7, page 8
Comparison of the structure of prokaryotic and eukaryotic cells.

MEDIA CONNECTIONS Life Cycle of a Cell

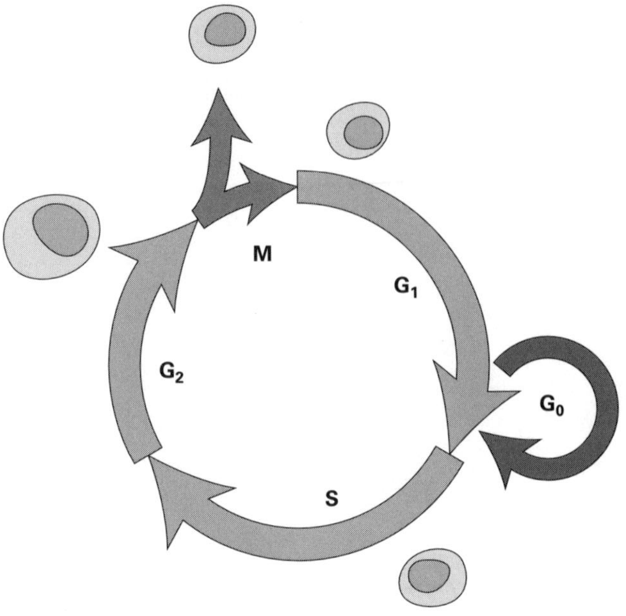

FIGURE 1-9, page 10
The eukaryotic cell cycle.

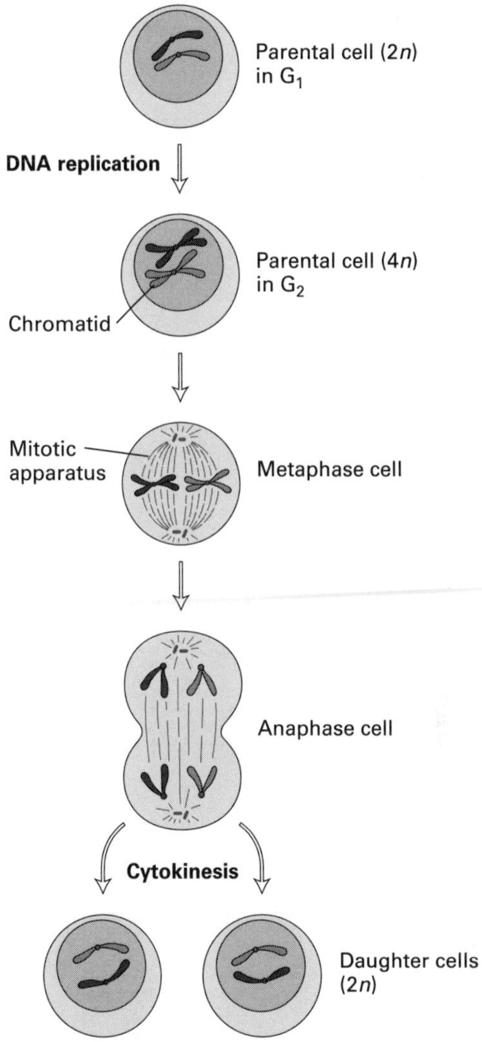

FIGURE 1-10, page 11
Cell division.

MEDIA CONNECTIONS Life Cycle of a Cell

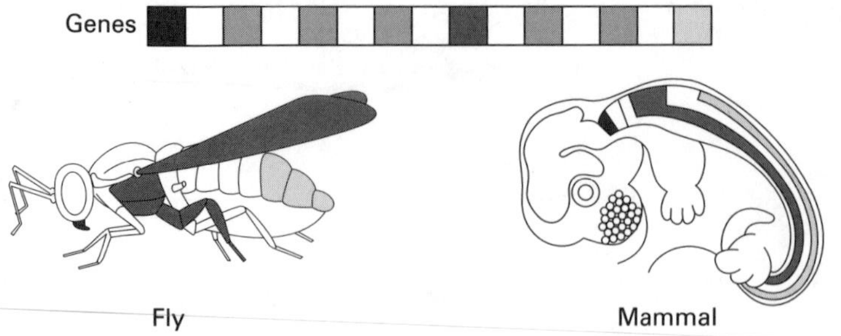

FIGURE 1-12, page 12
Common patterns of development are seen in animals as diverse as sea urchins, flies, mice, and humans.

Additional Notes

Chemical Foundations

(a) Water, ions, and small molecules (77%)

Water (70%): H—O
        H

Inorganic ions (1%): Na^+ Cl^- K^+ $H_2PO_4^-$

Small molecules (6%):

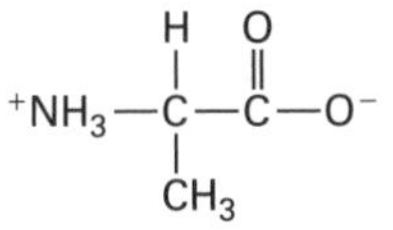

Amino acid
(alanine)

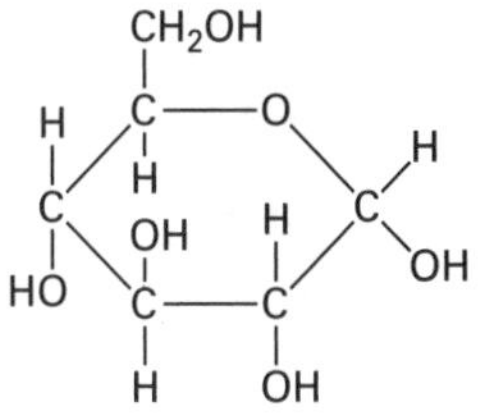

Sugar
(glucose)

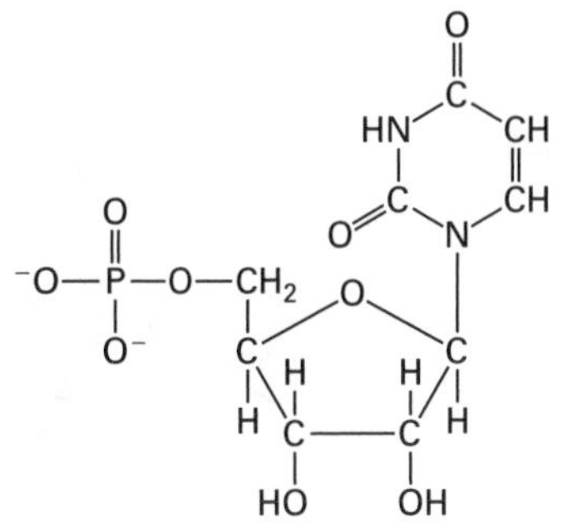

Nucleotide
(uradine monophosphate)

$CH_3—CH_2—CH_2—CH_2—CH_2—CH_2—CH_2—CH_2—CH_2—CH_2—CH_2—CH_2—CH_2—C—O^-$

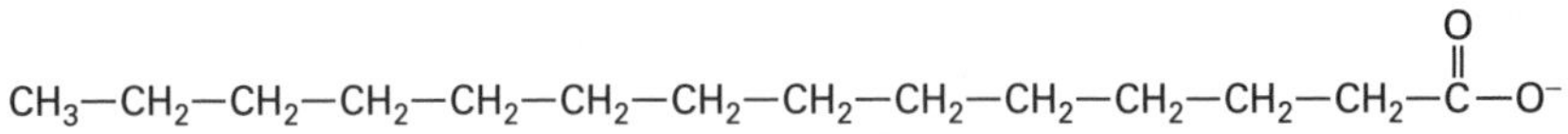

Fatty acid (myristic acid)

(b) Macromolecules (23%)

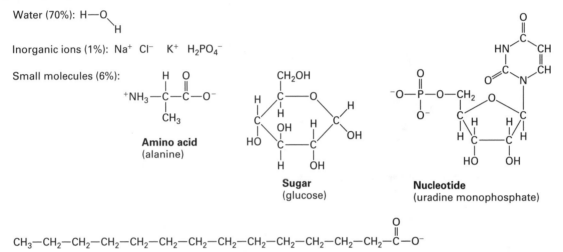

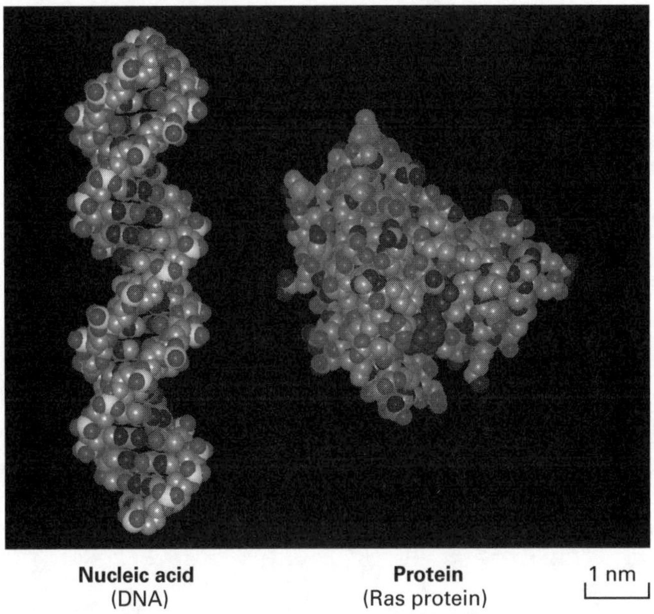

Nucleic acid
(DNA)

Protein
(Ras protein)

1 nm

FIGURE 2-1, page 15
The chemicals of life.

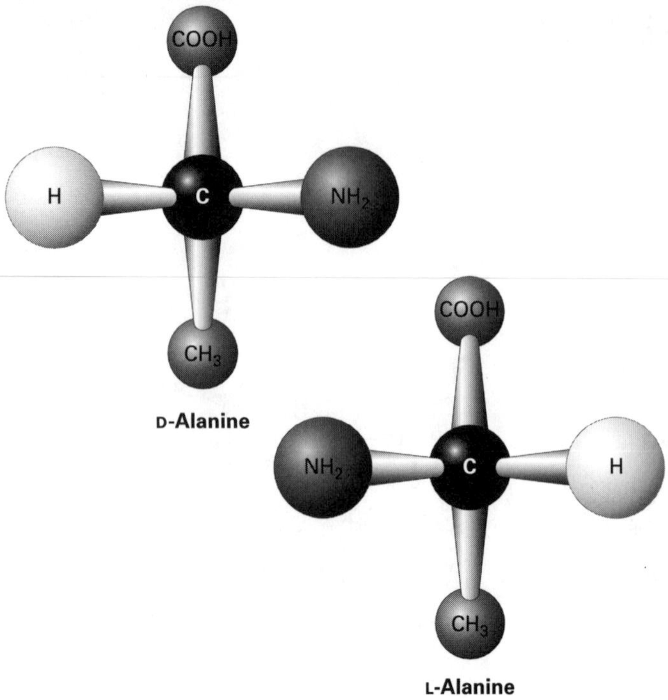

FIGURE 2-6, page 19
Stereoisomers of the amino acid alanine.

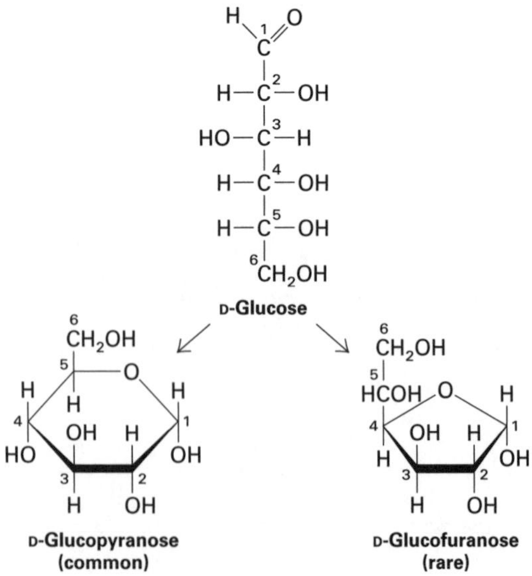

FIGURE 2-7, page 20
Three alternative configurations of D-glucose.

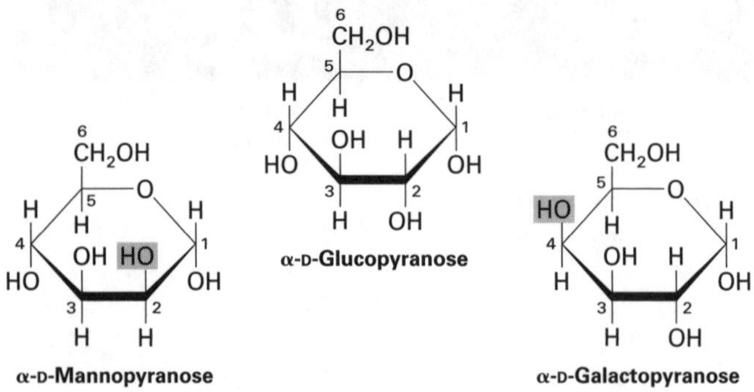

FIGURE 2-8, page 20
Haworth projections of the structures of glucose, mannose, and galactose in their pyranose forms.

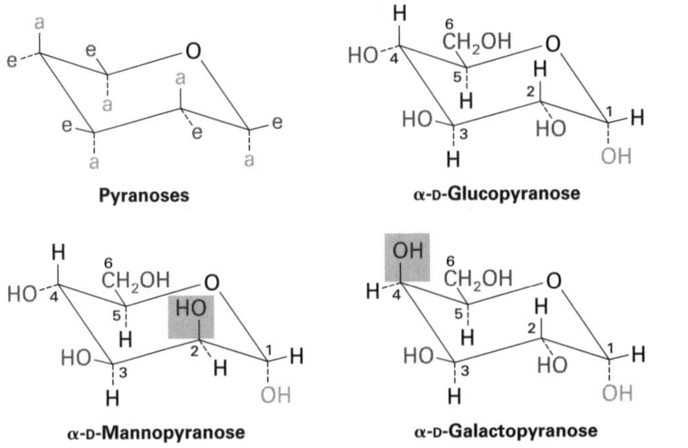

FIGURE 2-9, page 20
Chair conformations of glucose, mannose, and galactose in their pyranose forms.

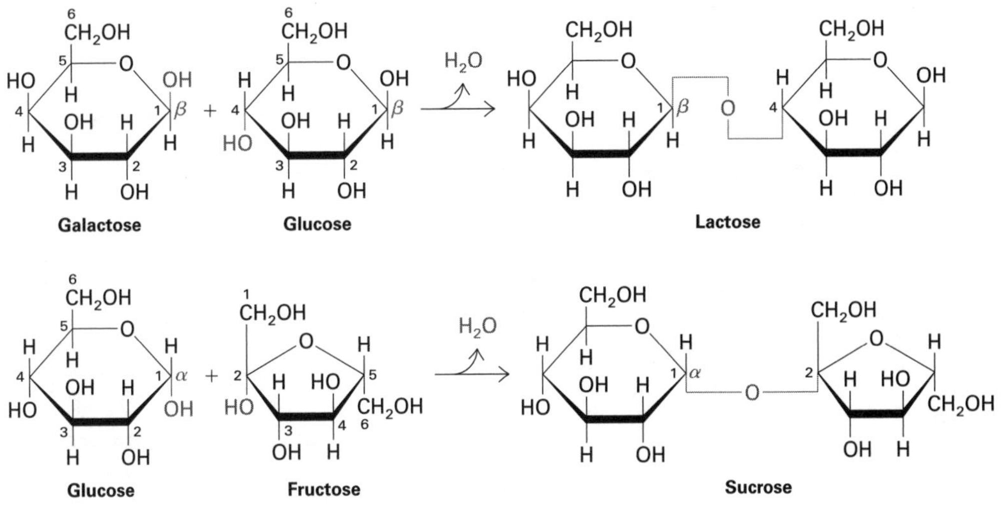

FIGURE 2-10, page 21
The formation of glycosidic linkages generate the disaccharides lactose and sucrose.

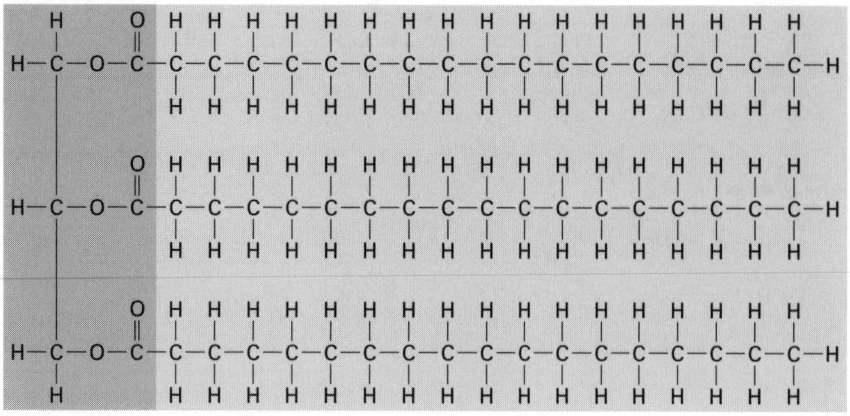

FIGURE 2-16, page 25
The chemical structure of tristearin, or tristearoyl glycerol, a component of natural fats.

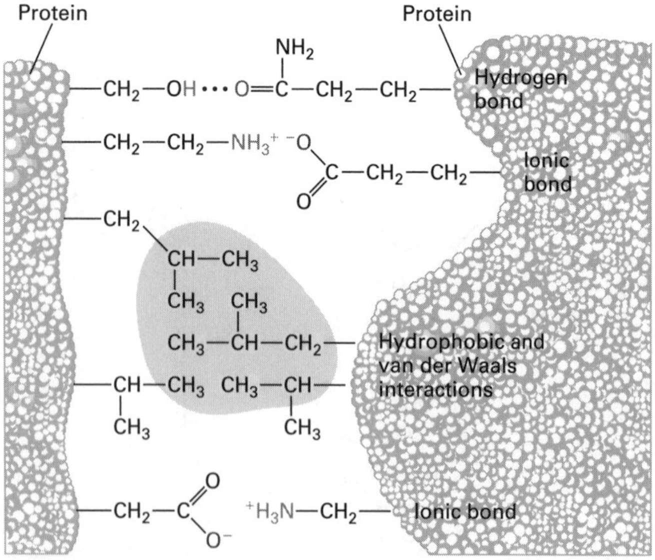

FIGURE 2-17, page 26
The binding of a hypothetical pair of proteins by two ionic bonds, one hydrogen bond, and one large combination of hydrophobic and van der Waals interactions.

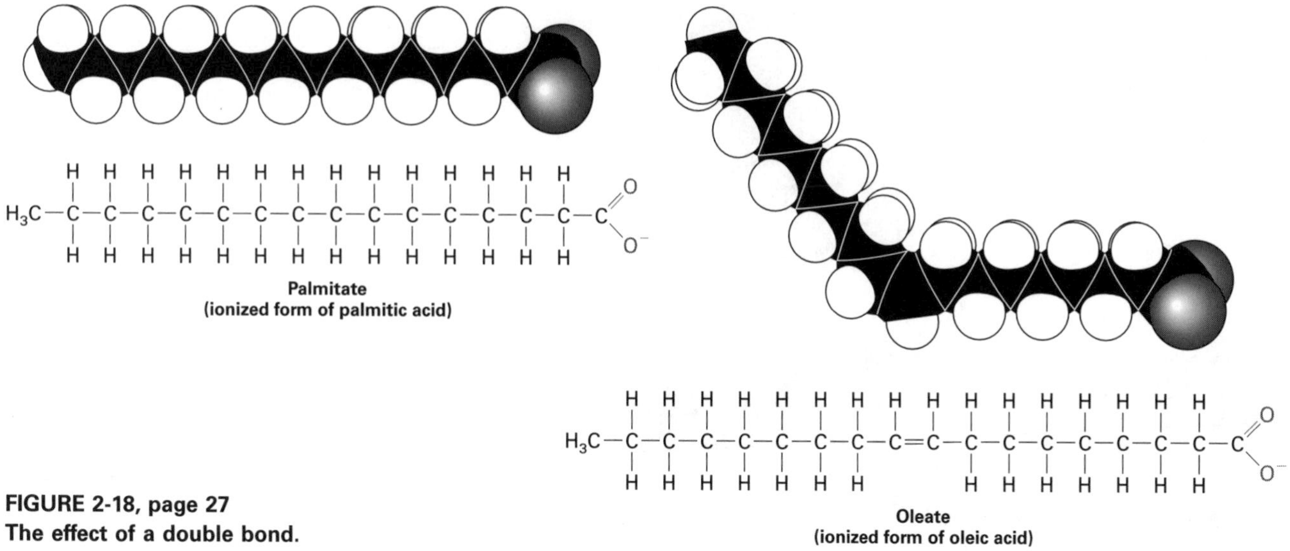

FIGURE 2-18, page 27
The effect of a double bond.

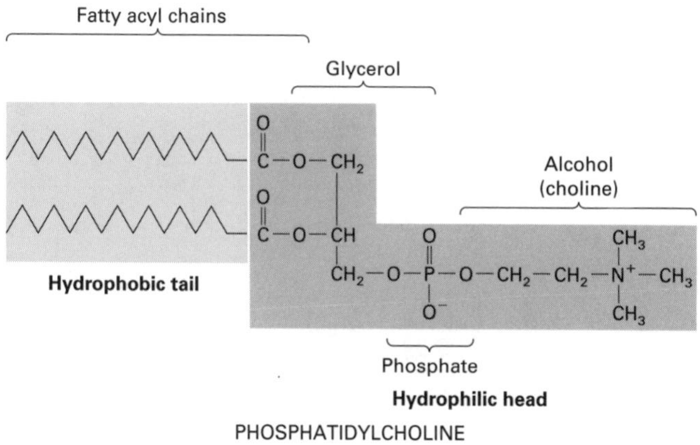

FIGURE 2-19, page 28
Phosphatidylcholine, a typical phosphoglyceride, has a hydrophobic tail and a hydrophilic head in which choline is linked to glycerol by phosphate.

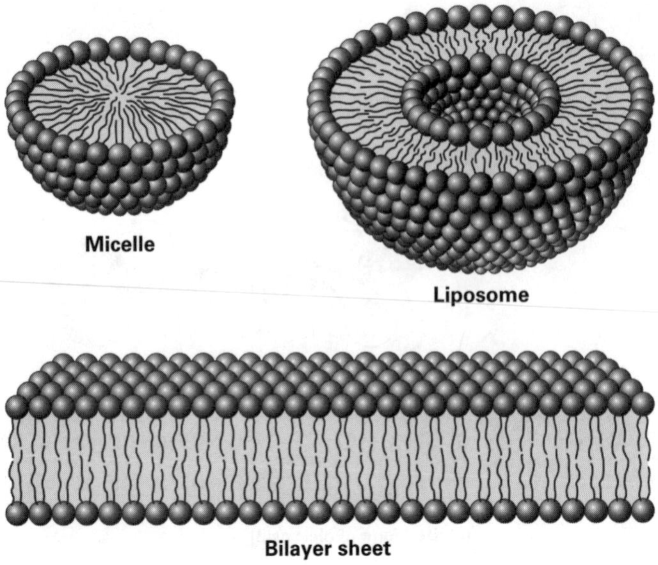

FIGURE 2-20, page 28
Cross-sectional views of the three structures that can be formed by mechanically dispersing a suspension of phospholipids in aqueous solutions.

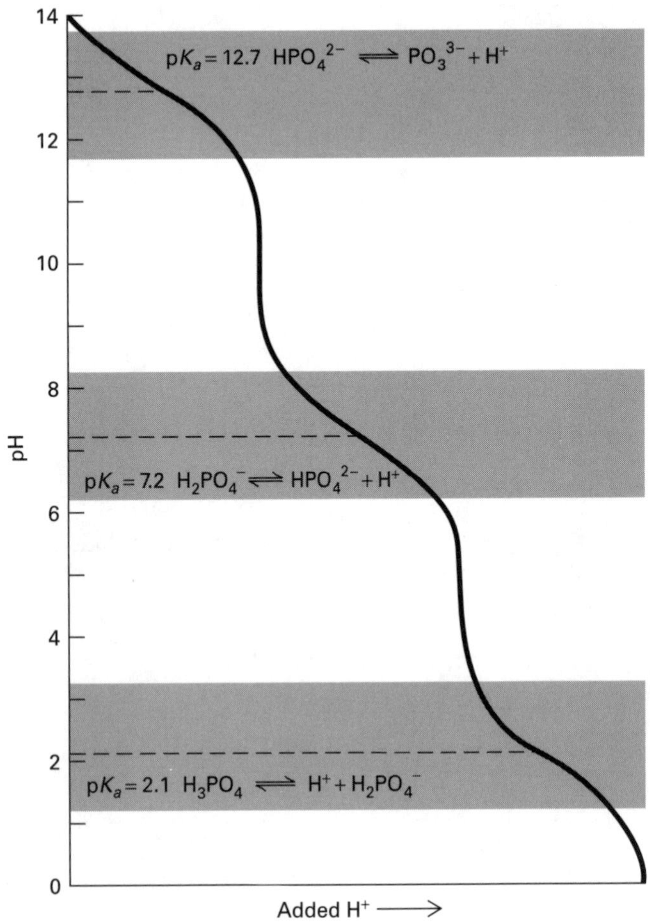

FIGURE 2-22, page 34
The titration curve of phosphoric acid (H_3PO_4).

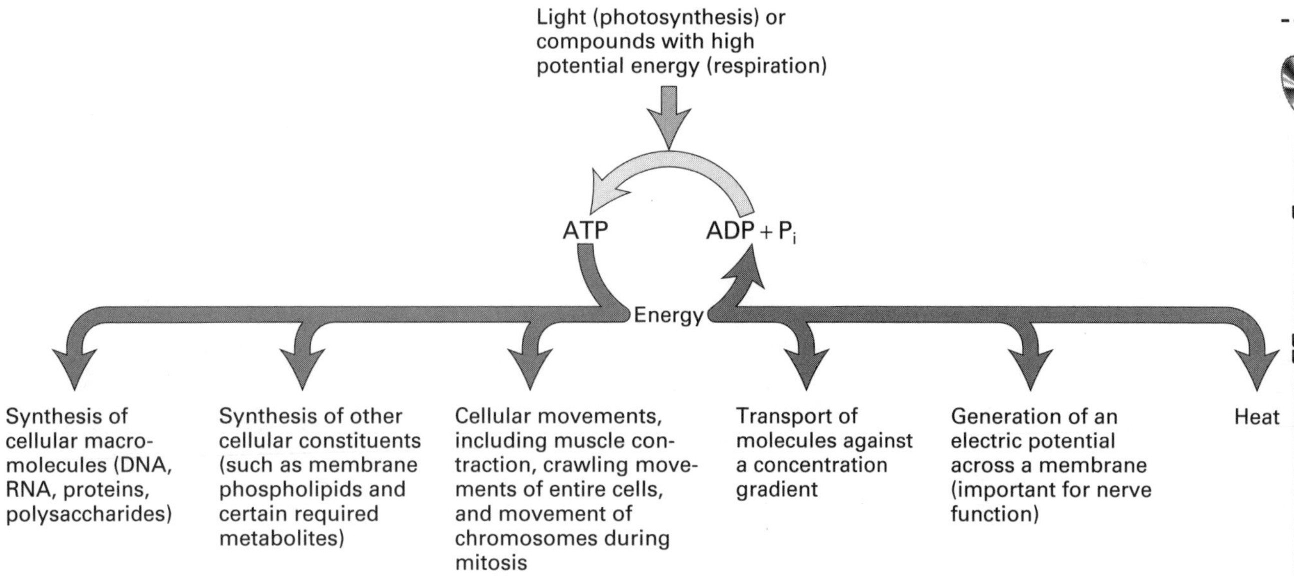

FIGURE 2-25, page 44
The ATP cycle.

MEDIA CONNECTIONS
Biological Energy Interconversions

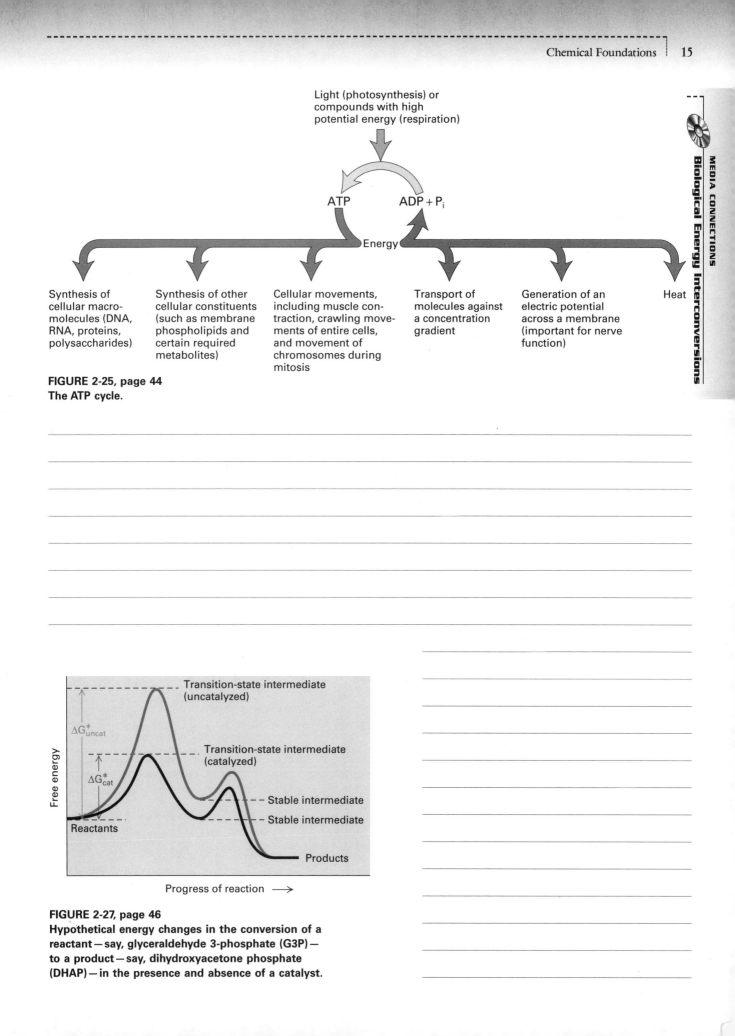

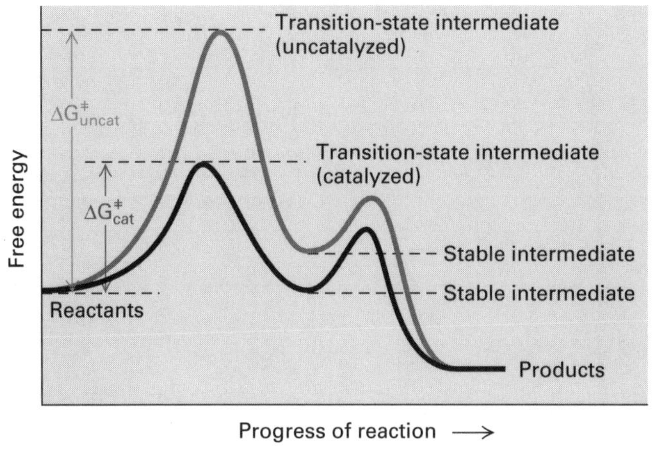

FIGURE 2-27, page 46
Hypothetical energy changes in the conversion of a reactant—say, glyceraldehyde 3-phosphate (G3P)—to a product—say, dihydroxyacetone phosphate (DHAP)—in the presence and absence of a catalyst.

Additional Notes

Protein Structure and Function

3

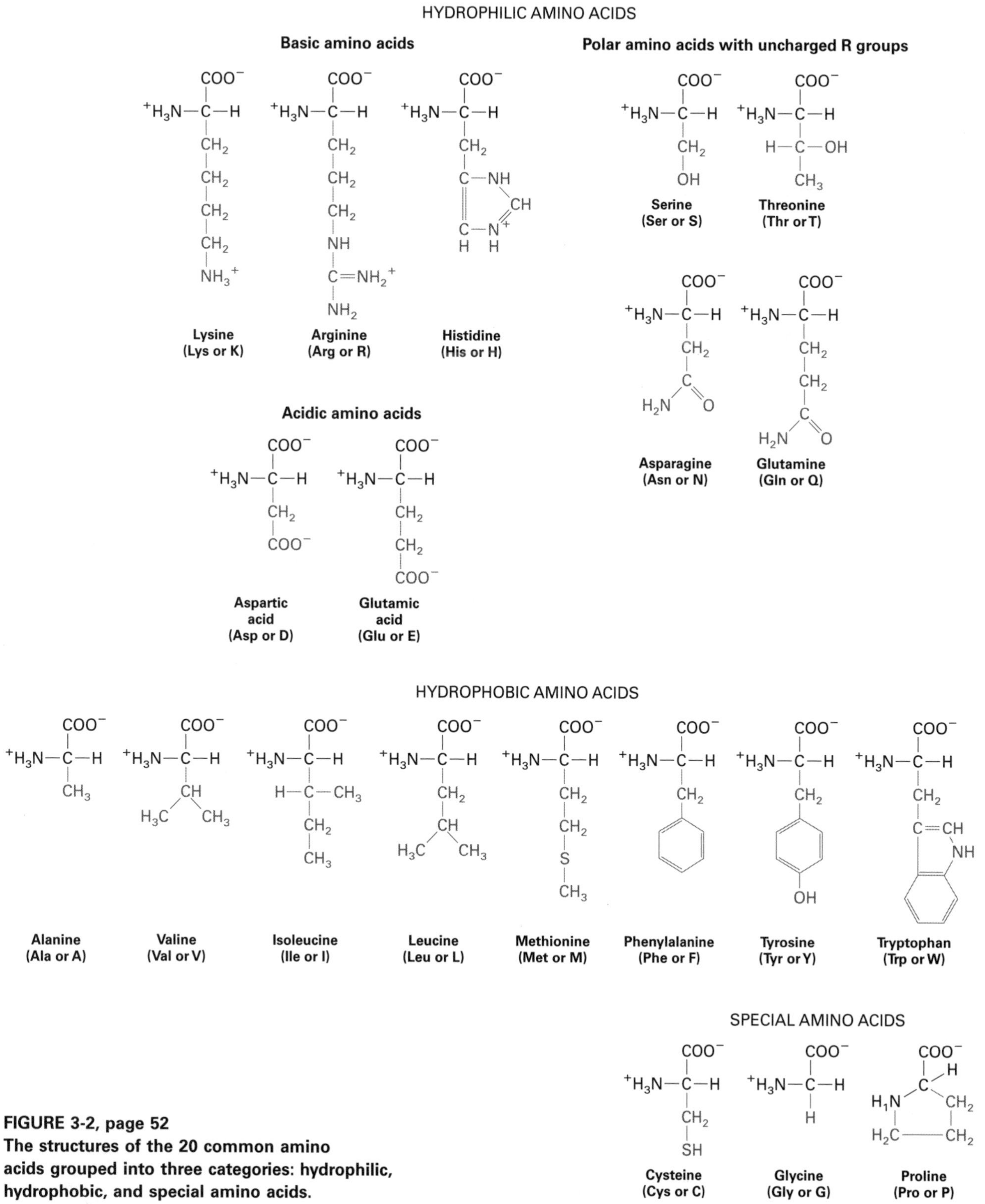

FIGURE 3-2, page 52
The structures of the 20 common amino acids grouped into three categories: hydrophilic, hydrophobic, and special amino acids.

(a)
H₂O
Peptide bond

(b)
Amino end
(N-terminus)
Carboxyl end
(C-terminus)

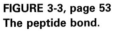

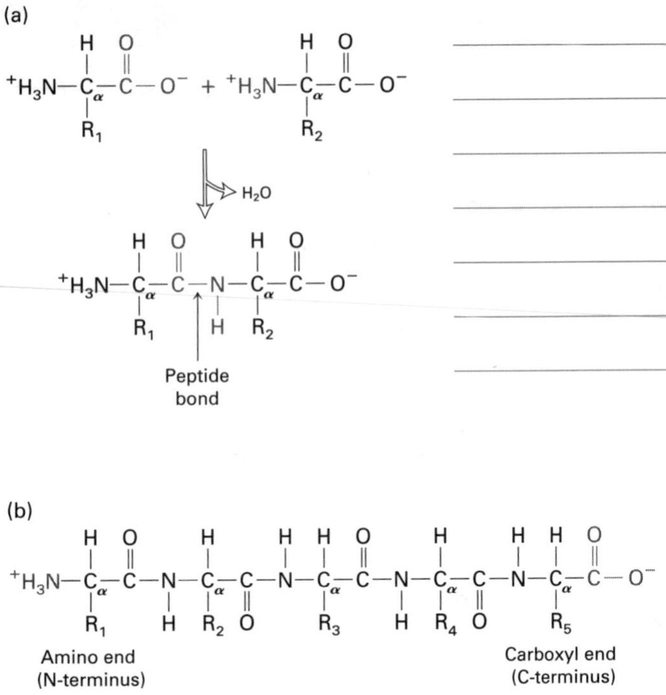

FIGURE 3-3, page 53
The peptide bond.

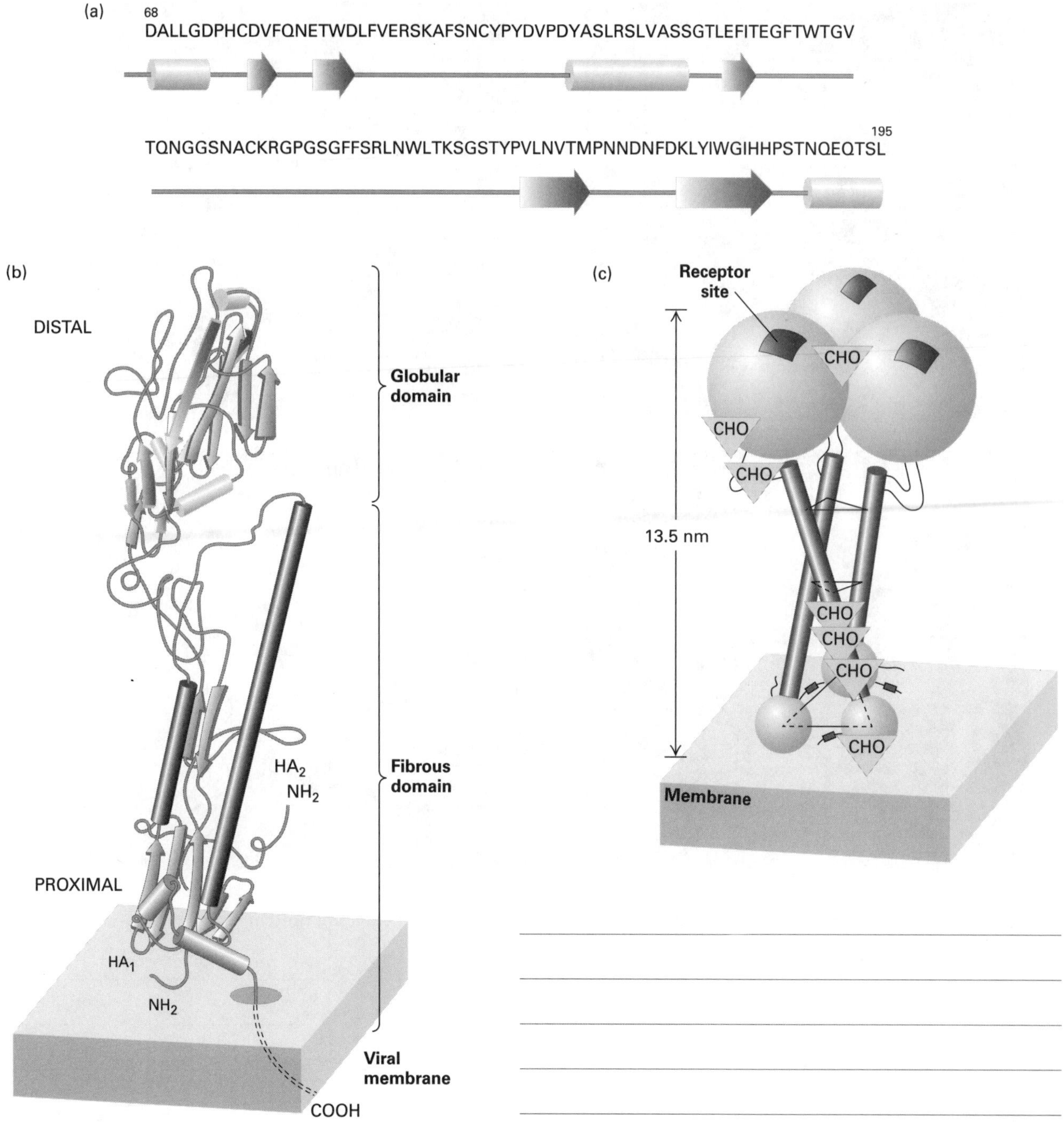

FIGURE 3-4, page 55
Four levels of structure in hemagglutinin, which is a long multimeric molecule whose three identical subunits are each composed of two chains, HA_1 and HA_2.

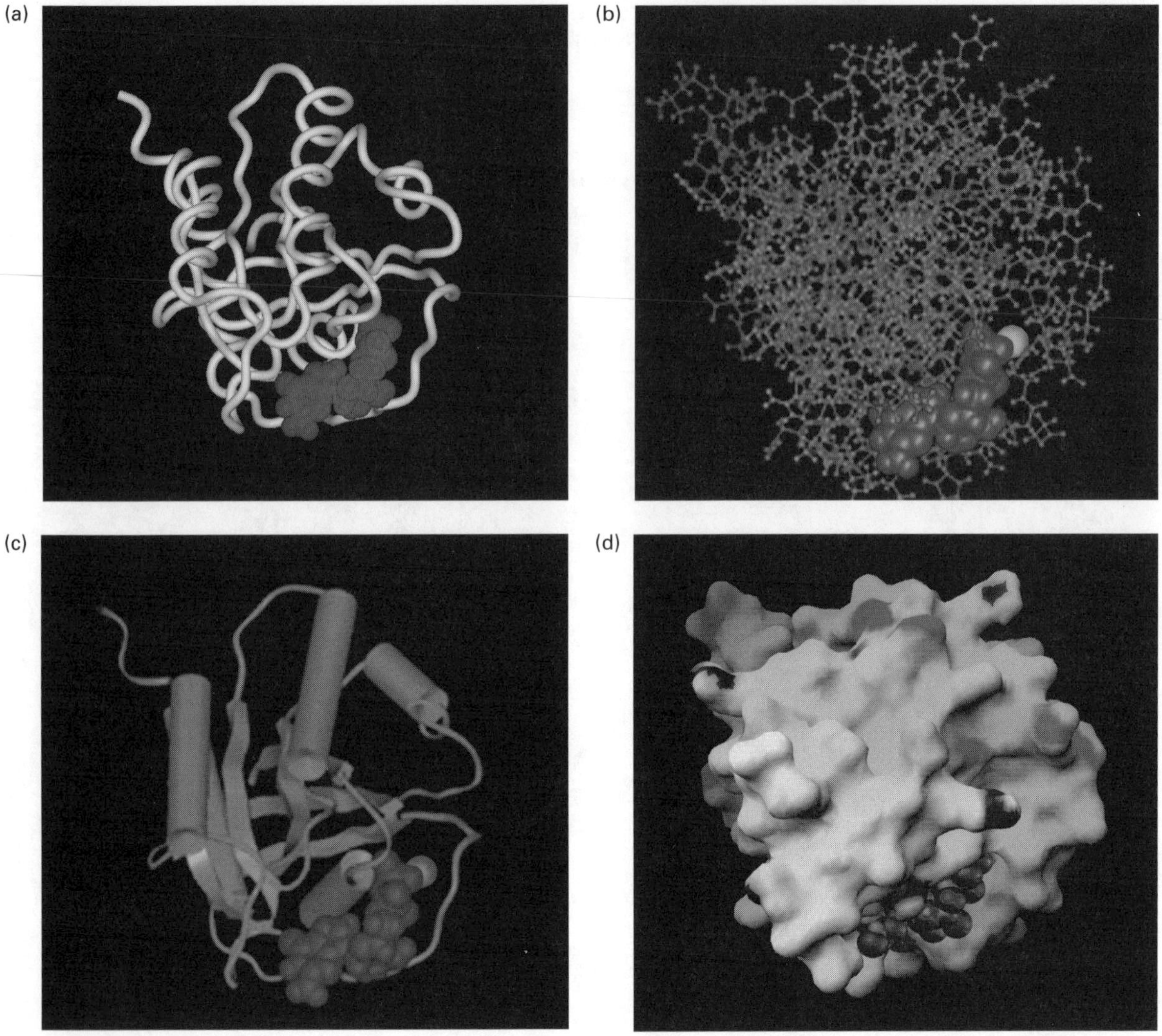

FIGURE 3-5, page 56
Various graphic representations of the structure of Ras, a guanine nucleotide–binding protein.

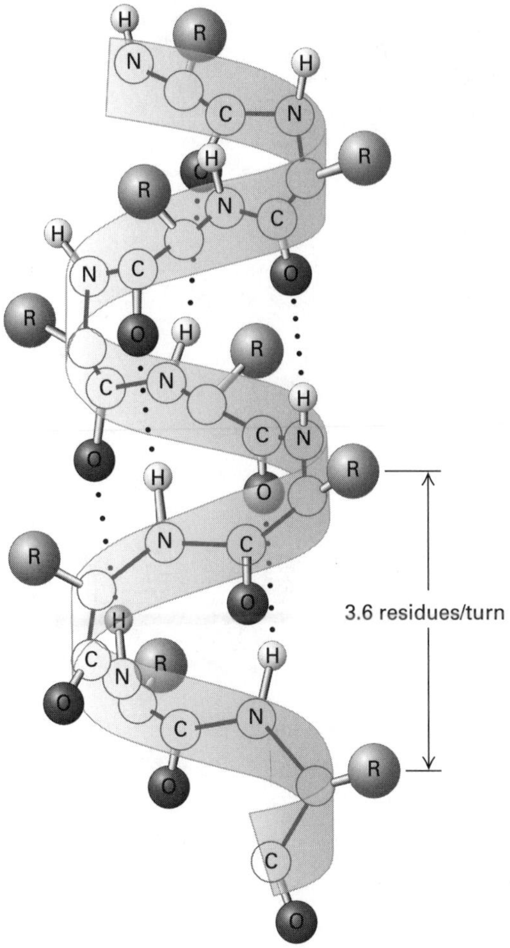

FIGURE 3-6, page 57
Model of the α helix.

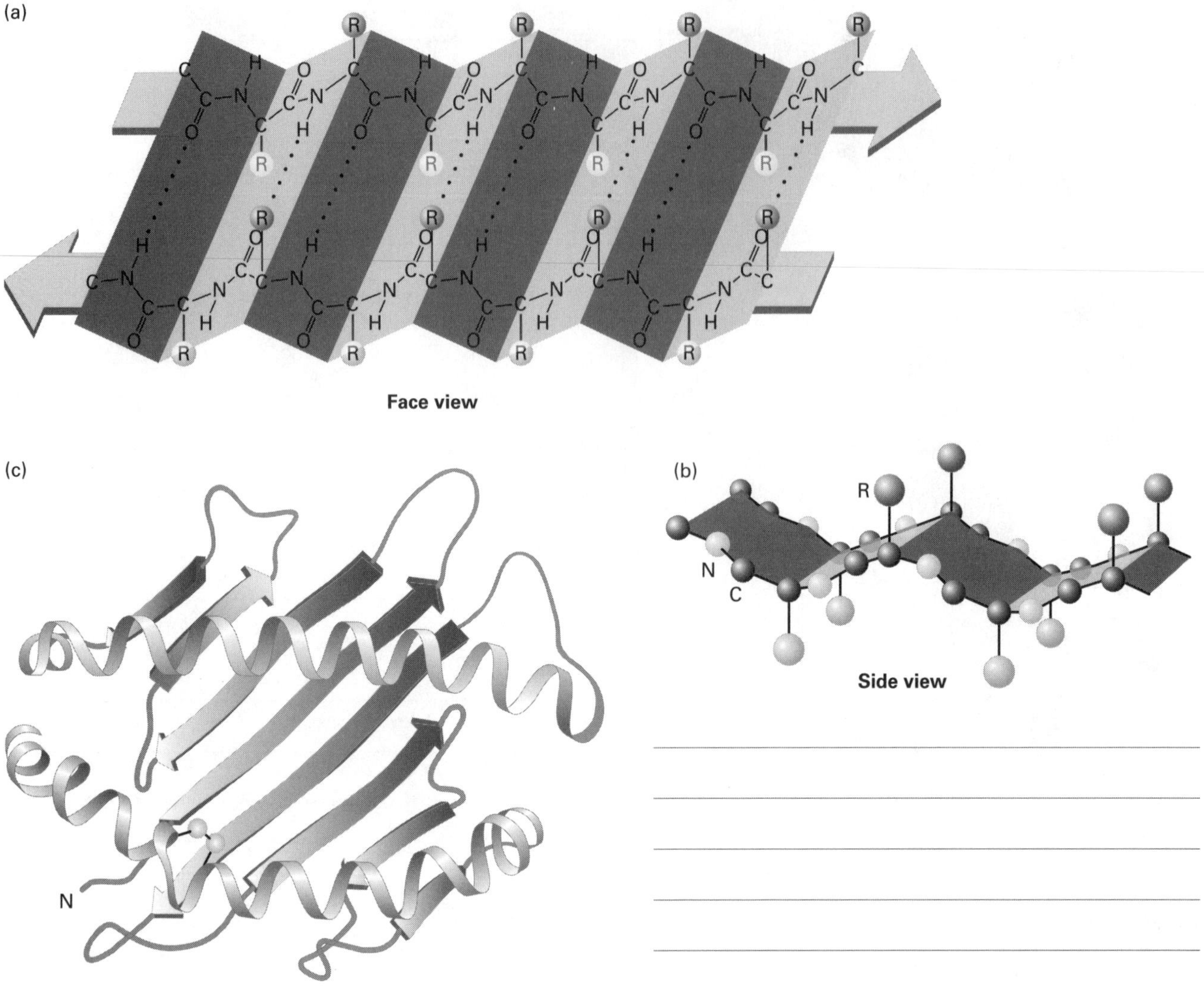
(a)
Face view
(c)
N
(b)
R
N
C
Side view

FIGURE 3-8, page 58
***β* sheets.**

(a)

(b) Helix–loop–helix motif
N
Helix
Loop
Ca2+
Helix
C
– Asp – Asp – Asp – – – Thr – – Glu –
Helix
Loop
Helix

(c) Zinc-finger motif
Cys
His
Zn2+
Cys
His
– – Cys – – Cys – – – – – – – – – – – – – – His – – – His –

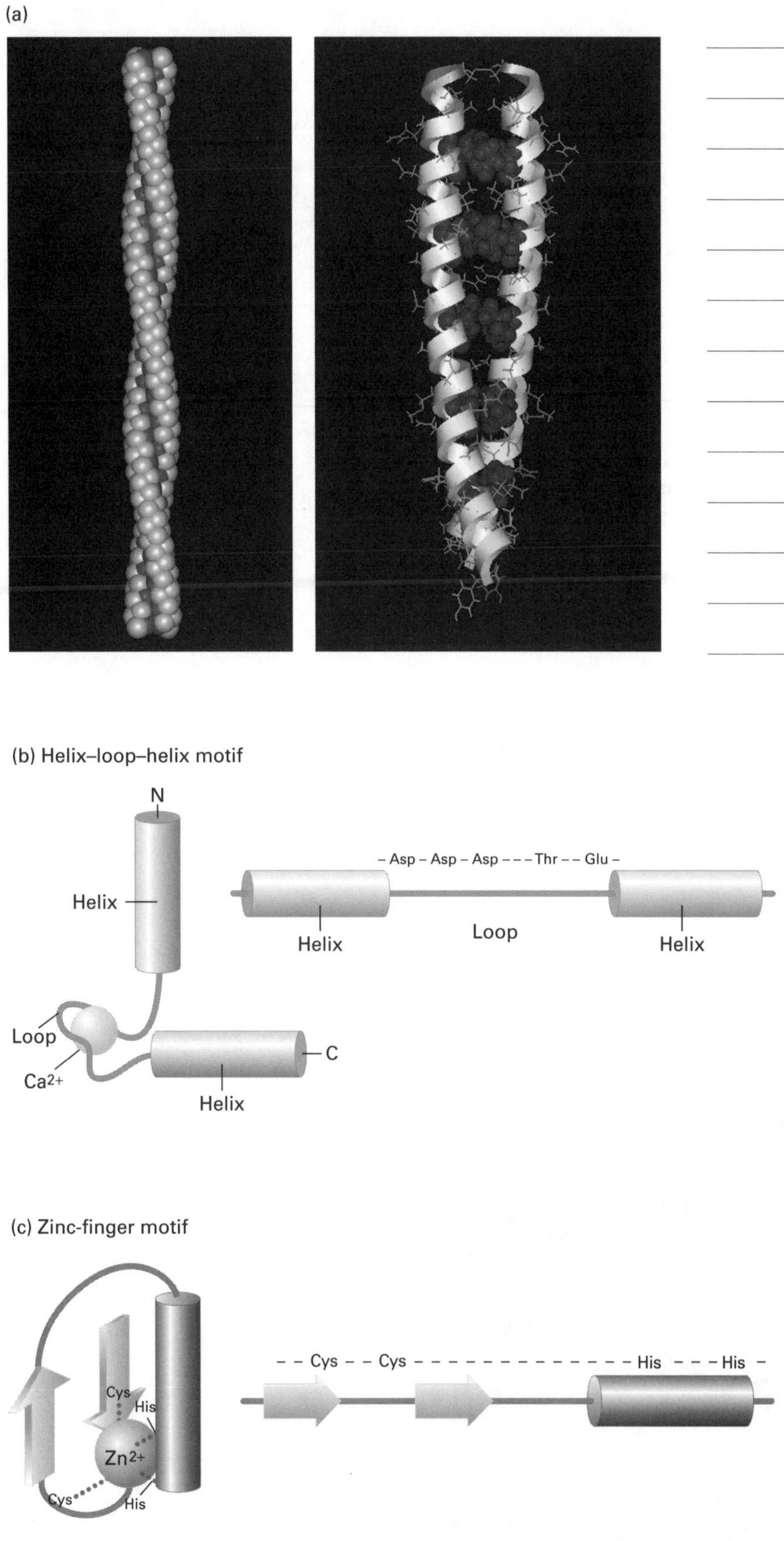

FIGURE 3-9, page 59
Secondary-structure motifs.

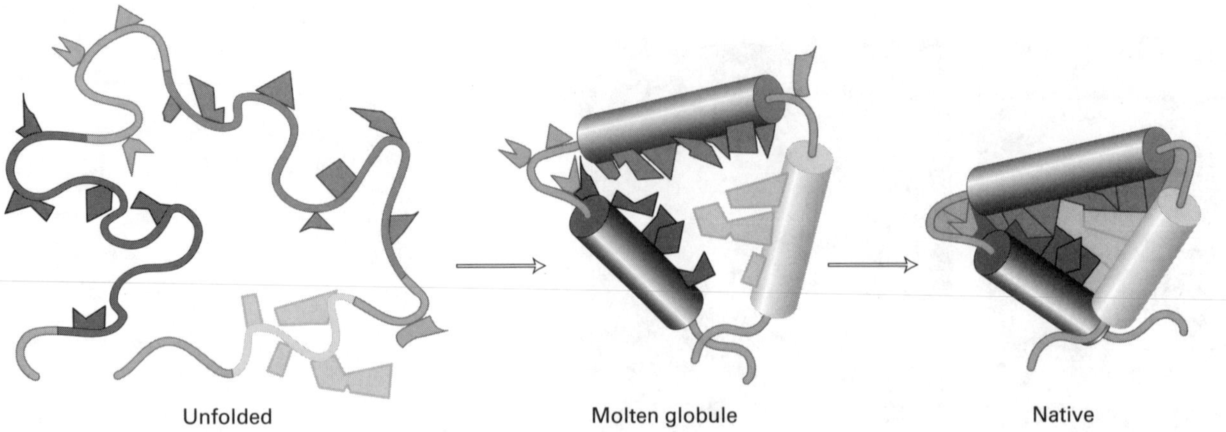

FIGURE 3-14, page 64
Three stages in unassisted protein folding. In its denatured state, the entire polypeptide chain assumes a random conformation.

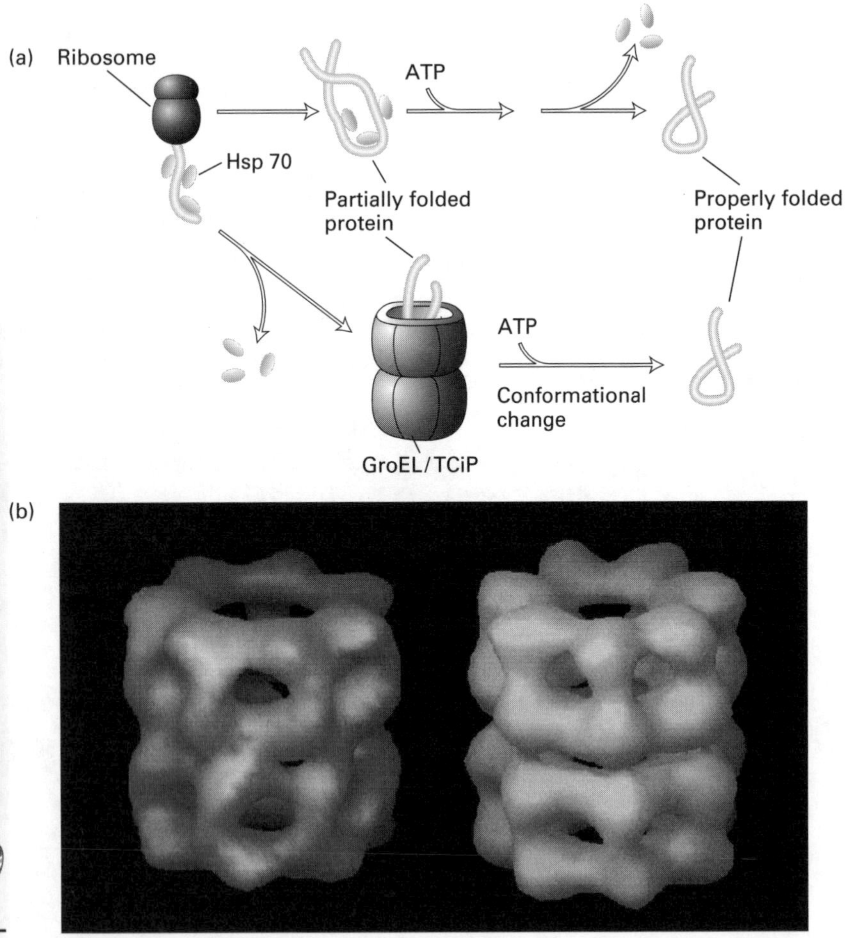

MEDIA CONNECTIONS
Chaperone-Mediated Folding

FIGURE 3-15, page 65
Chaperone-mediated protein folding.

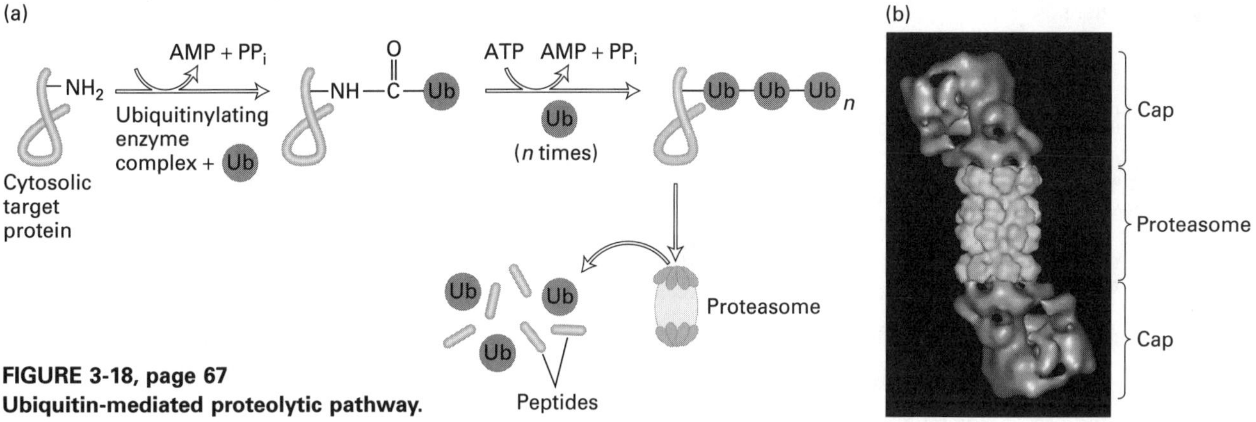

FIGURE 3-18, page 67
Ubiquitin-mediated proteolytic pathway.

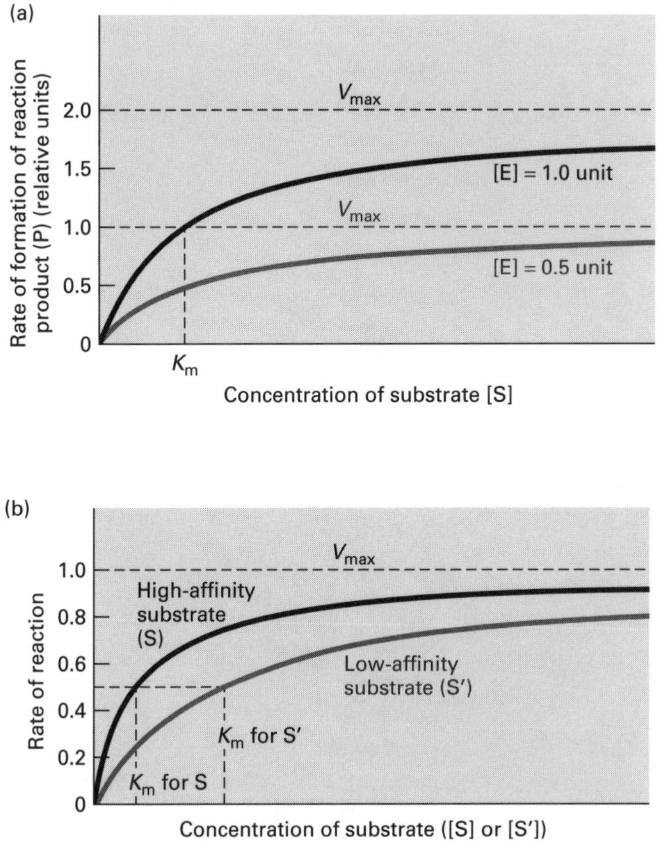

FIGURE 3-26, page 74
Dependence of the velocity of an enzyme-catalyzed reaction on substrate concentration.

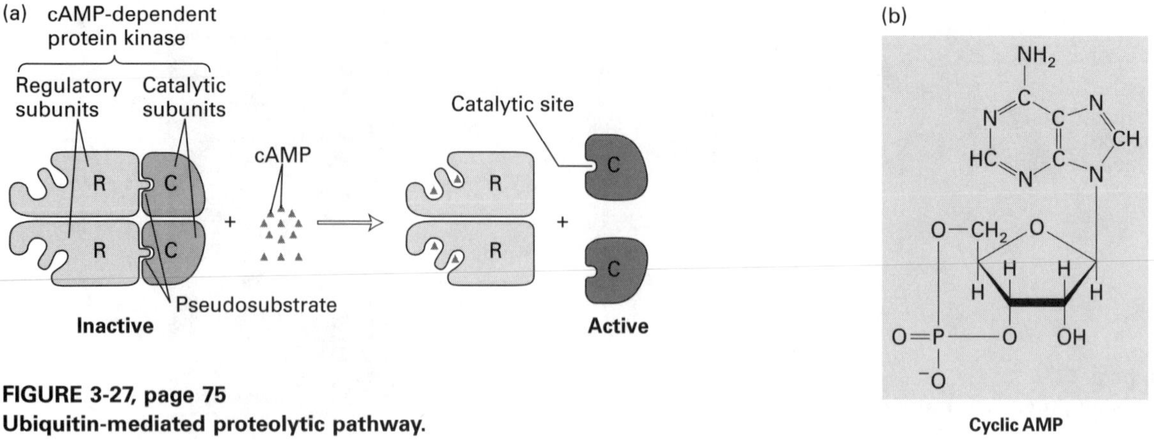

FIGURE 3-27, page 75
Ubiquitin-mediated proteolytic pathway.

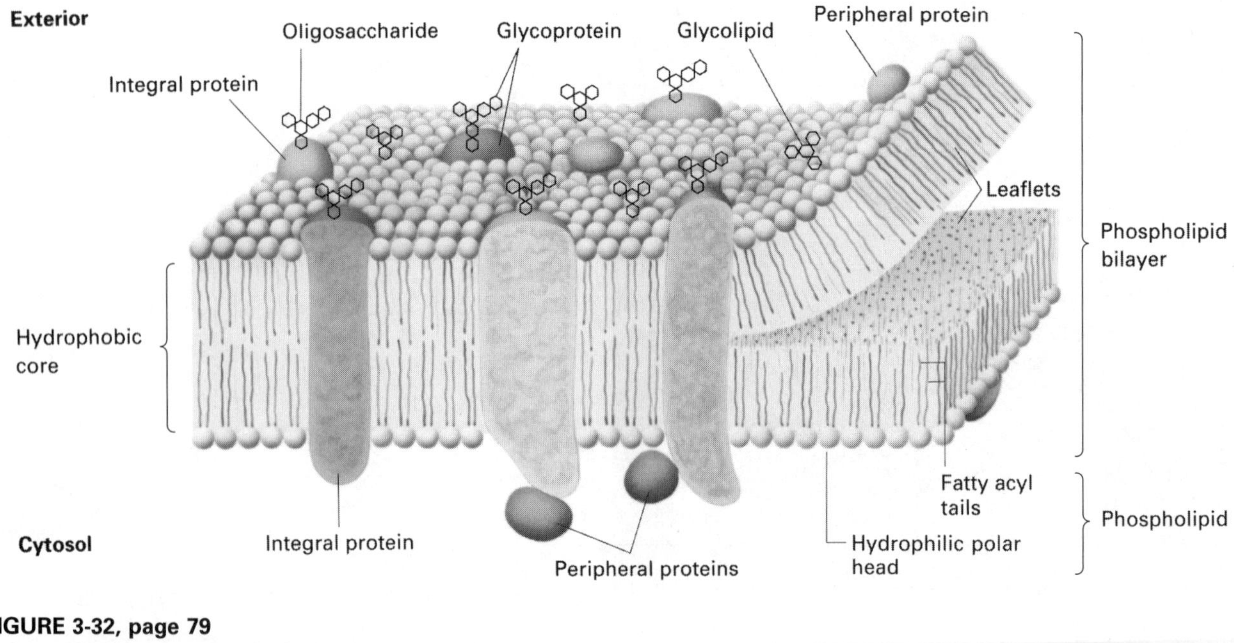

FIGURE 3-32, page 79
Schematic diagram of typical membrane proteins in a biological membrane.

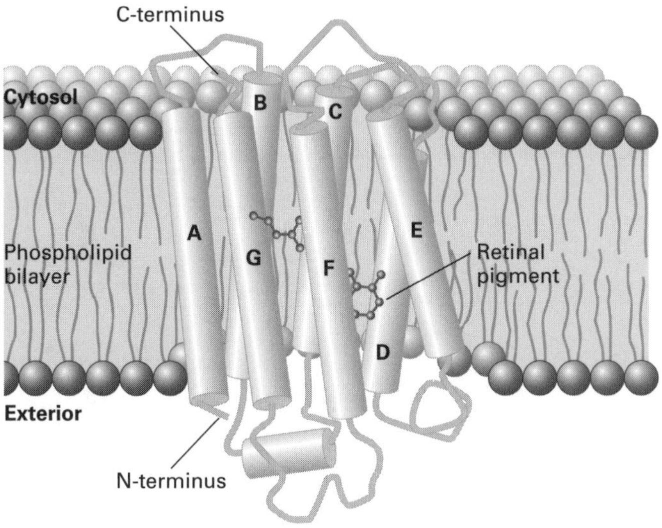

FIGURE 3-34, page 81
Overall structure of bacteriorhodopsin as deduced from electron diffraction analyses of two-dimensional crystals of the protein in the bacterial membrane.

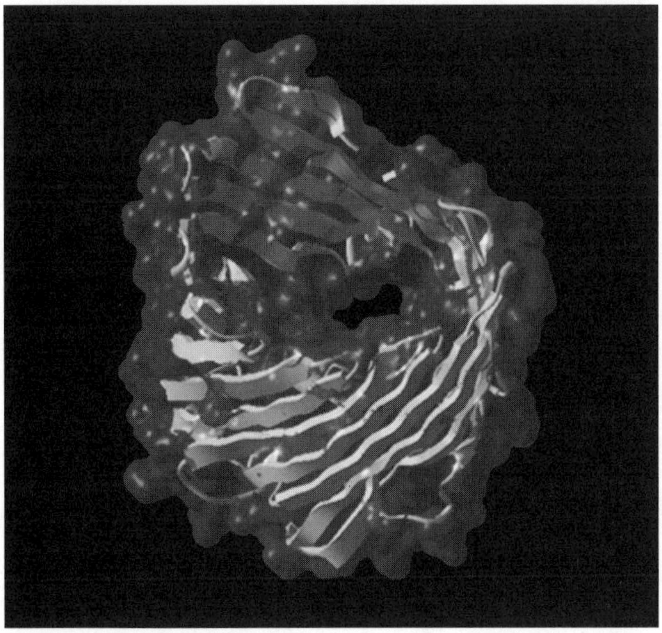

FIGURE 3-35, page 81
Model of the three-dimensional structure of a subunit of OmpF, a porin found in the *E. coli* outer membrane.

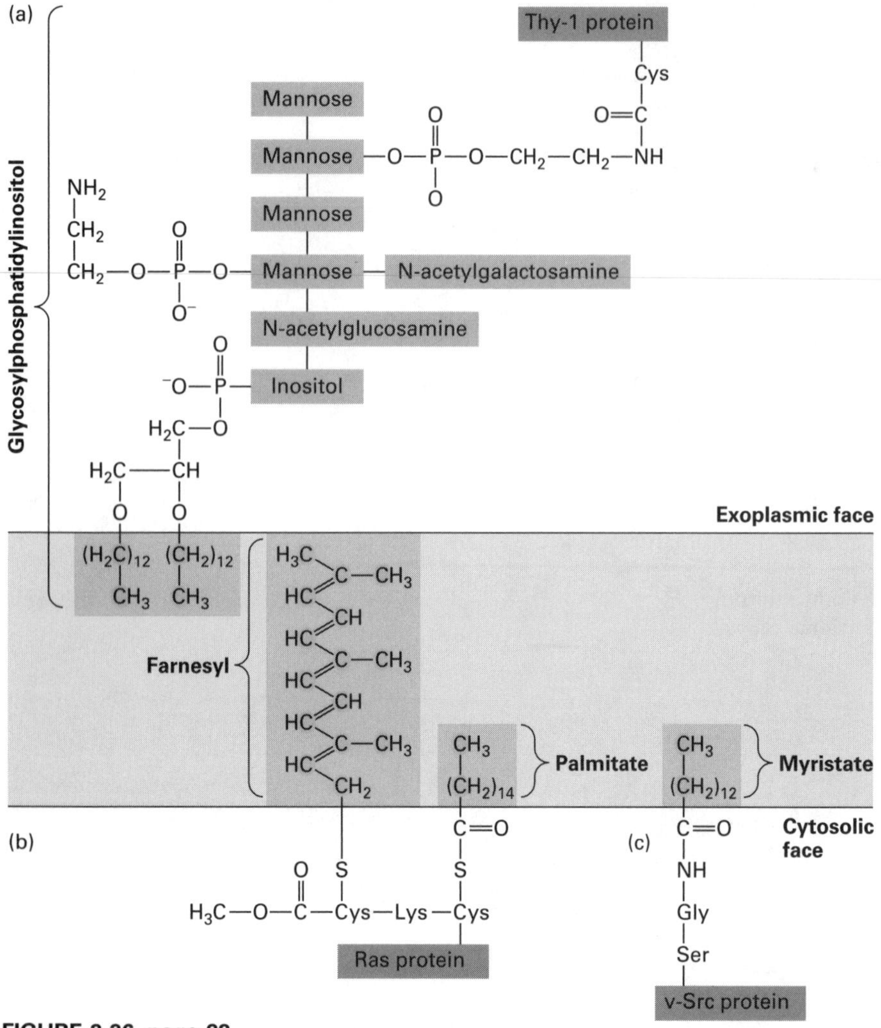

FIGURE 3-36, page 82
Anchoring of integral proteins to the plasma membrane by membrane-embedded hydrocarbon groups (highlighted in red).

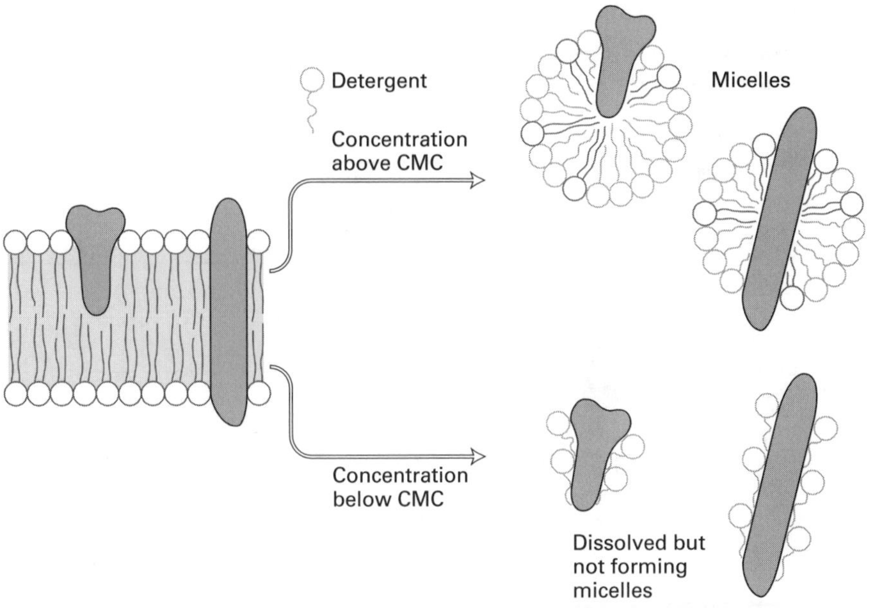

FIGURE 3-39, page 84
Solubilization of integral membrane proteins by nonionic detergents.

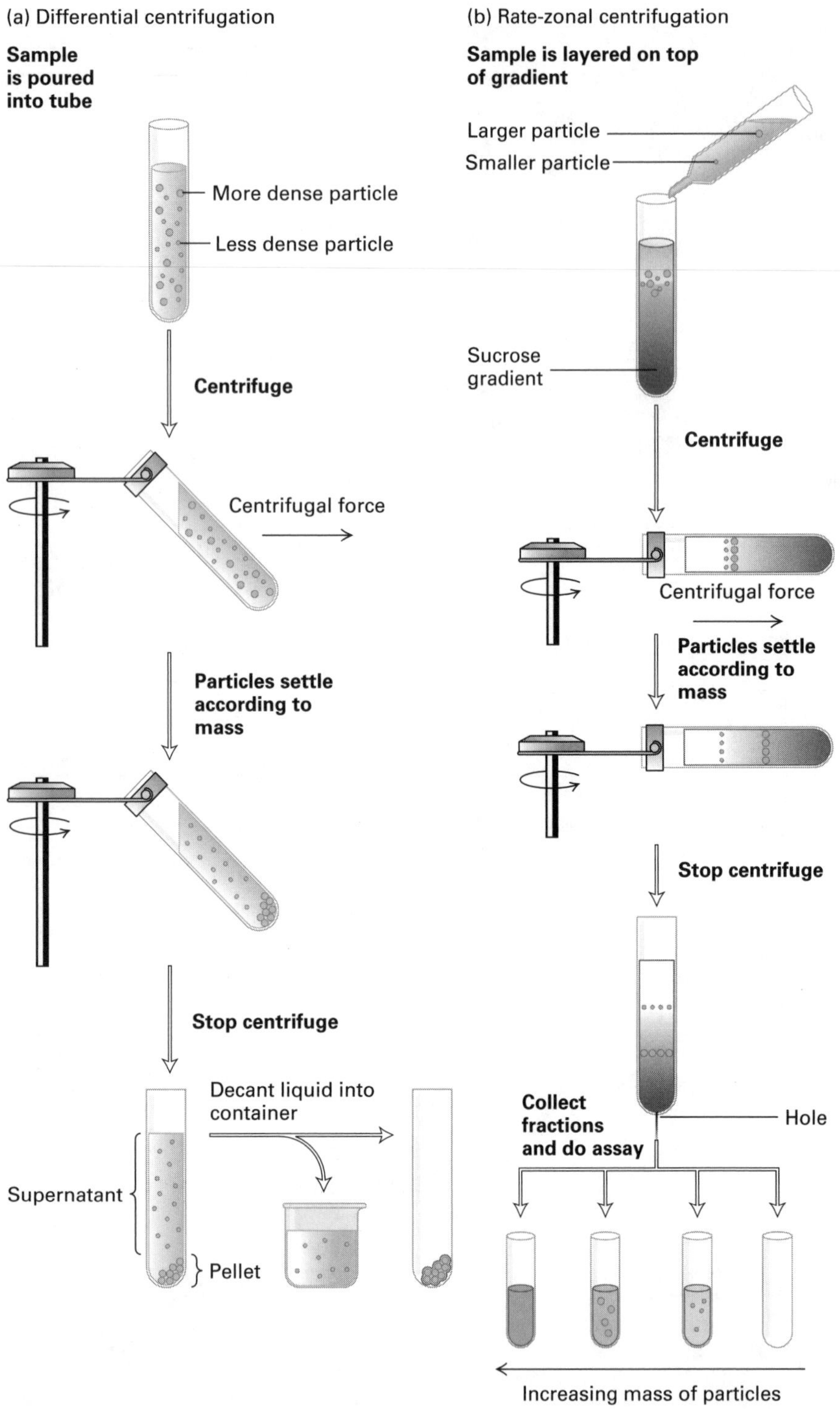

FIGURE 3-40, page 86
Two common centrifugation techniques for separating particles.

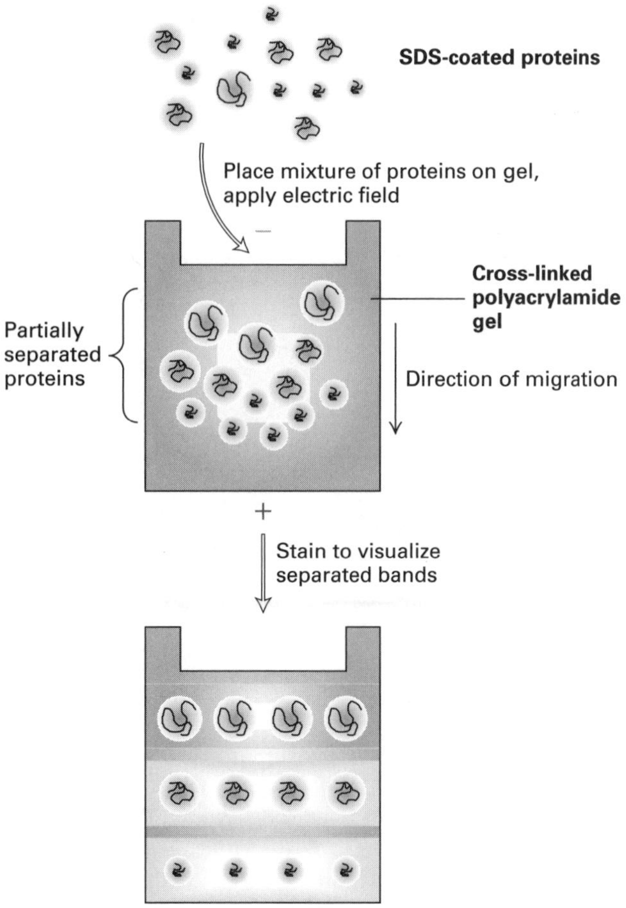

FIGURE 3-41, page 87
SDS-polyacrylamide gel electrophoresis, a common technique for separating proteins at good resolution.

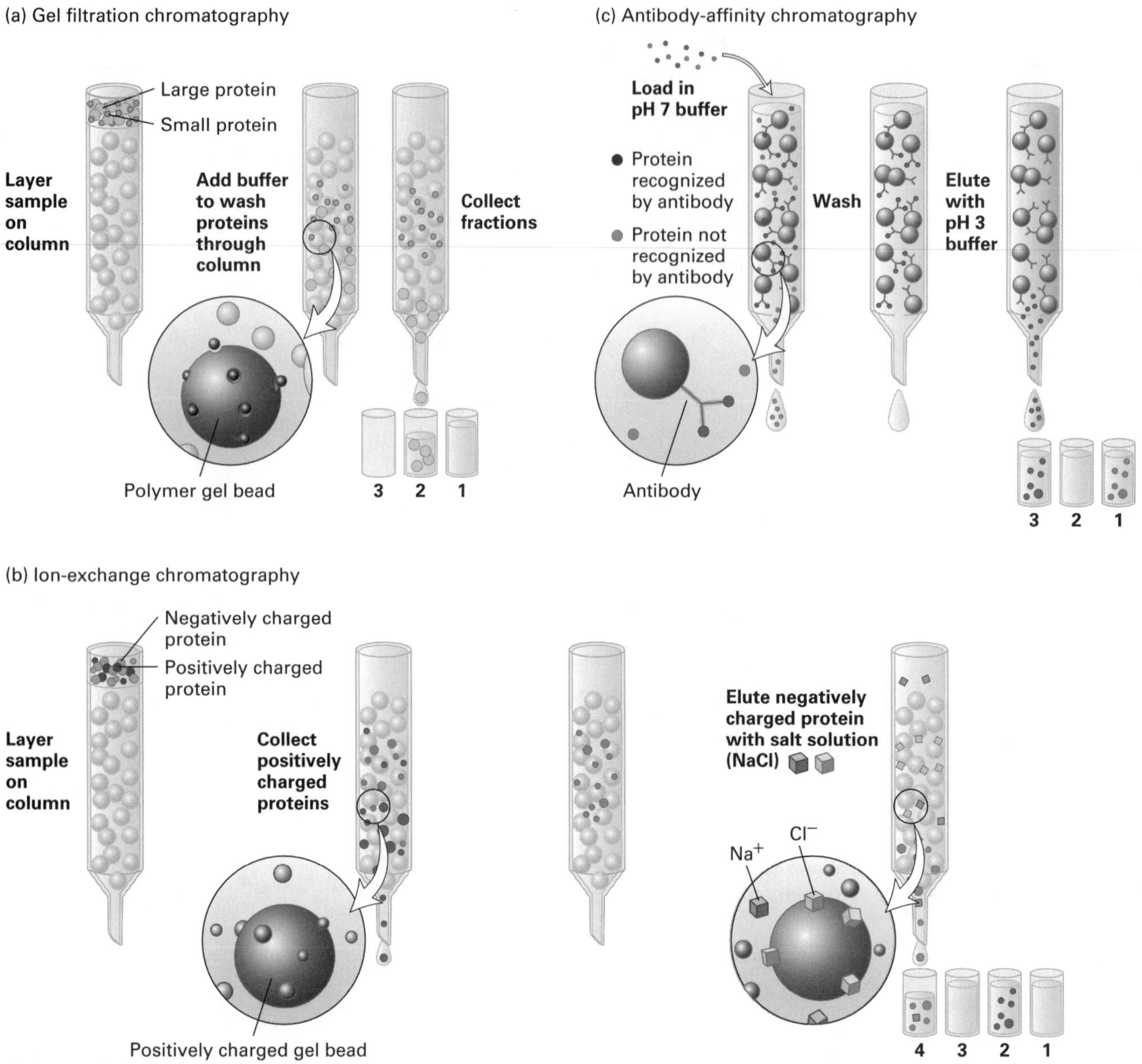

FIGURE 3-43, page 89
Three commonly used liquid chromatographic techniques.

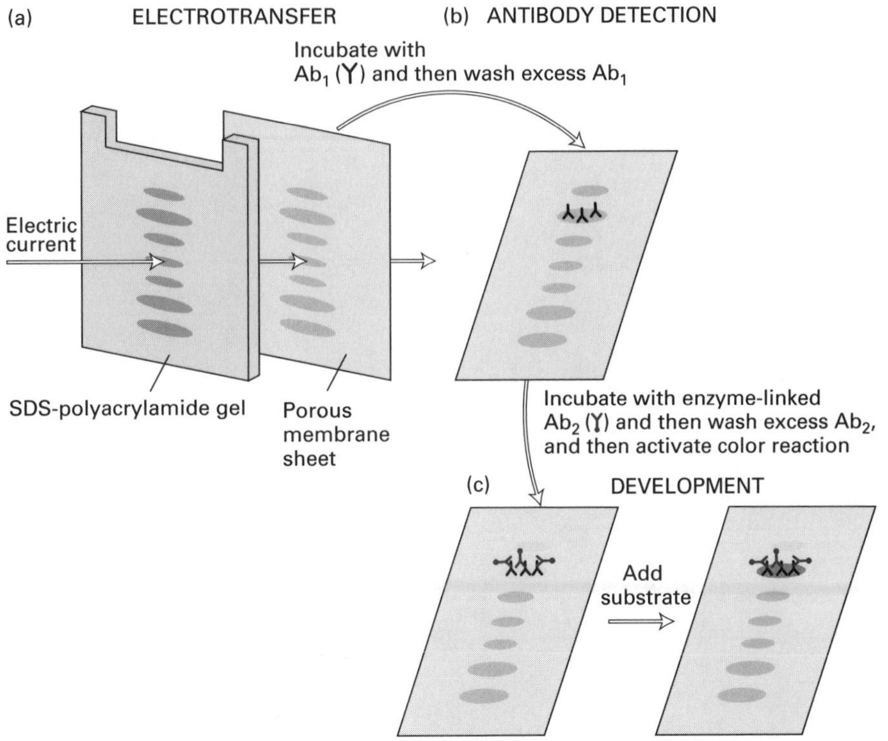
(a) ELECTROTRANSFER
(b) ANTIBODY DETECTION
Incubate with
Ab1 (Y) and then wash excess Ab1
Electric current
SDS-polyacrylamide gel
Porous membrane sheet
Incubate with enzyme-linked
Ab2 (Y) and then wash excess Ab2,
and then activate color reaction
(c) DEVELOPMENT
Add substrate

FIGURE 3-44, page 91
Western blotting, or immunoblotting.

MEDIA CONNECTIONS
Immunoblotting

Additional Notes

Nucleic Acids, the Genetic Code, and the Synthesis of Macromolecules

4

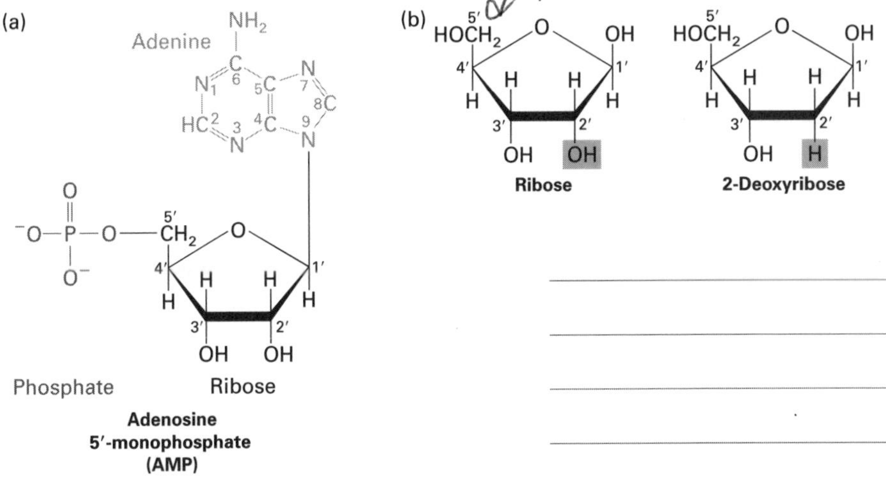

FIGURE 4-1, page 101
All nucleotides have a common structure.

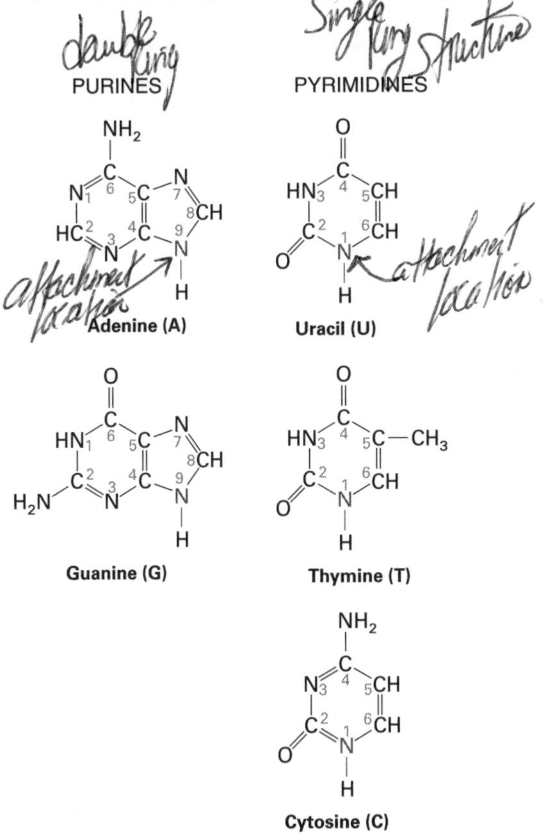

FIGURE 4-2, page 102
The chemical structures of the principal bases in nucleic acids.

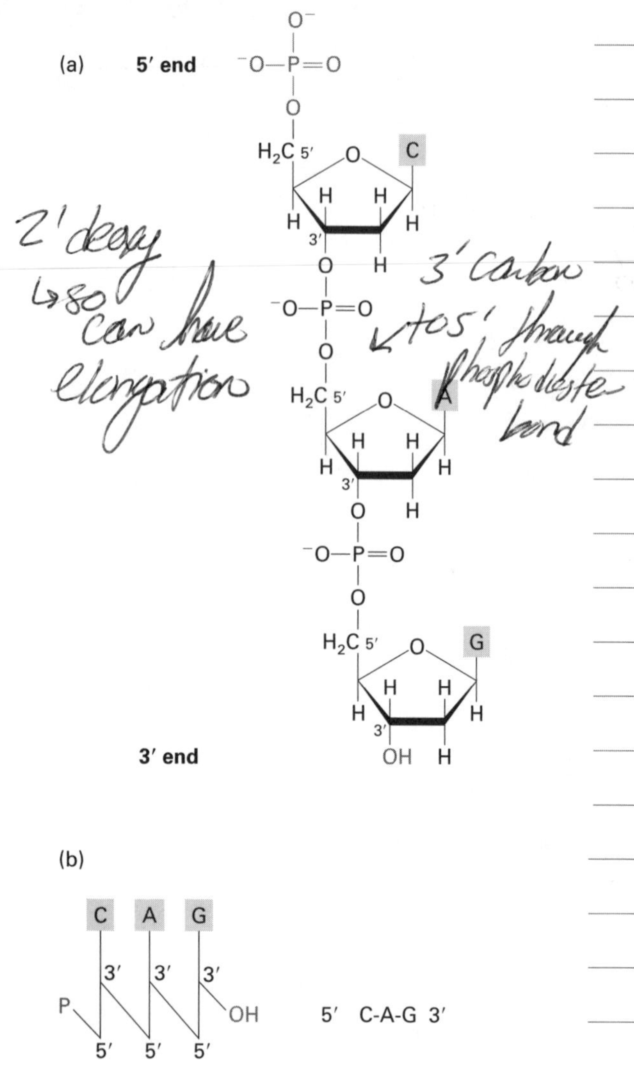

FIGURE 4-3, page 102
Alternative ways of representing nucleic acid chains, in this case a single strand of DNA containing only three bases: cytosine (C), adenine (A), and guanine (G).

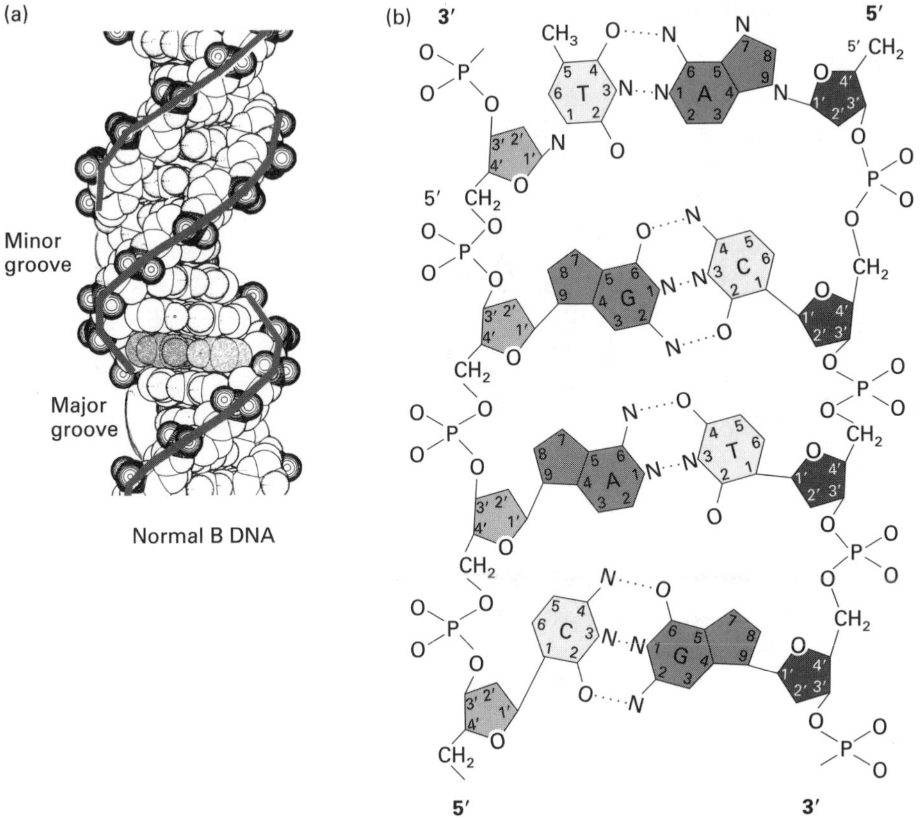

FIGURE 4-4, page 104
Two representations of contacts within the DNA double helix.

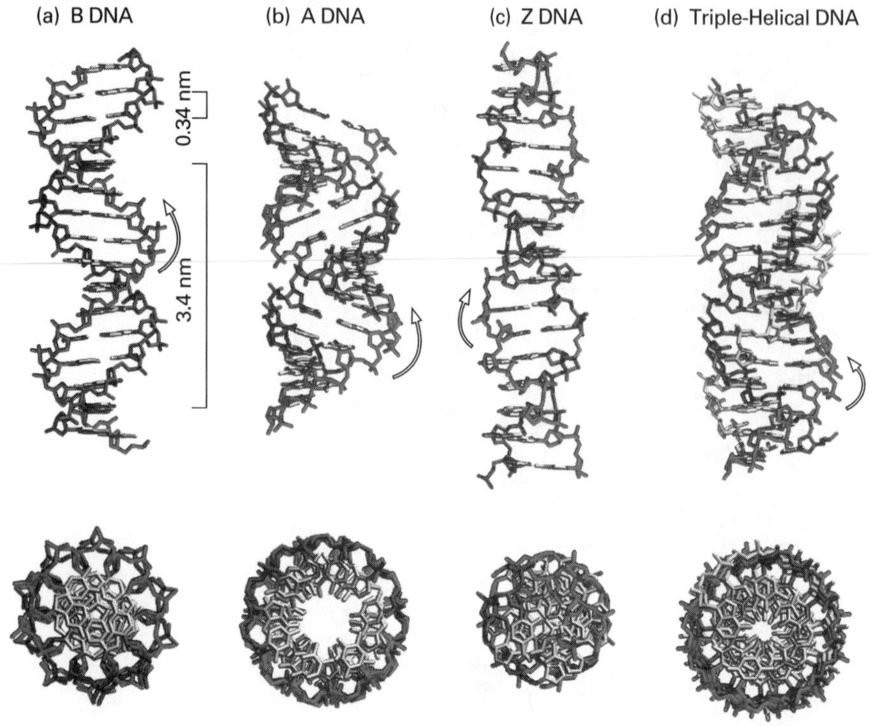

FIGURE 4-6, page 105
Models of various DNA structures that are known to exist.

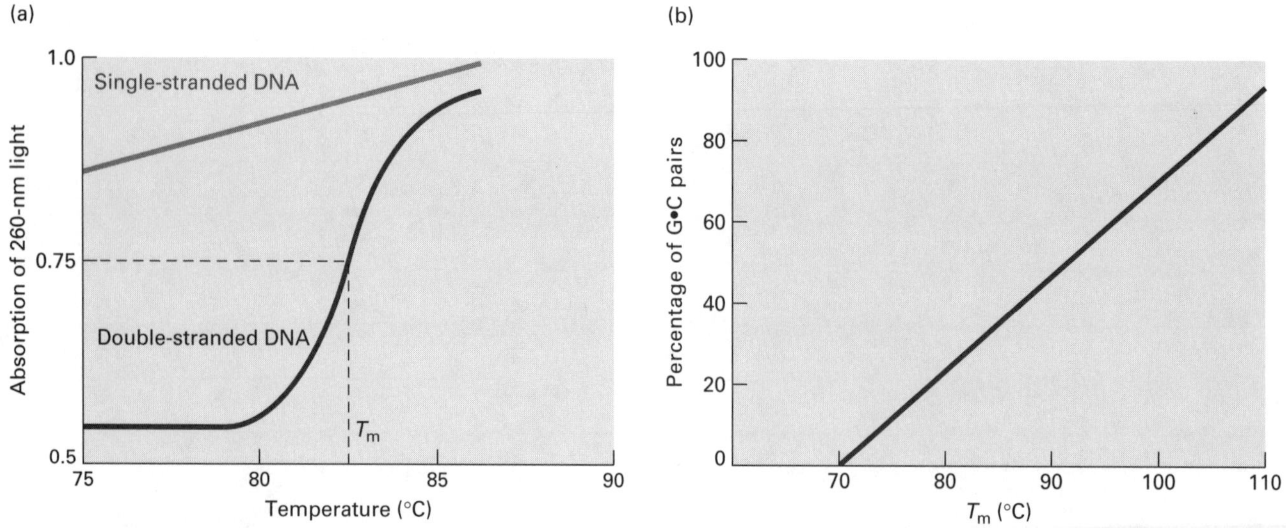

FIGURE 4-9, page 107
Light absorption and temperature in DNA denaturation.

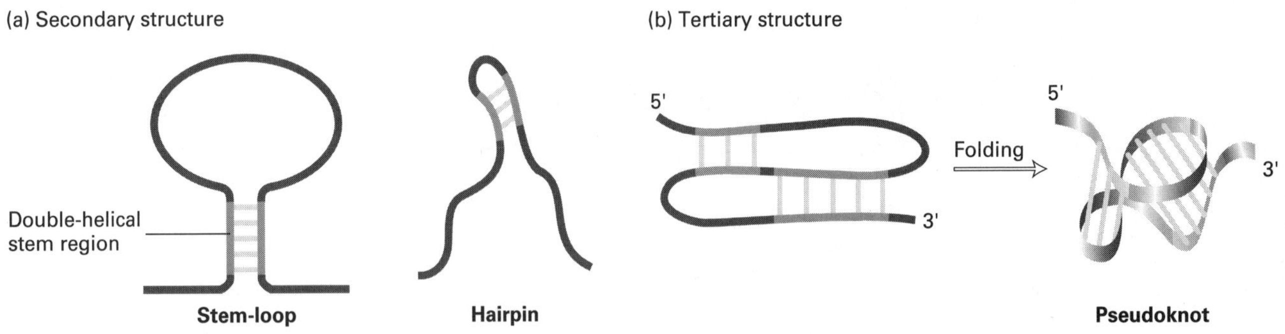

FIGURE 4-12, page 109
RNA secondary and tertiary structures.

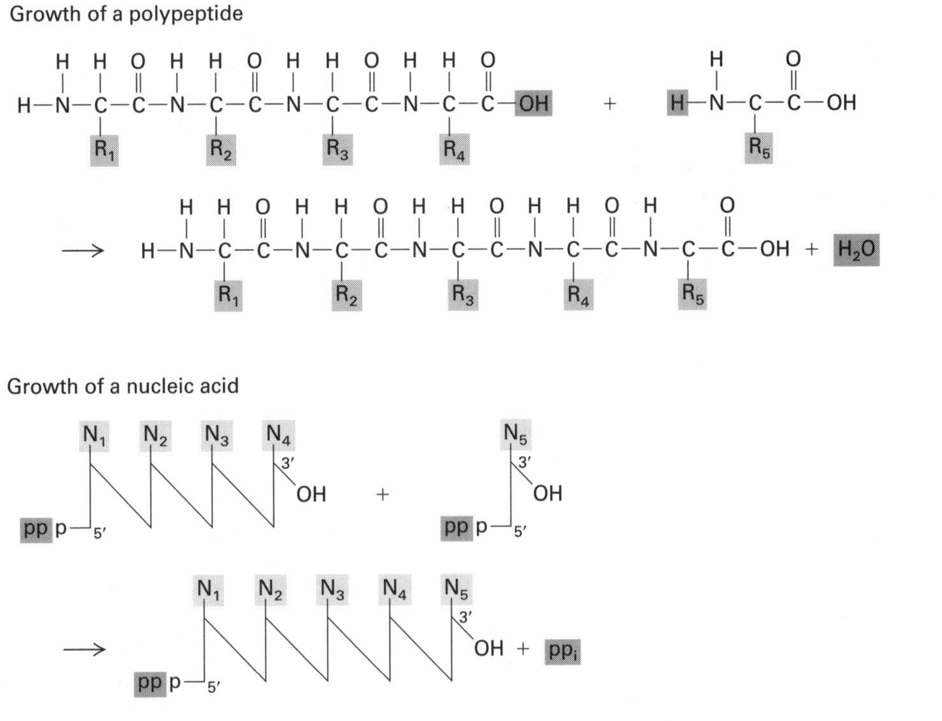

FIGURE 4-13, page 110
Chain elongation in the in vivo synthesis of both proteins and nucleic acids proceeds by sequential addition of monomeric units—amino acids and nucleotides, respectively.

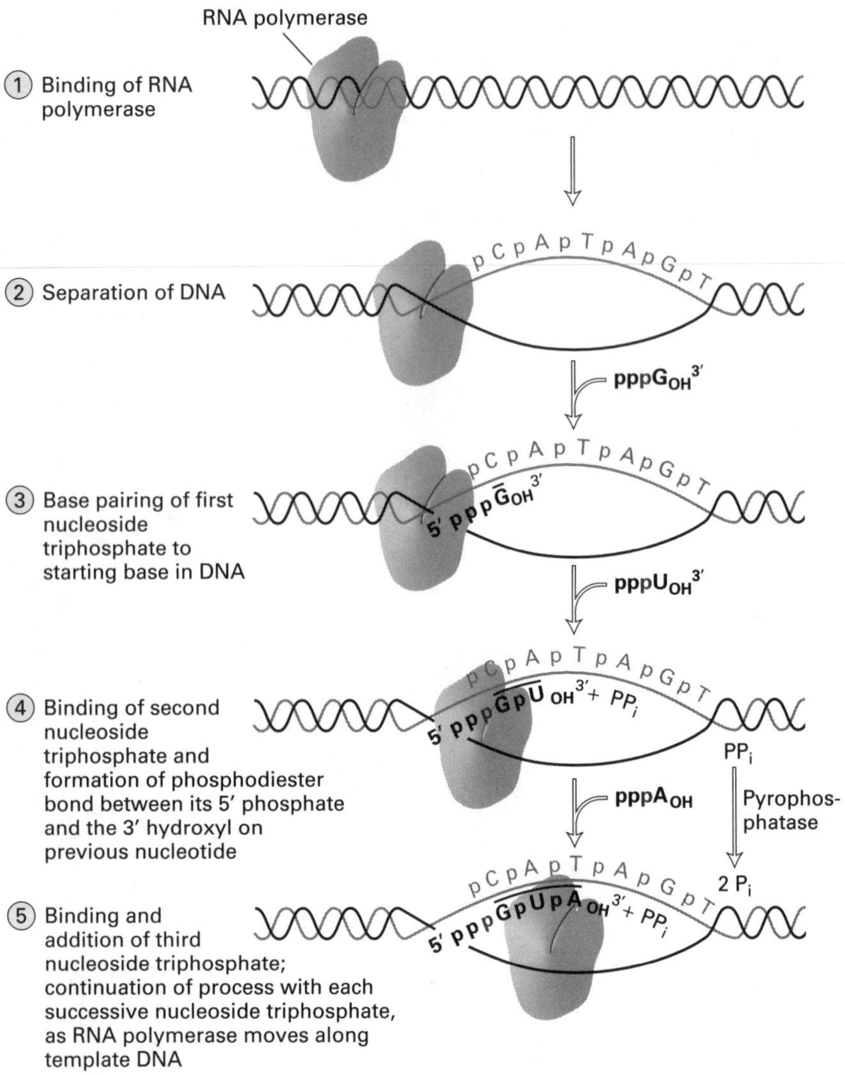
RNA polymerase
1 Binding of RNA polymerase
2 Separation of DNA
pCpApTpApGpT
pppG OH 3′
3 Base pairing of first nucleoside triphosphate to starting base in DNA
5′ pppG OH 3′
pppU OH 3′
4 Binding of second nucleoside triphosphate and formation of phosphodiester bond between its 5′ phosphate and the 3′ hydroxyl on previous nucleotide
5′ pppGpU OH 3′ + PPi
PPi
pppA OH
Pyrophos-phatase
2 Pi
5 Binding and addition of third nucleoside triphosphate; continuation of process with each successive nucleoside triphosphate, as RNA polymerase moves along template DNA
5′ pppGpUpA OH 3′ + PPi

FIGURE 4-15, page 112
Transcription of DNA into RNA is catalyzed by RNA polymerase, which can initiate the synthesis of strands de novo on DNA templates.

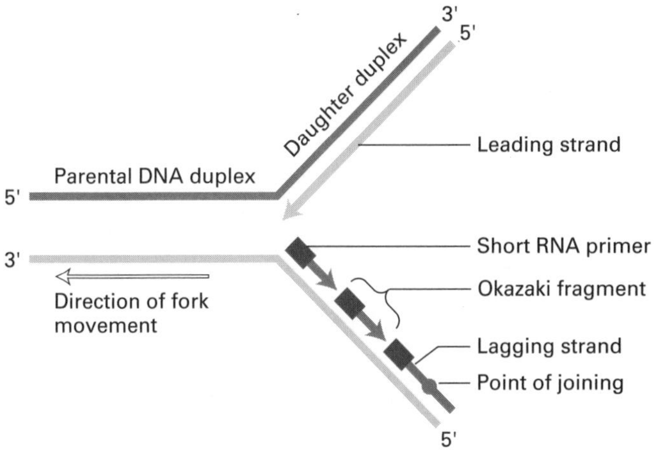

FIGURE 4-16, page 113
Schematic diagram of DNA replication at a growing fork.

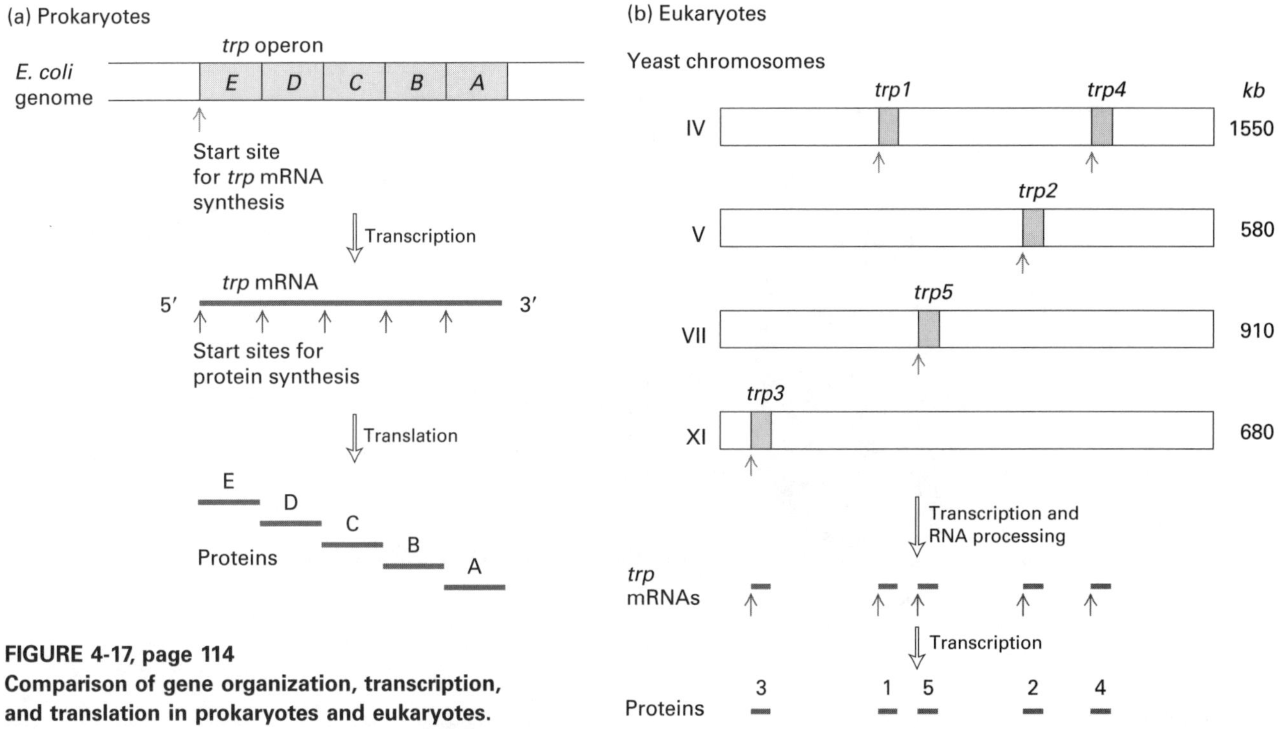

FIGURE 4-17, page 114
Comparison of gene organization, transcription, and translation in prokaryotes and eukaryotes.

MEDIA CONNECTIONS Life Cycle of an mRNA

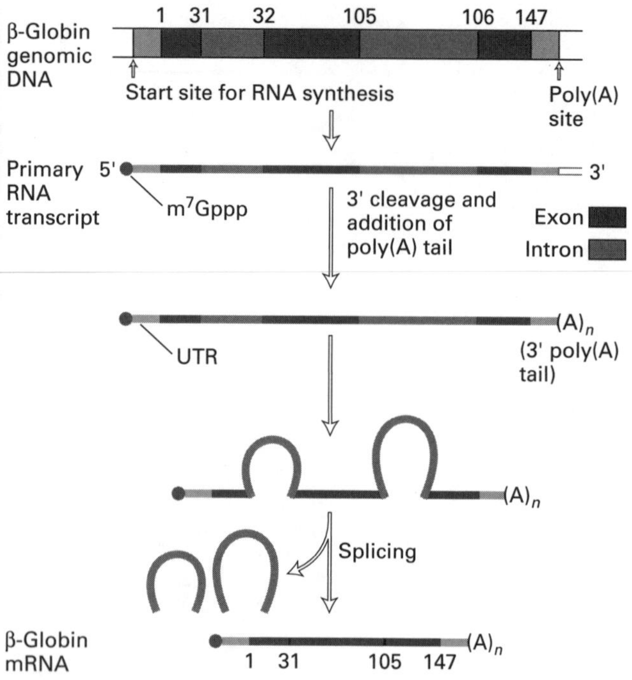

FIGURE 4-19, page 115
Overview of RNA processing in eukaryotes using β-globin gene as an example.

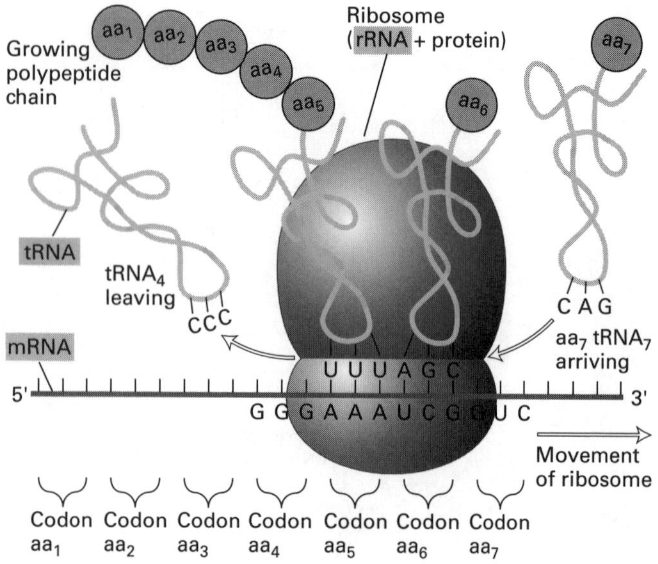

FIGURE 4-20, page 117
The three roles of RNA in protein synthesis.

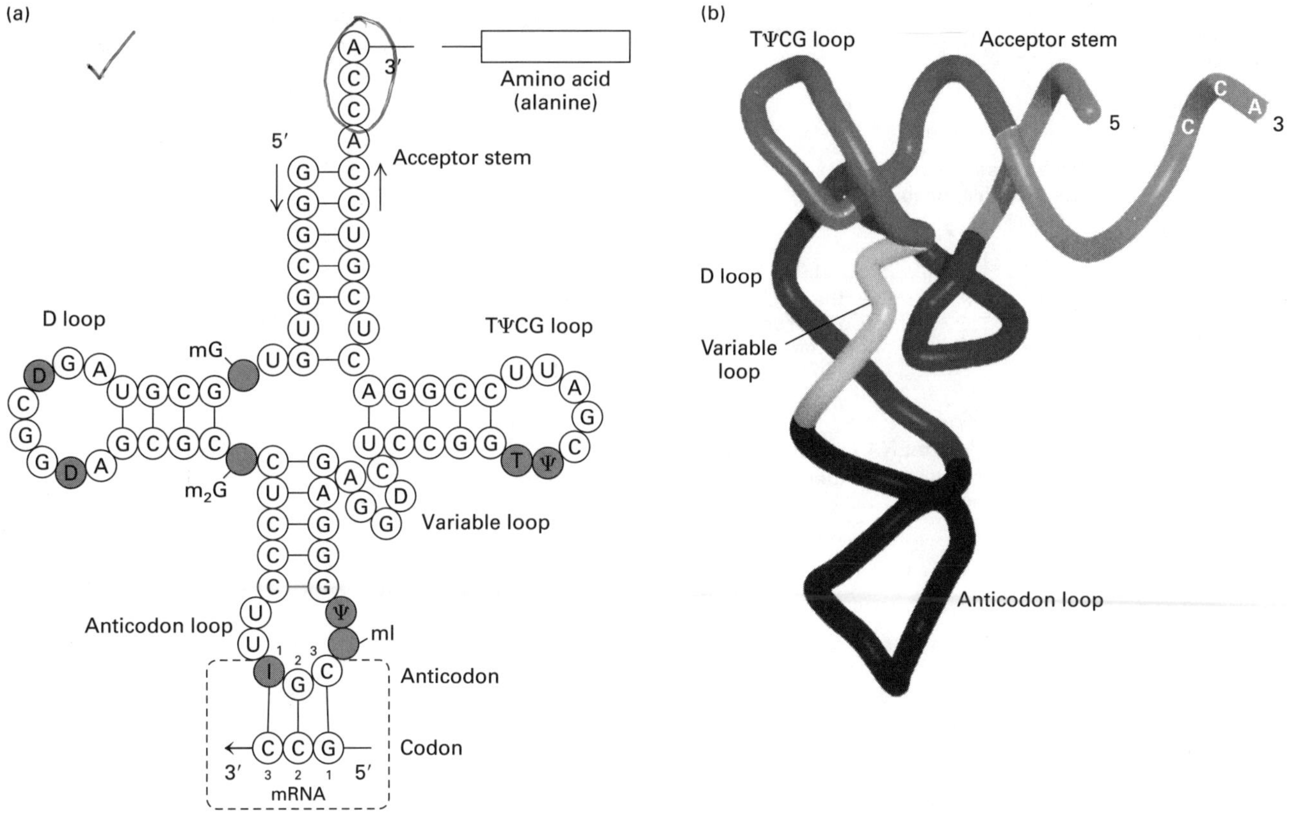
(a)
Amino acid (alanine)
3′
5′
Acceptor stem
D loop
TΨCG loop
mG
m_2G
Variable loop
Anticodon loop
mI
Anticodon
Codon
3′ 3 2 1 5′
mRNA
(b)
TΨCG loop
Acceptor stem
5
3
D loop
Variable loop
Anticodon loop

FIGURE 4-26, page 121
Structure of tRNAs.

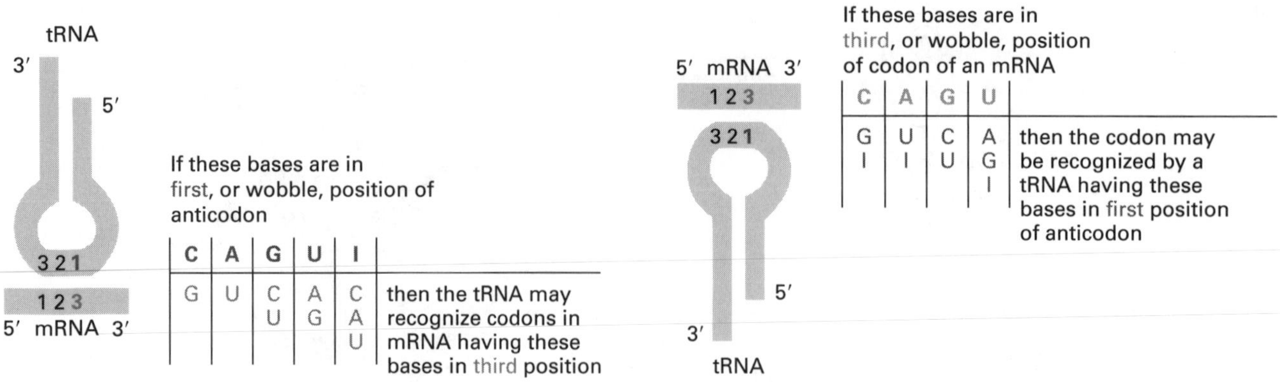

FIGURE 4-27, page 122
The first and second bases in an mRNA codon form Watson-Crick base pairs with the third and second bases, respectively, of a tRNA anticodon.

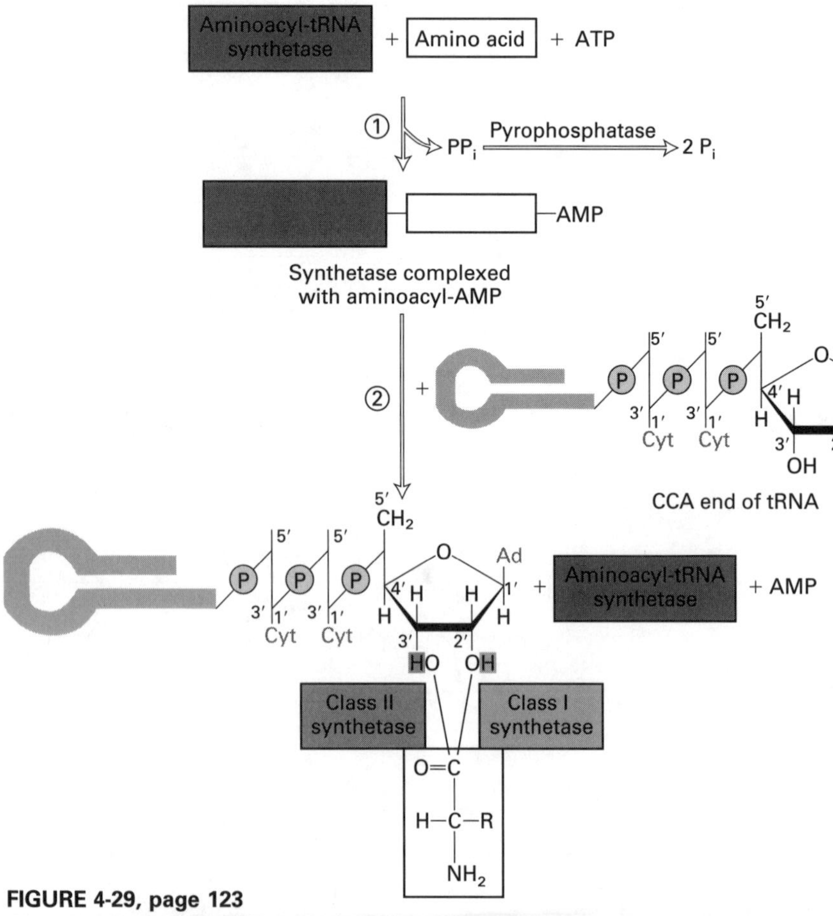

FIGURE 4-29, page 123
Aminoacylation of tRNA. Amino acids are covalently linked to tRNAs by aminoacyl-tRNA synthetases.

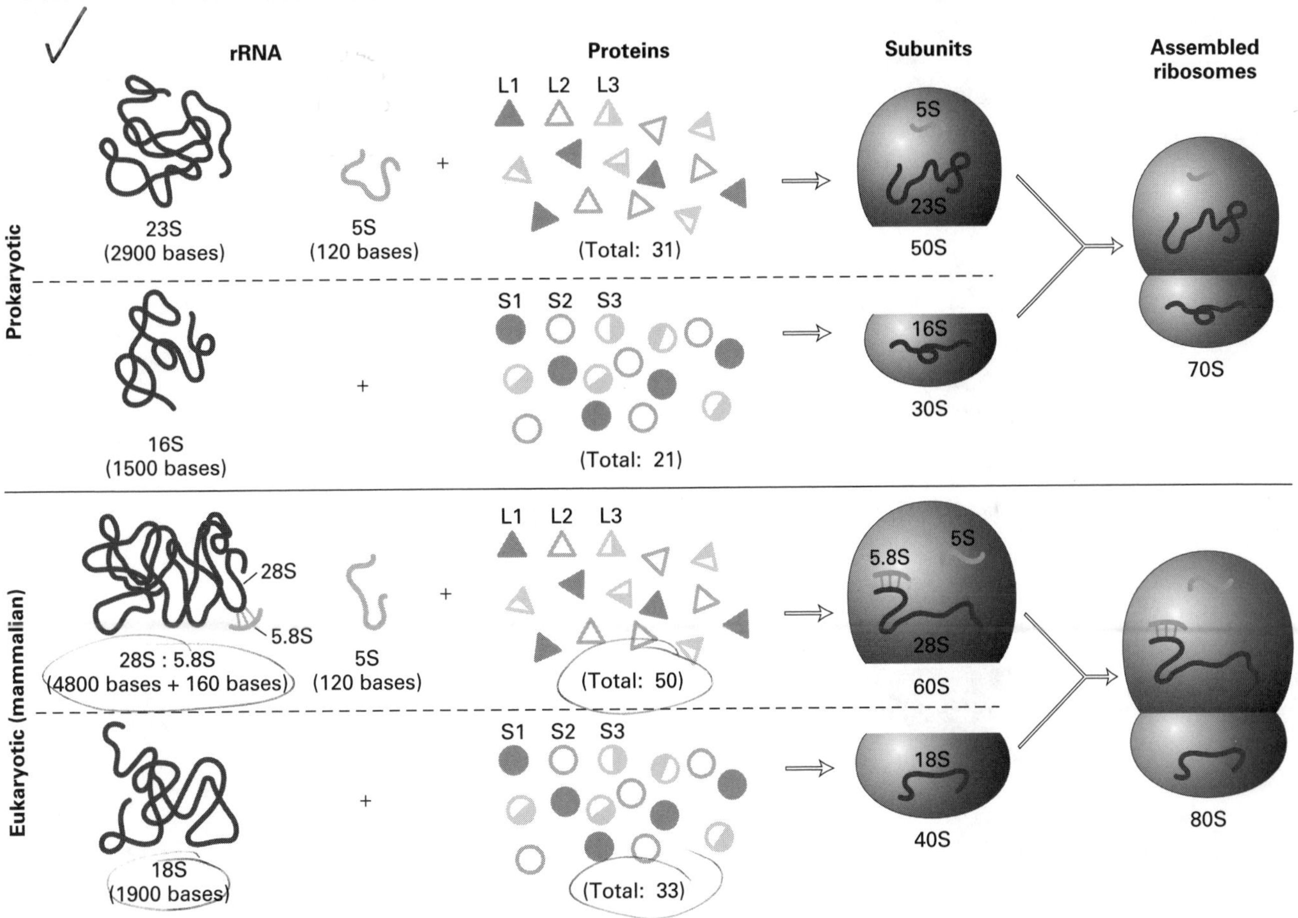

FIGURE 4-32, page 126
The general structure of ribosomes in prokaryotes and eukaryotes.

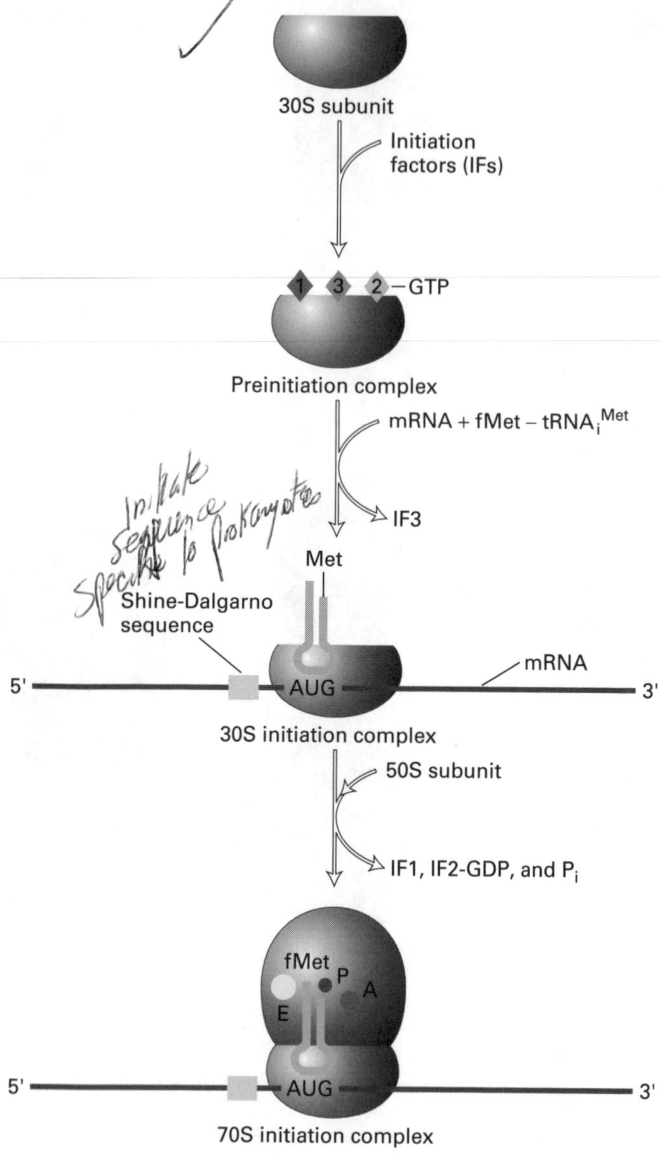

FIGURE 4-36, page 129
Bacterial initiation of protein synthesis.

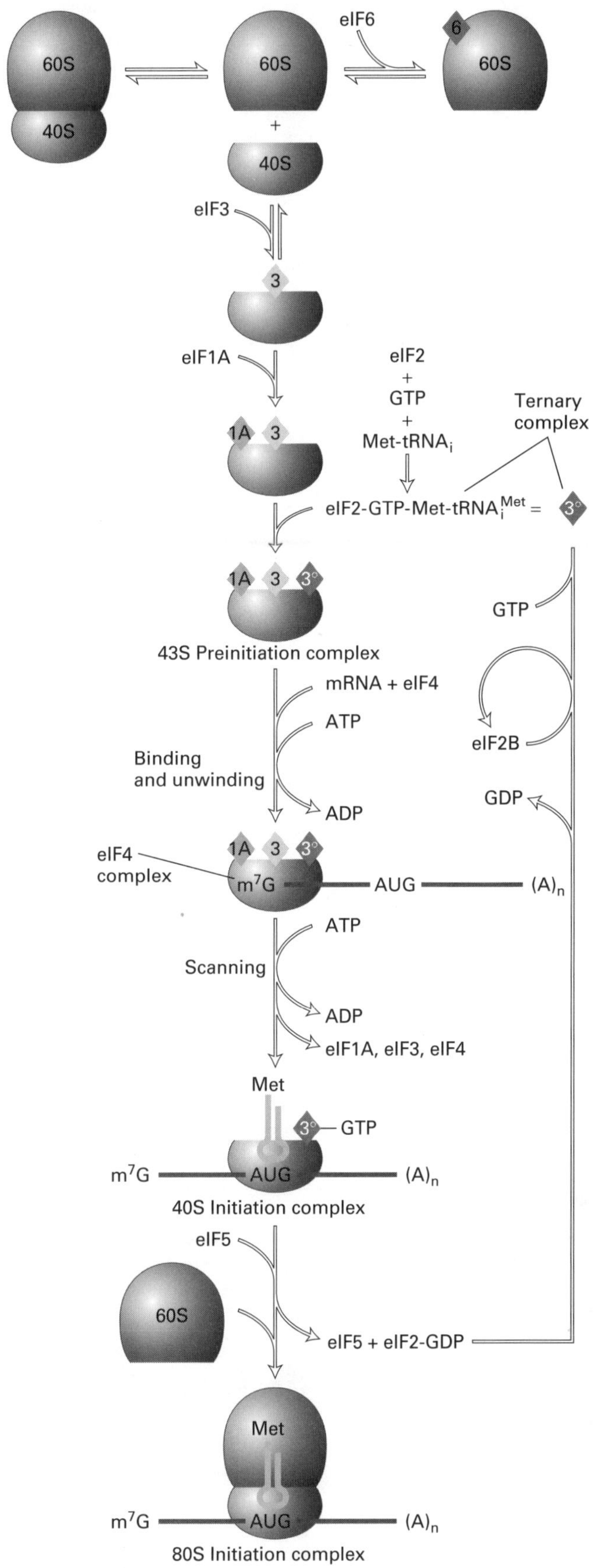

FIGURE 4-37, page 130
Eukaryotic initiation of protein synthesis.

MEDIA CONNECTIONS **Protein Synthesis**

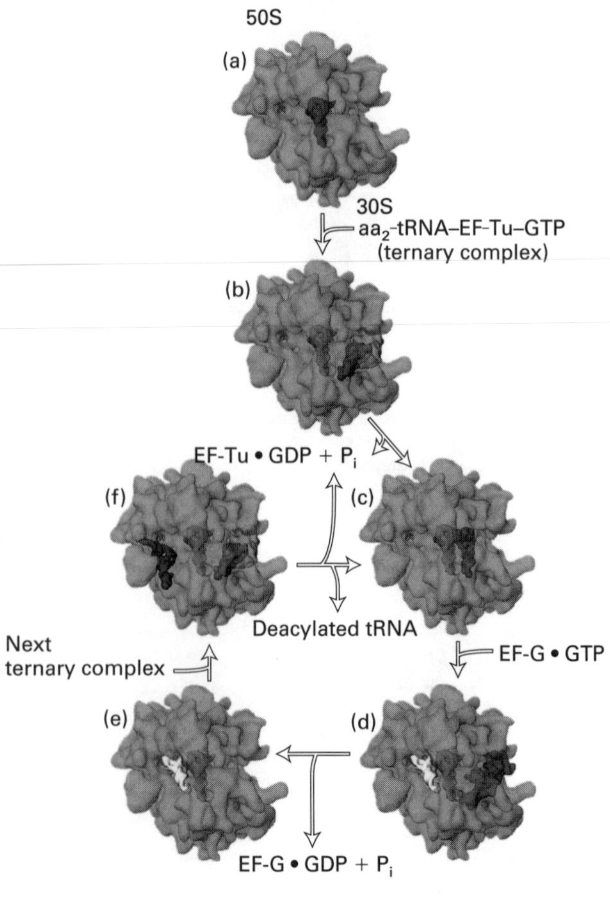

FIGURE 4-39, page 132
The elongation cycle in protein synthesis visualized for *E. coli* ribosomes.

MEDIA CONNECTIONS **Life Cycle of an mRNA**

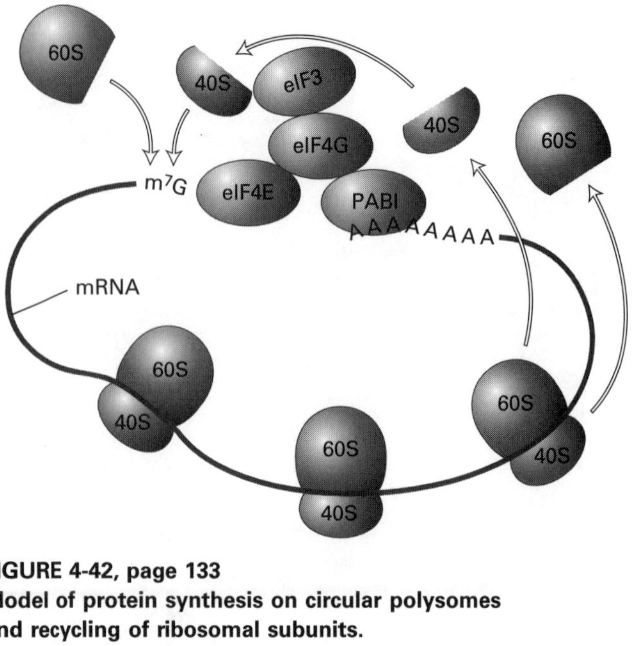

FIGURE 4-42, page 133
Model of protein synthesis on circular polysomes and recycling of ribosomal subunits.

Additional Notes

5

Biomembranes and the Subcellular Organization of Eukaryotic Cells

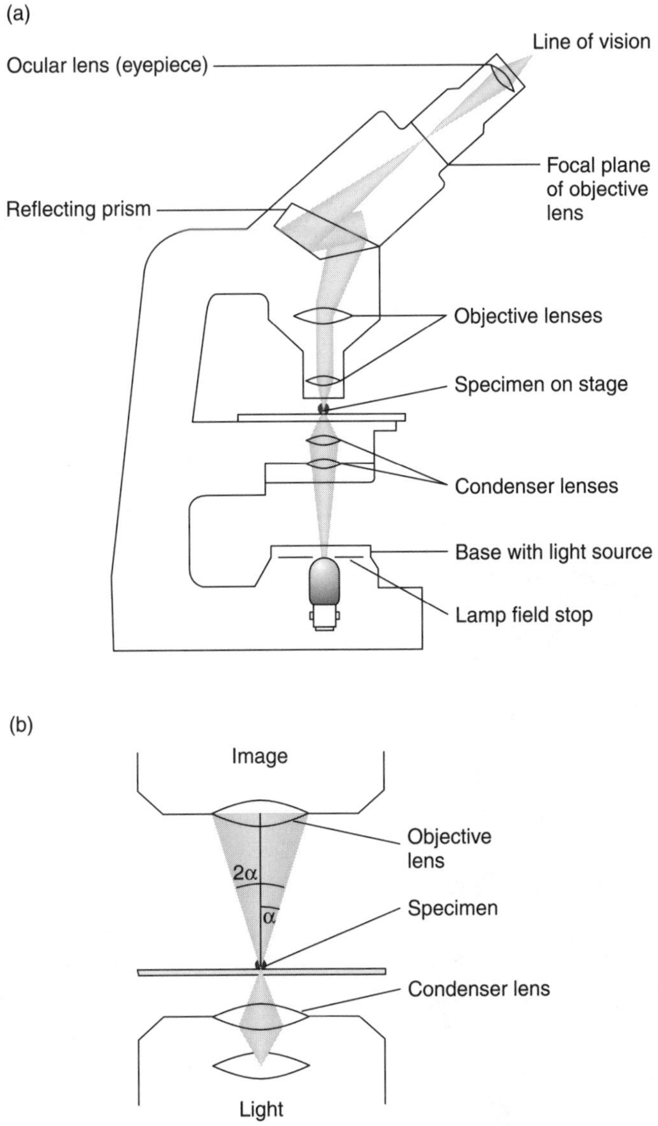

FIGURE 5-2, page 140
The optical pathway in a modern compound optical microscope.

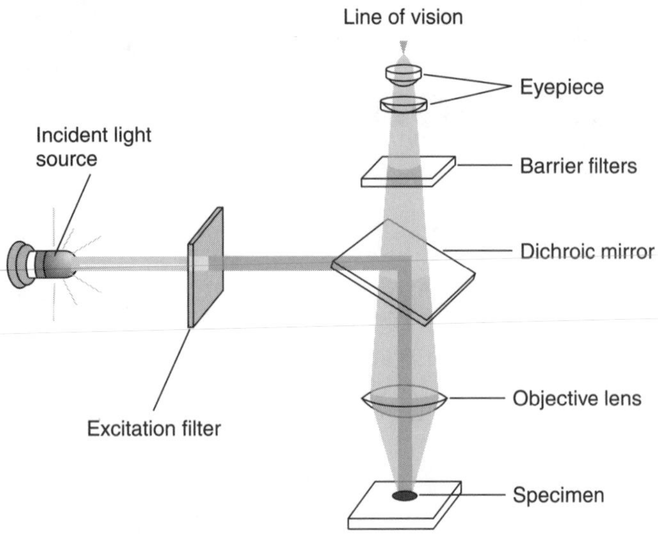

FIGURE 5-5, page 142
The optical pathway in an epi-flourescence microscope.

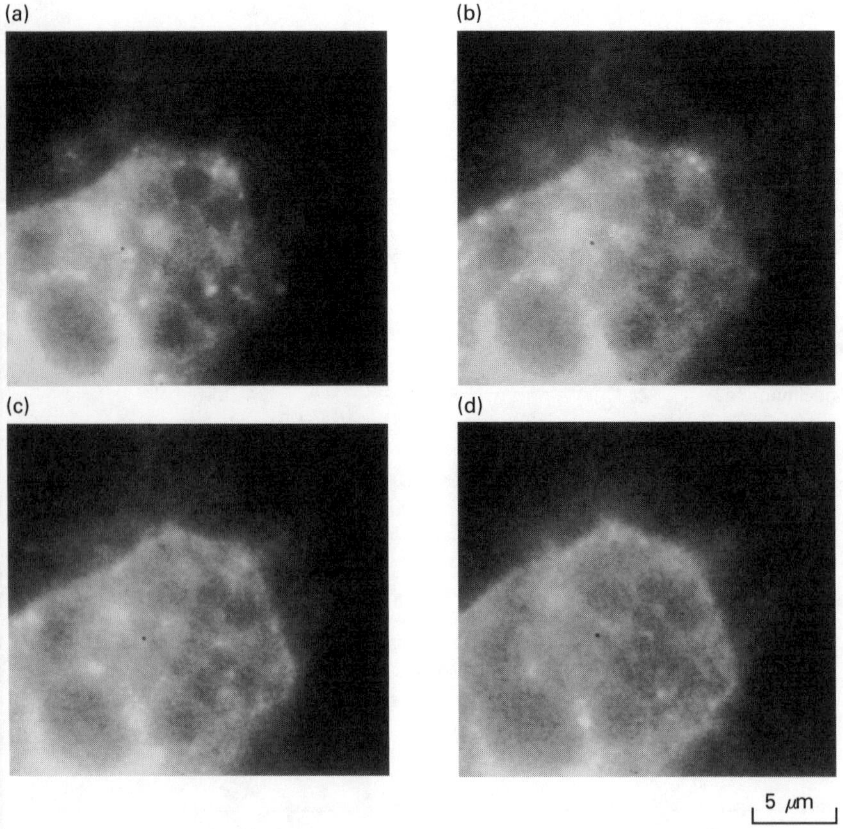

FIGURE 5-7, page 143
Use of green flourescent protein (GFP) to localize GLUT4, a glucose transport protein, within living fat cells.

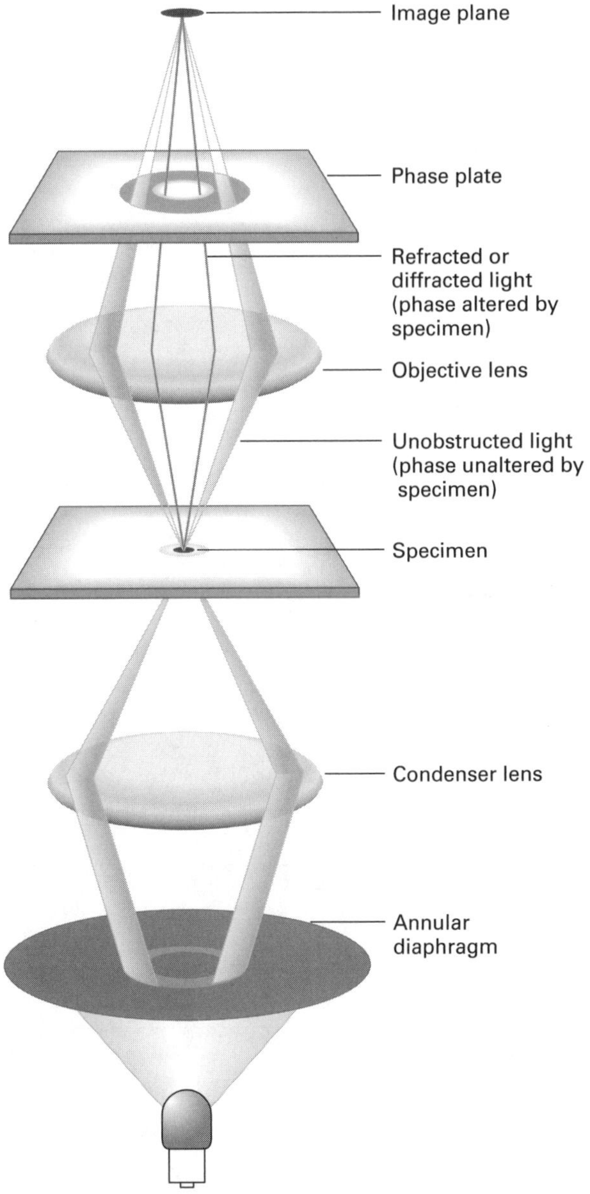

FIGURE 5-12, page 147
The optical pathway of the phase-contrast microscope.

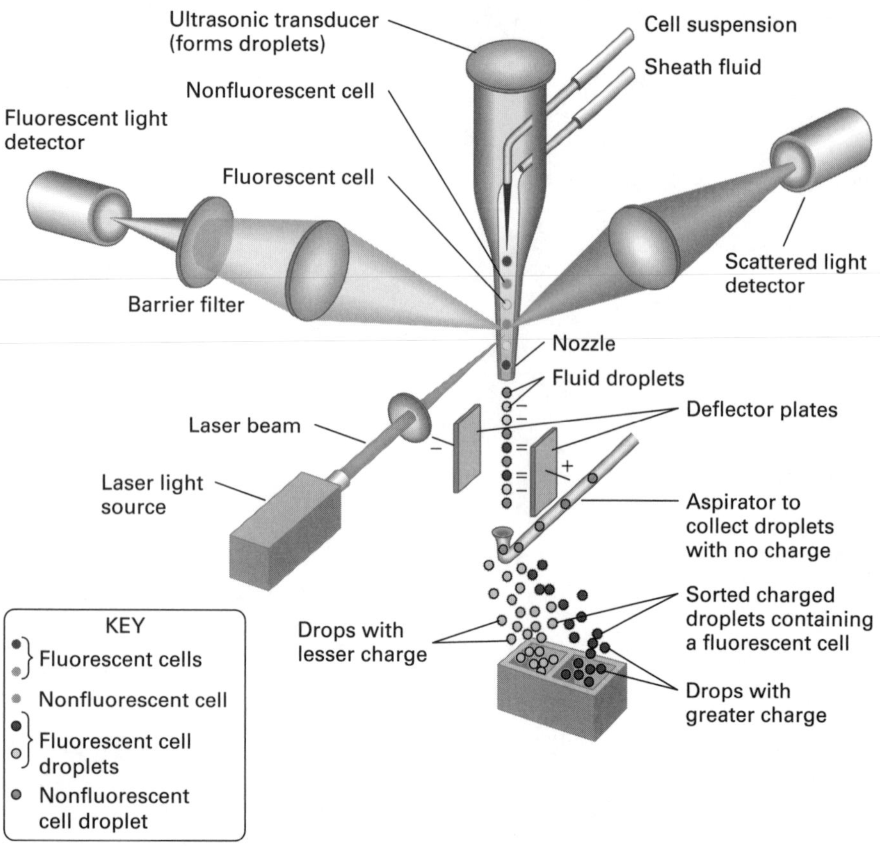

FIGURE 5-21, page 153
Flourescence-activated cell sorter (FACS).

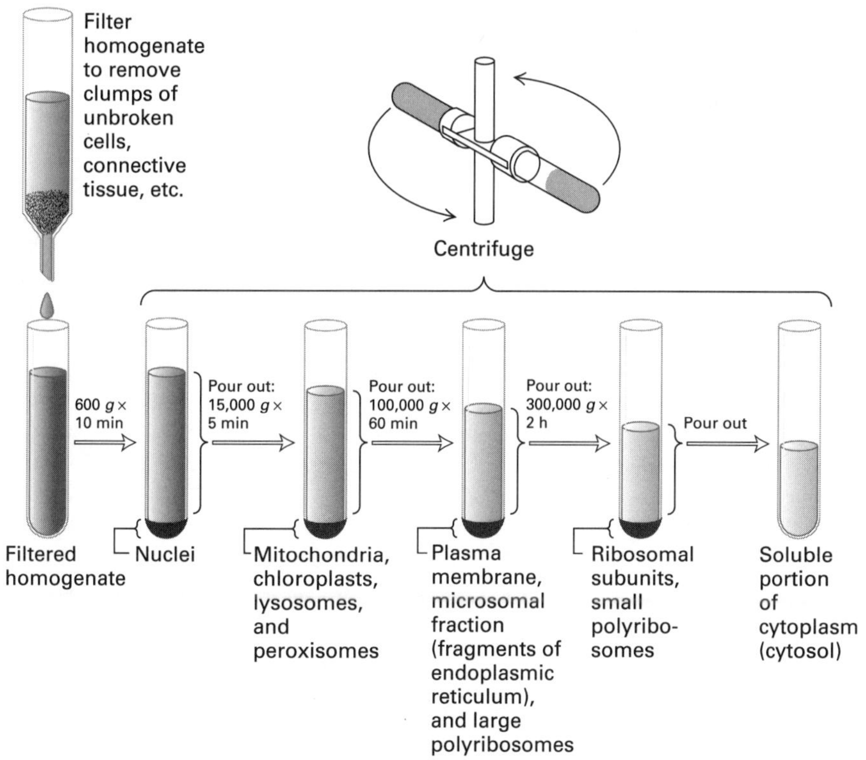

FIGURE 5-23, page 155
Cell fractionation by differential centrifugation.

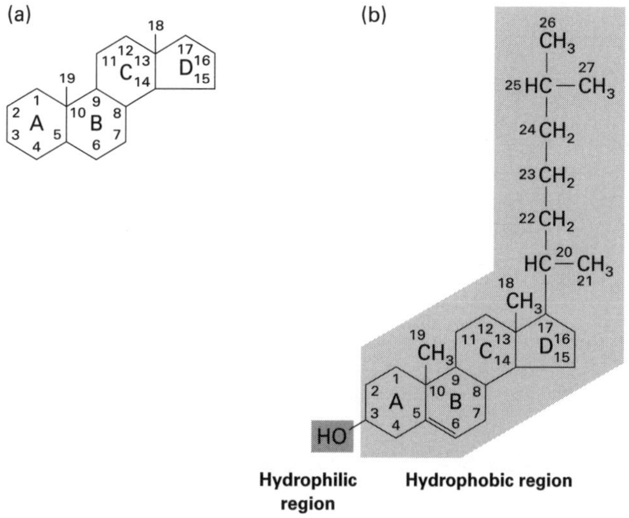

FIGURE 5-29, page 159
(a) The general structure of a steriod and (b) the structure of cholesterol.

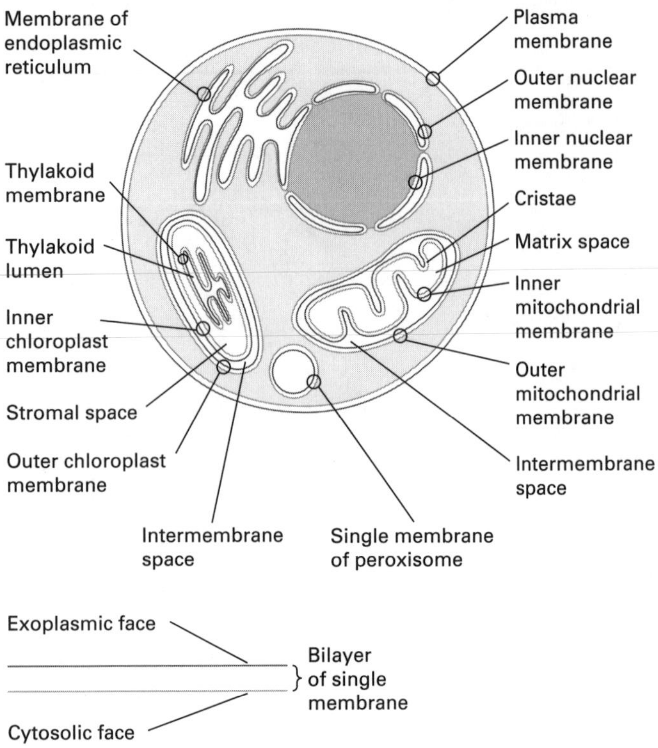

FIGURE 5-31, page 160
Faces of cellular membranes.

(a)
Intercellular space
Plasma membrane
Endoplasmic reticulum
Golgi vesicles
Secretory vesicle
Nuclear membrane
Nucleus
Mitochondrion
2 μm
(b)
Plasma membrane
Cilia
Mitochondrion
Peroxisome
Nucleus
Nucleolus
Rough ER
Smooth ER
Golgi complex
Nuclear pore
Cytoskeletal fibers
Lysosome

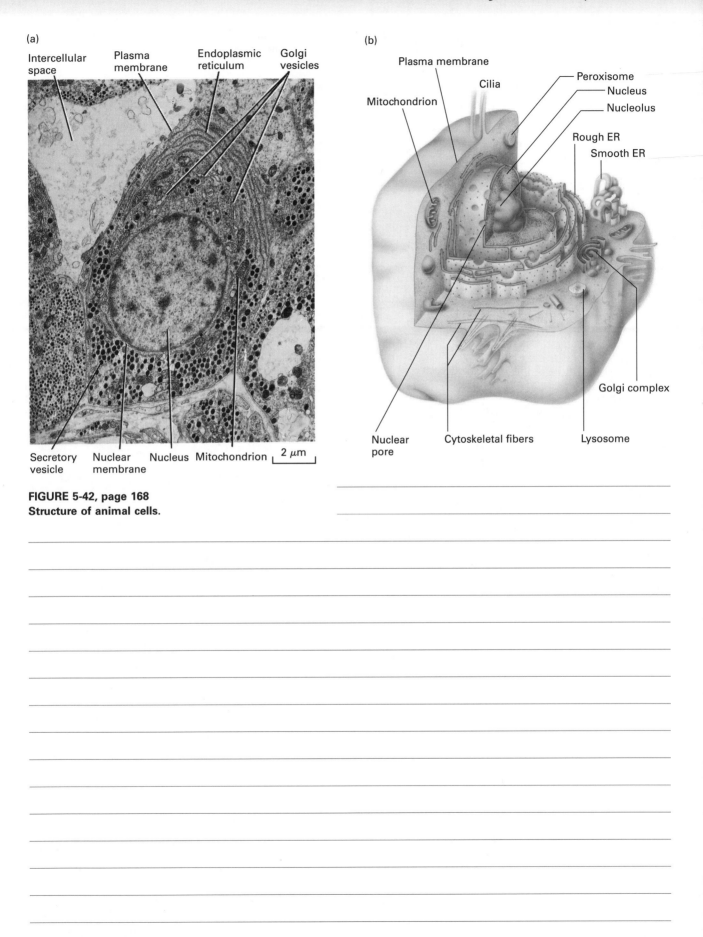

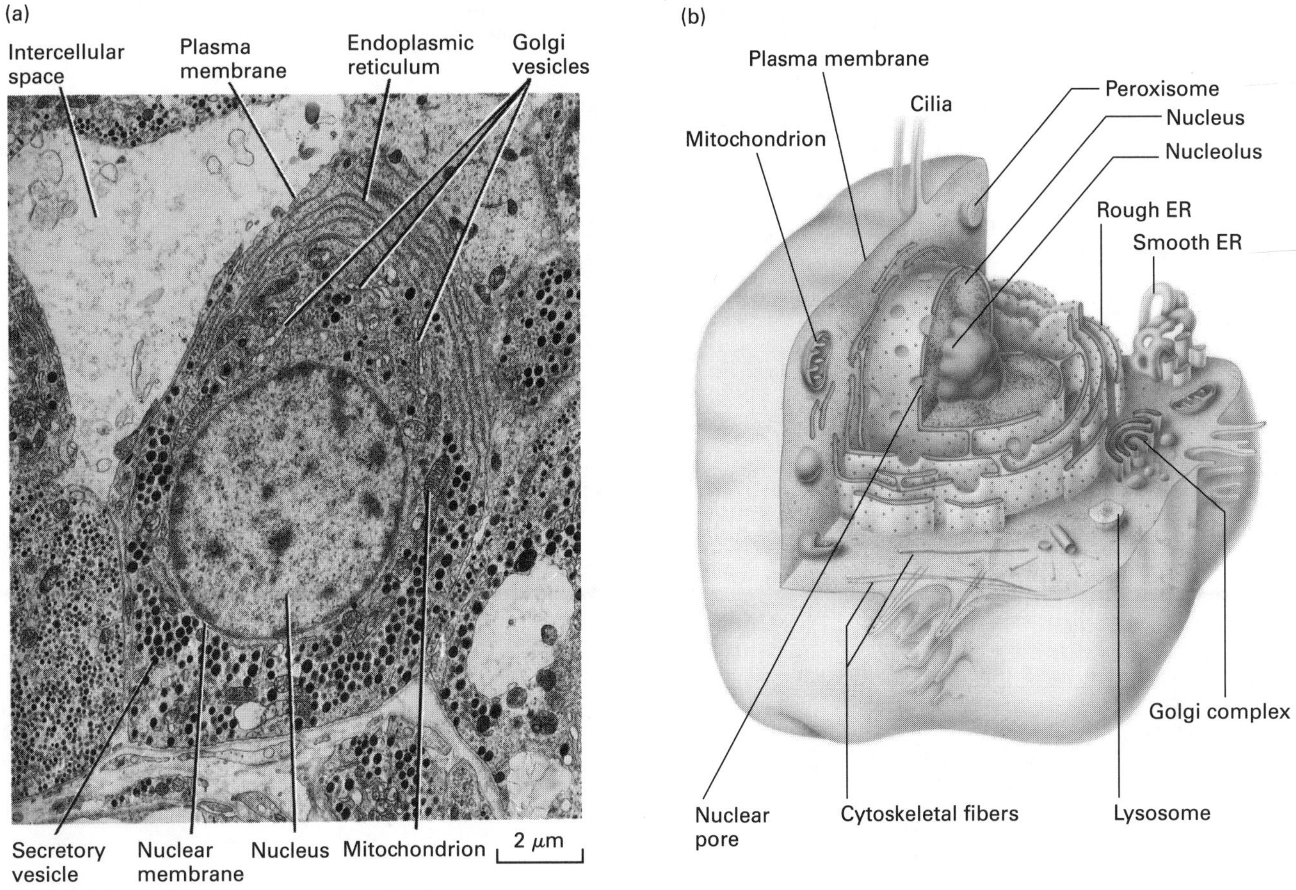

FIGURE 5-42, page 168
Structure of animal cells.

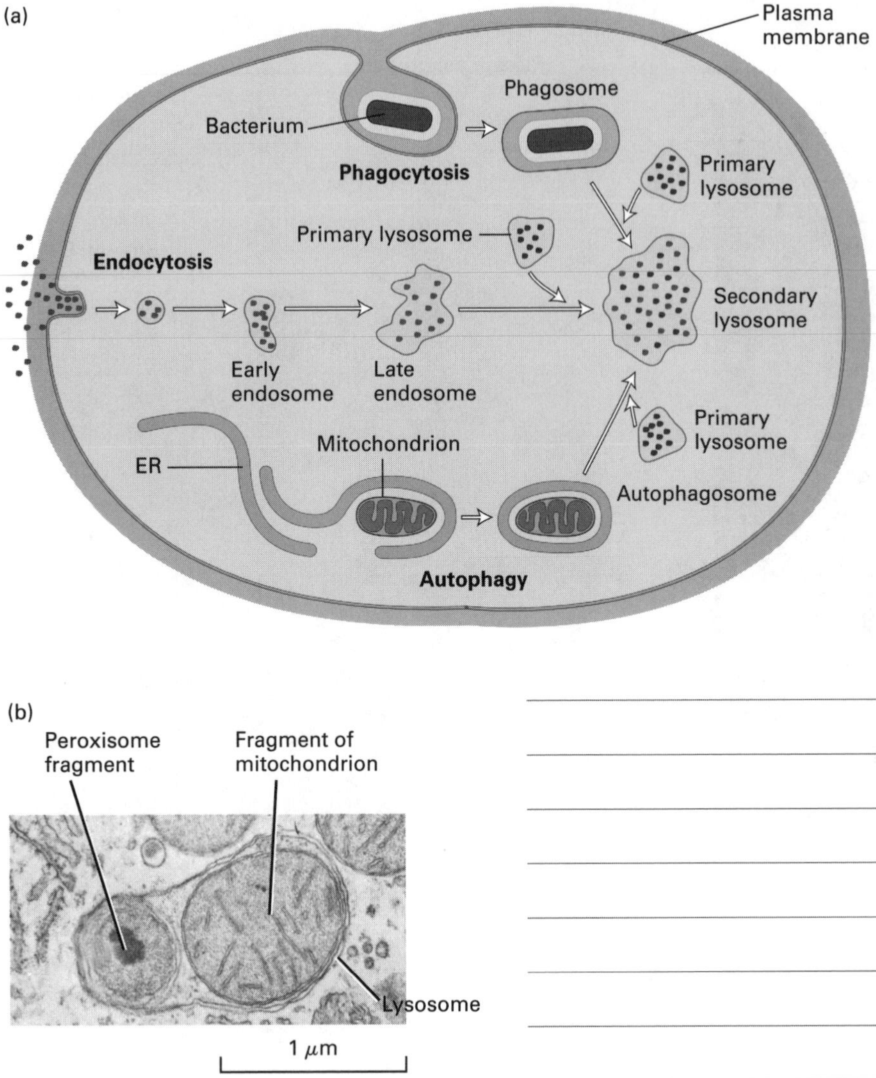
(a)
Plasma membrane
Phagosome
Bacterium
Phagocytosis
Primary lysosome
Primary lysosome
Endocytosis
Secondary lysosome
Early endosome
Late endosome
Primary lysosome
Mitochondrion
ER
Autophagosome
Autophagy
(b)
Peroxisome fragment
Fragment of mitochondrion
Lysosome
1 μm

FIGURE 5-44, page 170
Lysosomal degradation.

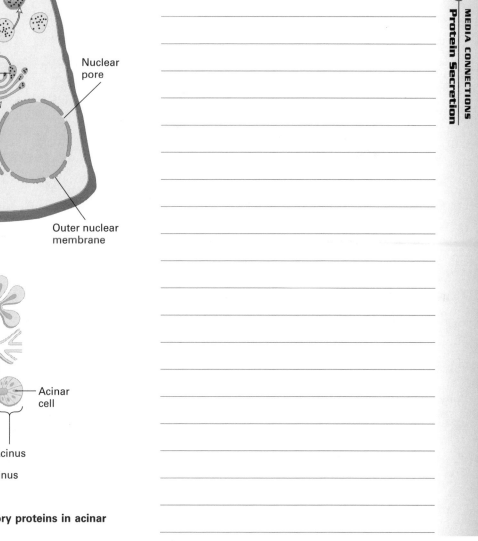

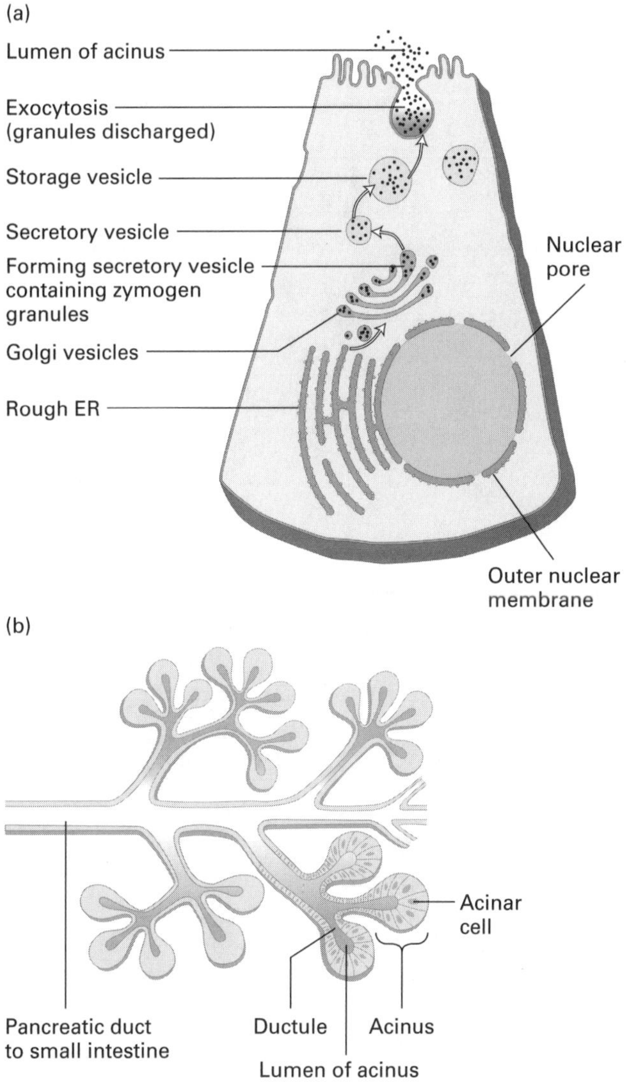

FIGURE 5-48, page 173
The synthesis and release of secretory proteins in acinar cells of the rat pancreas.

MEDIA CONNECTIONS
Protein Secretion

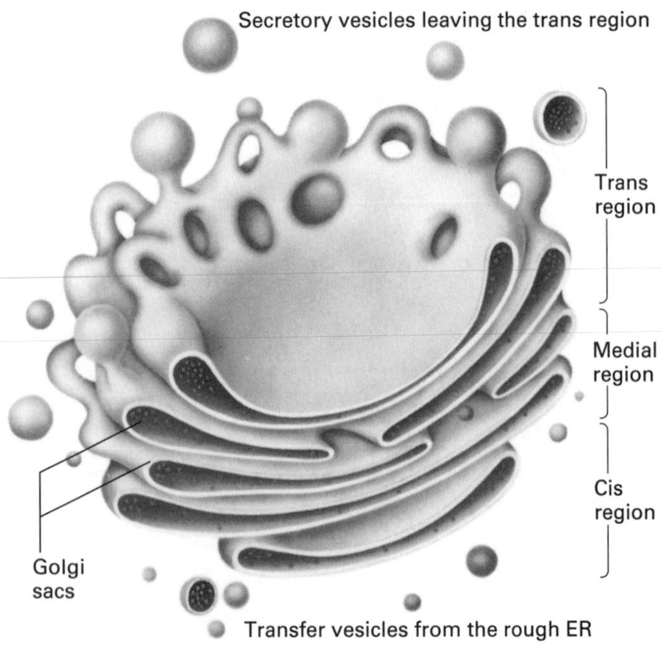

FIGURE 5-49, page 174
Three-dimensional model of the Golgi complex built by analyzing micrographs of serial sections through a secretory cell.

Additional Notes

Manipulating Cells and Viruses in Culture

6

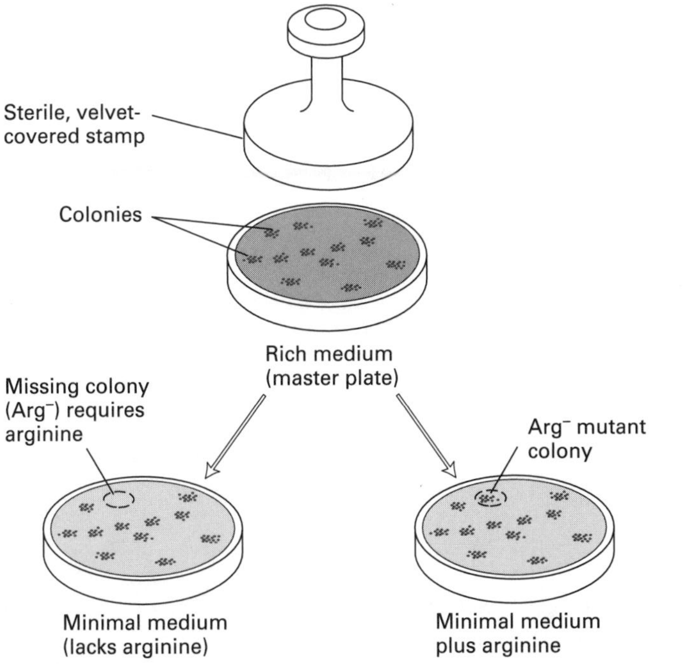

FIGURE 6-2, page 182
Replica plating is used to detect random mutation in bacterial and yeast cultures that produce cells differing in a single genetic trait from other cells in the culture.

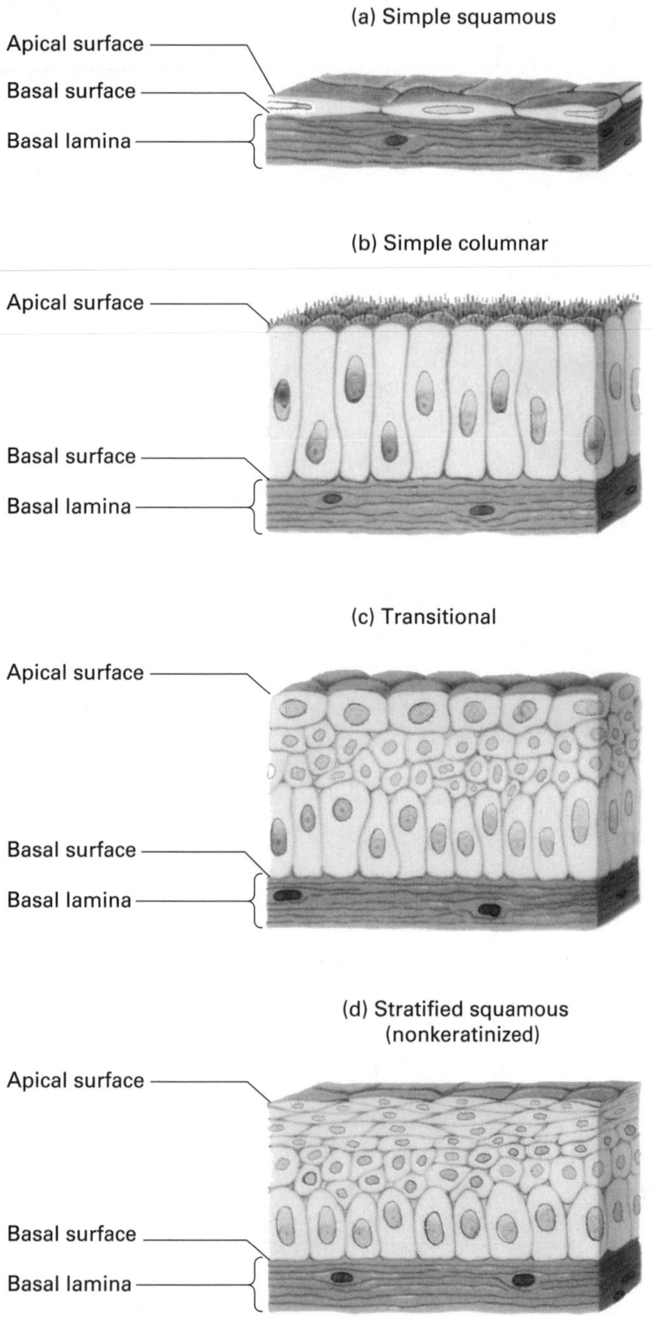

FIGURE 6-4, page 185
Principal types of epithelium.

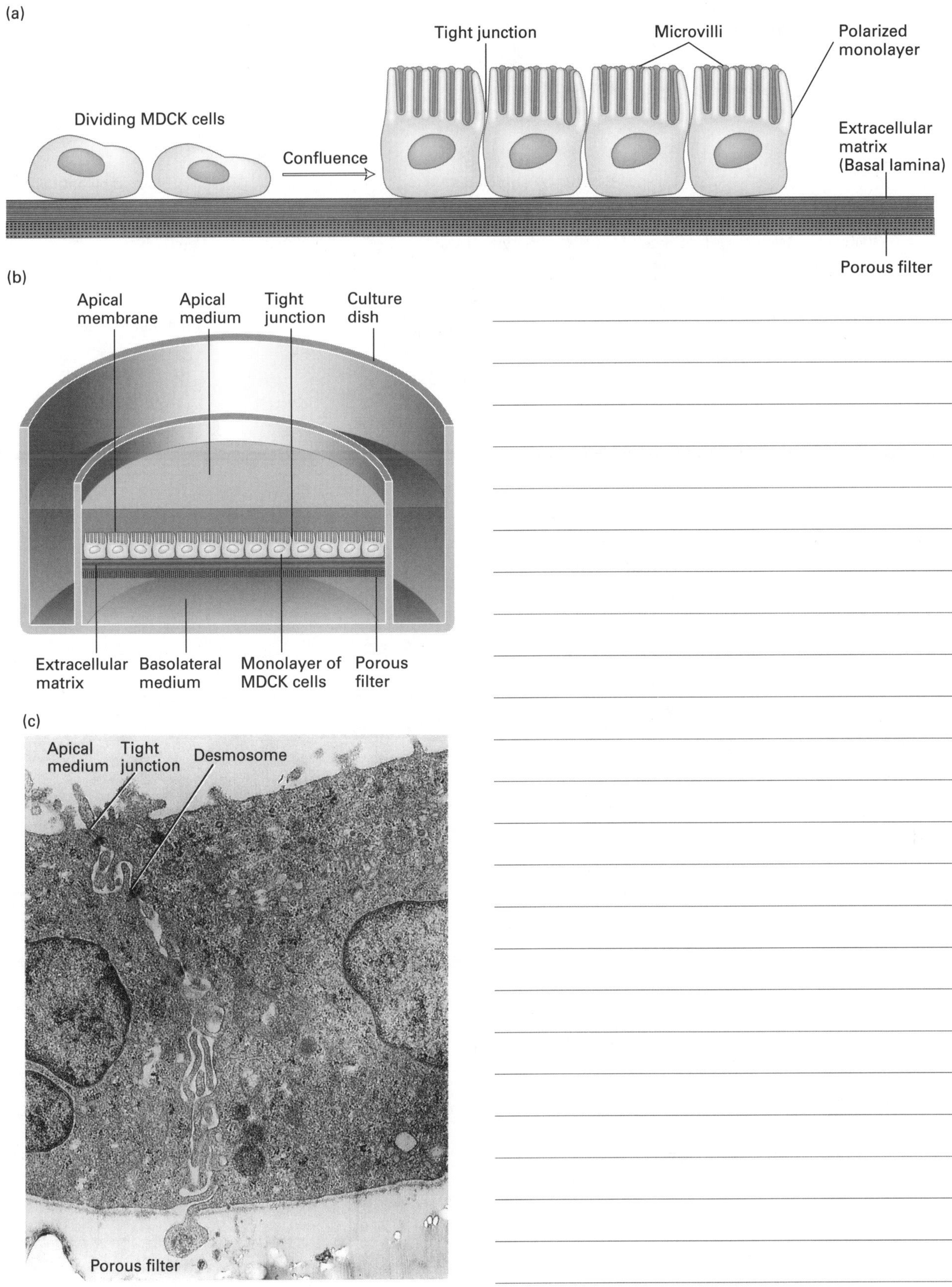

FIGURE 6-7, page 188
Culture of Madin-Darby canine kidney (MDCK) cells, a line of differentiated epithelial cells.

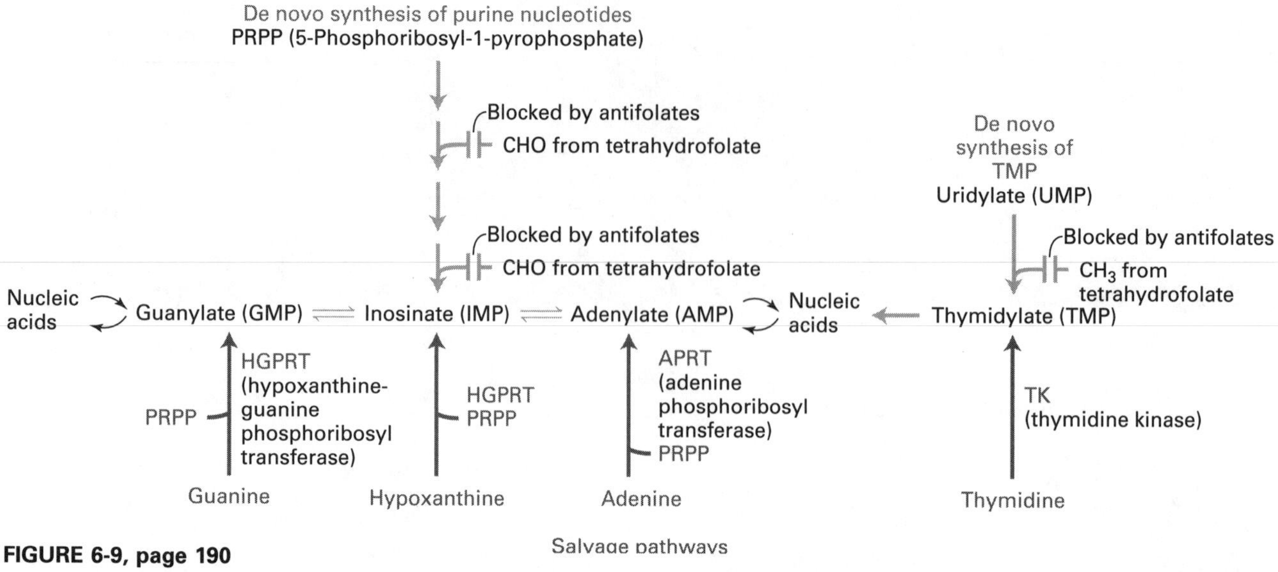

FIGURE 6-9, page 190
De novo and salvage pathways for nucleotide synthesis.

MEDIA CONNECTIONS
Preparing Monoclonal Antibodies

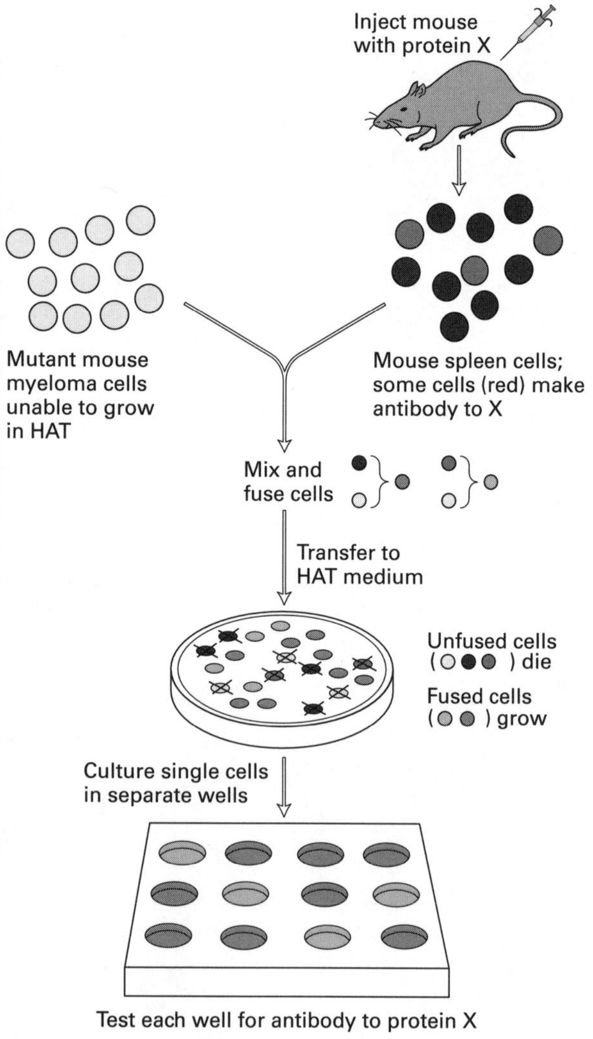

FIGURE 6-10, page 191
Procedure for producing a monoclonal antibody to protein X.

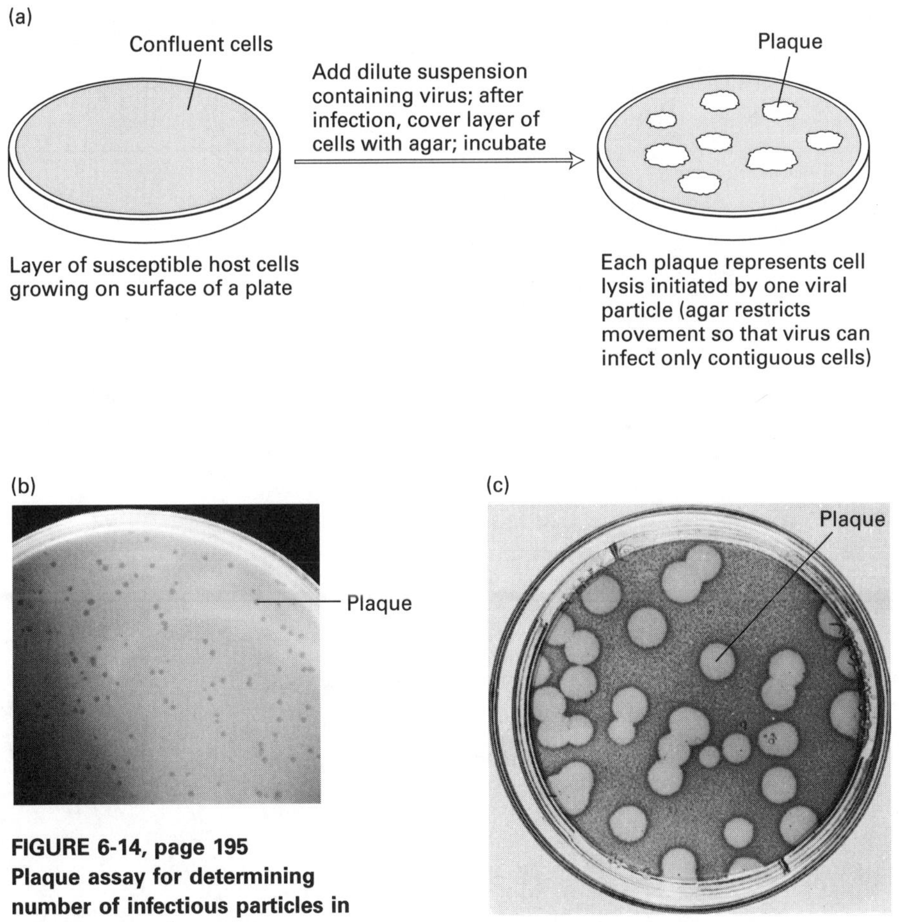
(a)
Confluent cells
Add dilute suspension containing virus; after infection, cover layer of cells with agar; incubate
Plaque
Layer of susceptible host cells growing on surface of a plate
Each plaque represents cell lysis initiated by one viral particle (agar restricts movement so that virus can infect only contiguous cells)
(b)
Plaque
(c)
Plaque

FIGURE 6-14, page 195
Plaque assay for determining number of infectious particles in a viral suspension.

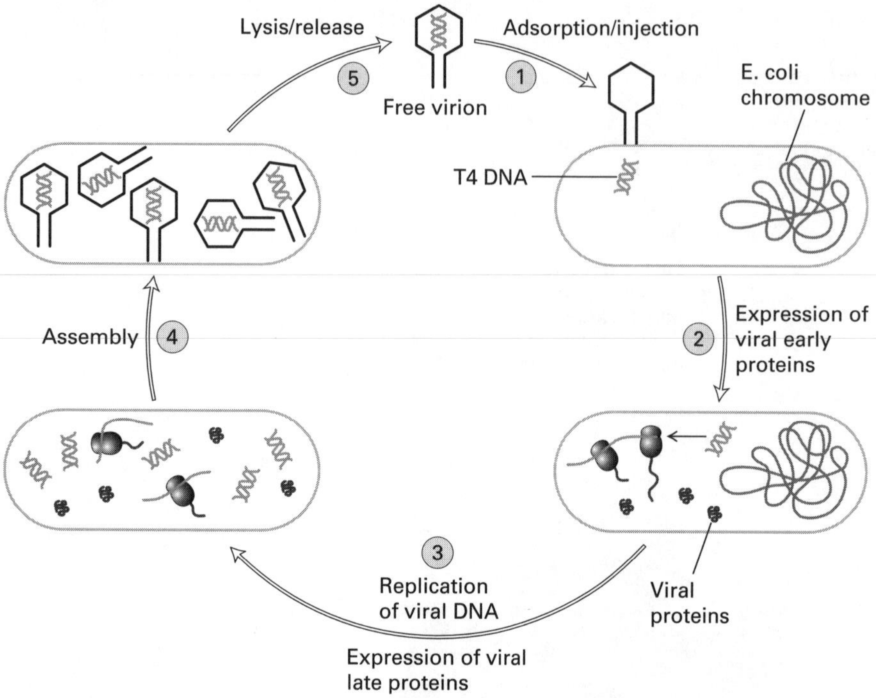

FIGURE 6-16, page 196
The steps in the lytic replication cycle of a nonenveloped virus are illustrated for *E. coli* bacteriophage T4, which has a double-stranded DNA genome.

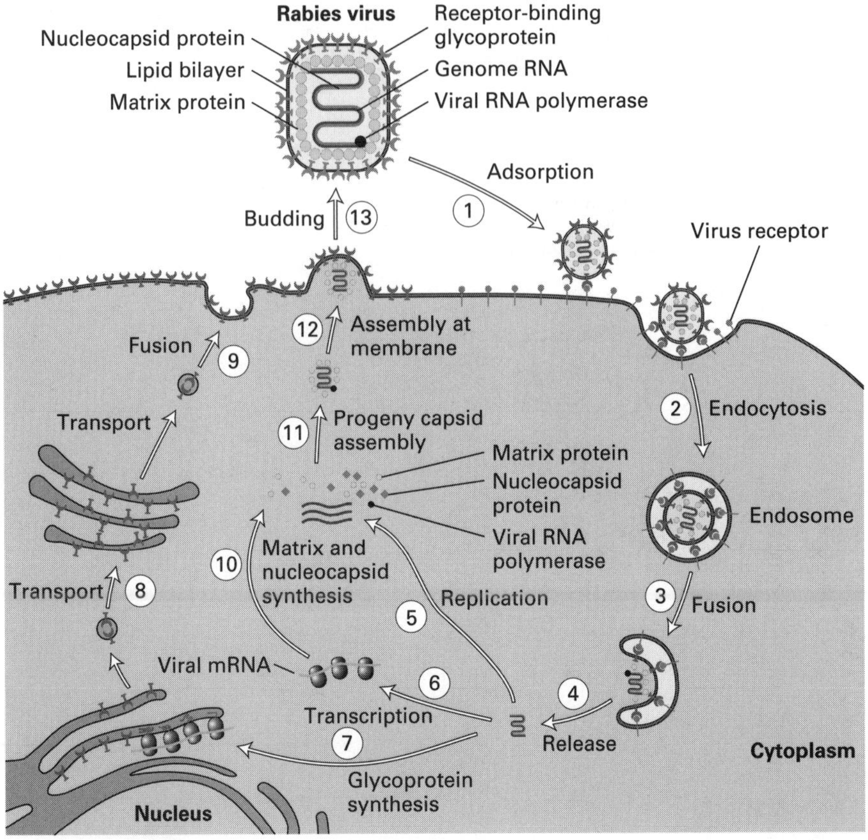

FIGURE 6-17, page 197
The steps in the lytic replication cycle of an enveloped virus are illustrated for rabies virus, which has a single-stranded RNA genome.

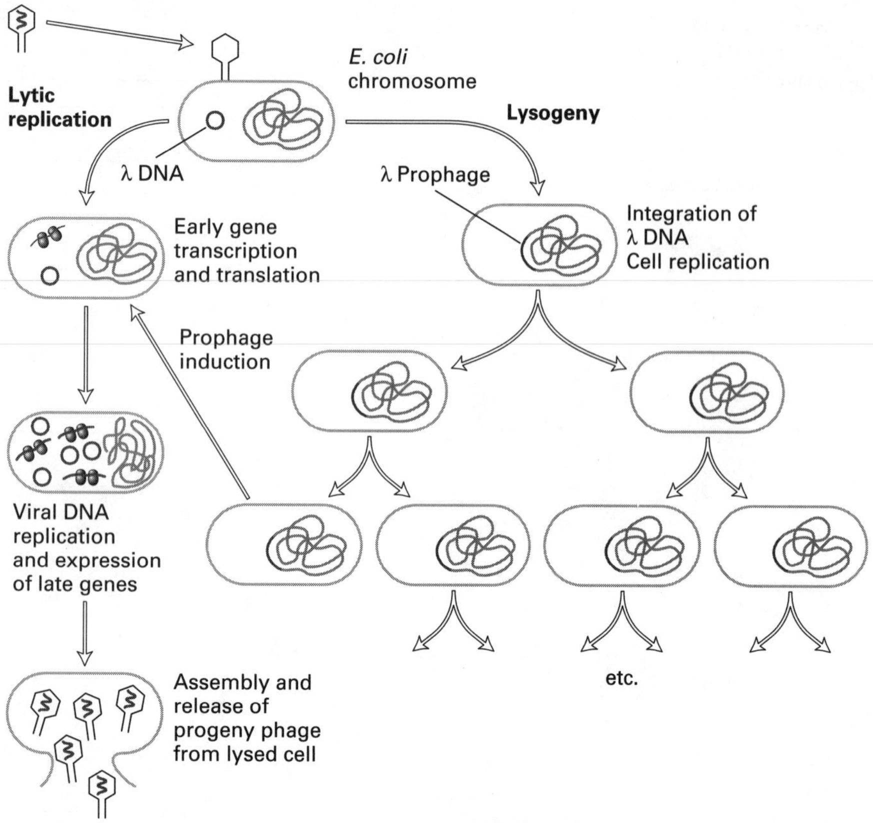

FIGURE 6-19, page 198
λ bacteriophage undergoes either lytic replication or lysogeny following infection of *E. coli.*

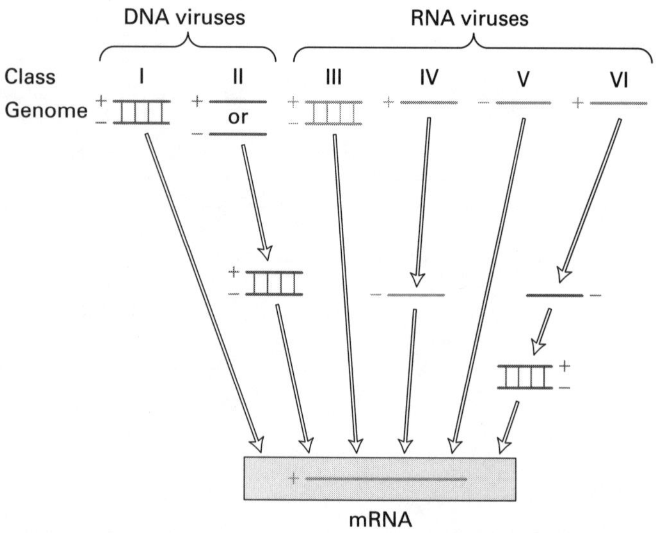

FIGURE 6-20, page 199
Classification of animal viruses based on the composition of their genomes and pathway of mRNA formation.

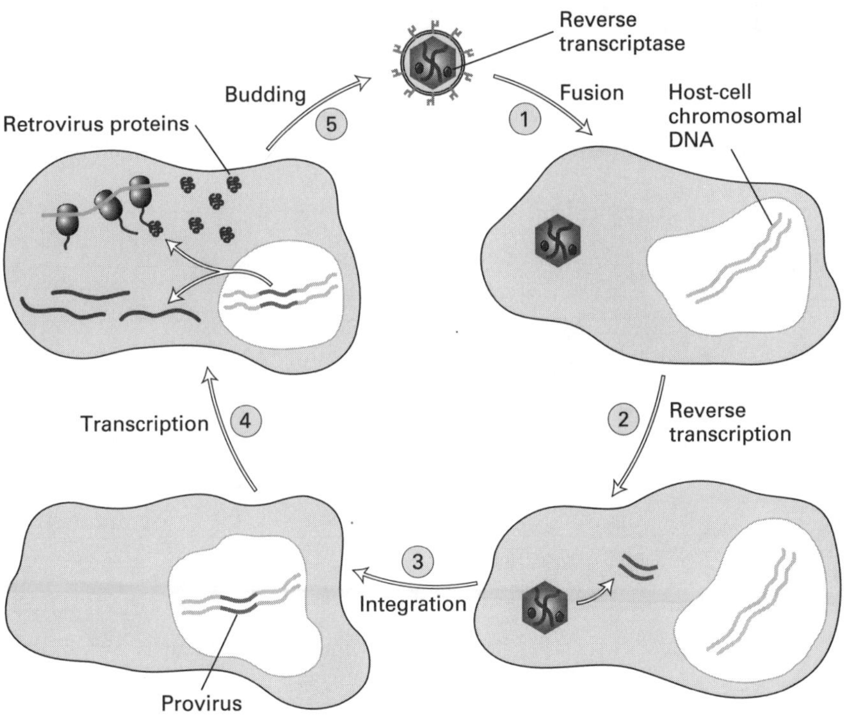

FIGURE 6-22, page 202
Retroviral life cycle.

MEDIA CONNECTIONS
Life Cycle of a Retrovirus

Additional Notes

Recombinant DNA and Genomics

7

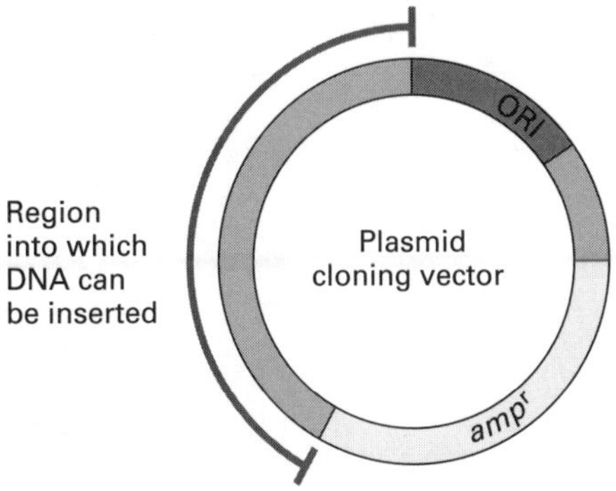

FIGURE 7-1, page 209
Diagram of a simple cloning vector derived from a plasmid, a circular, double-stranded DNA molecule that can replicate within an *E. coli* cell.

MEDIA CONNECTIONS
Plasmid Cloning

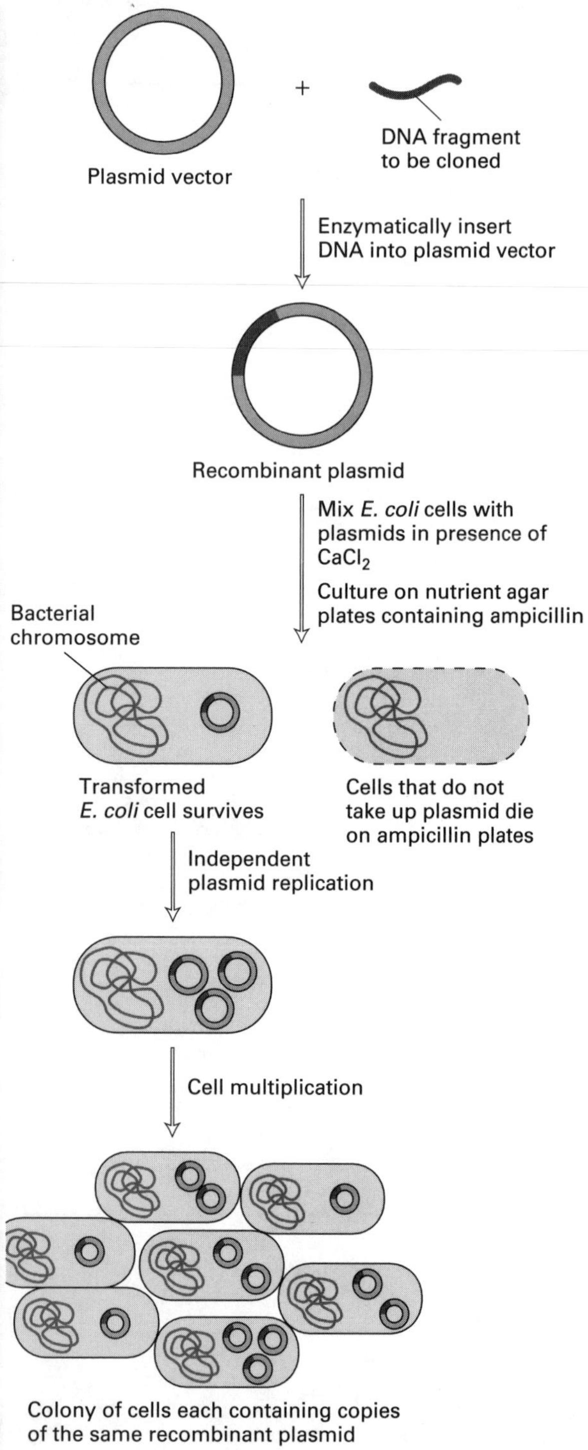
Plasmid vector
+
DNA fragment
to be cloned
Enzymatically insert
DNA into plasmid vector
Recombinant plasmid
Mix E. coli cells with
plasmids in presence of
$CaCl_2$
Culture on nutrient agar
plates containing ampicillin
Bacterial
chromosome
Transformed
E. coli cell survives
Cells that do not
take up plasmid die
on ampicillin plates
Independent
plasmid replication
Cell multiplication
Colony of cells each containing copies
of the same recombinant plasmid

FIGURE 7-3, page 210
General procedure for cloning a DNA fragment in a plasmid vector.

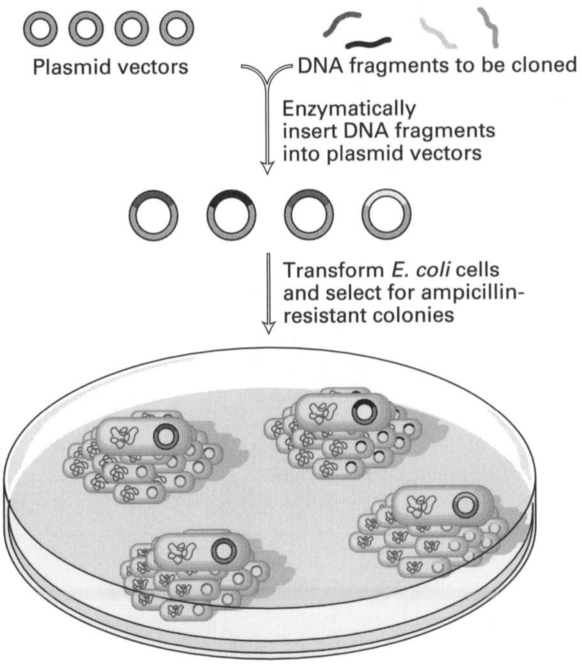

FIGURE 7-4, page 211
Isolation of DNA fragments from a mixture by cloning in a plasmid vector.

MEDIA CONNECTIONS
Plasmid Cloning

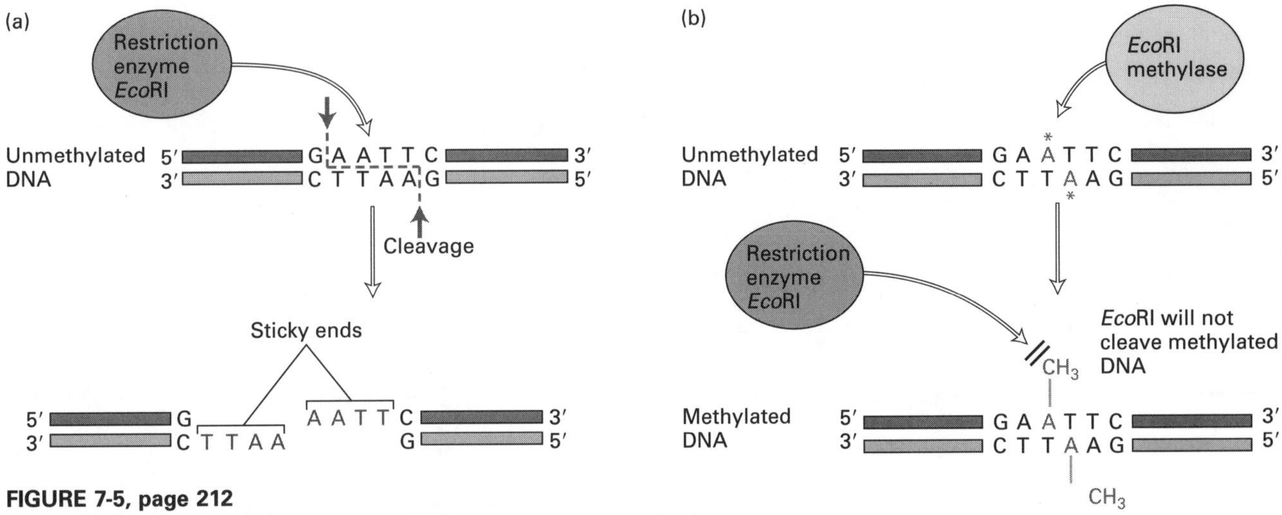

FIGURE 7-5, page 212
Restriction-recognition sites are short DNA sequences recognized and cleaved by various restriction endonucleases.

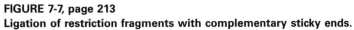

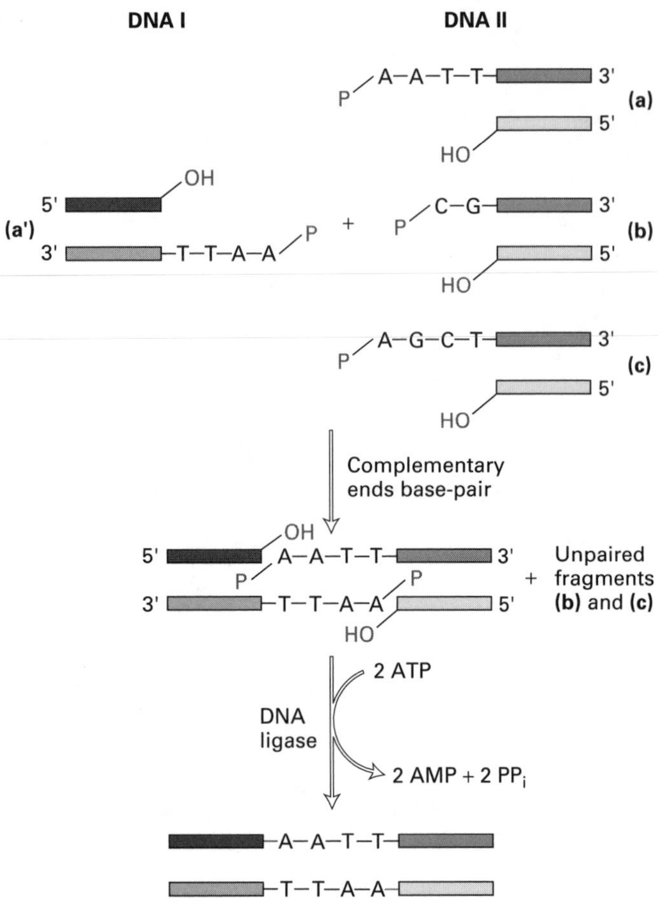

FIGURE 7-7, page 213
Ligation of restriction fragments with complementary sticky ends.

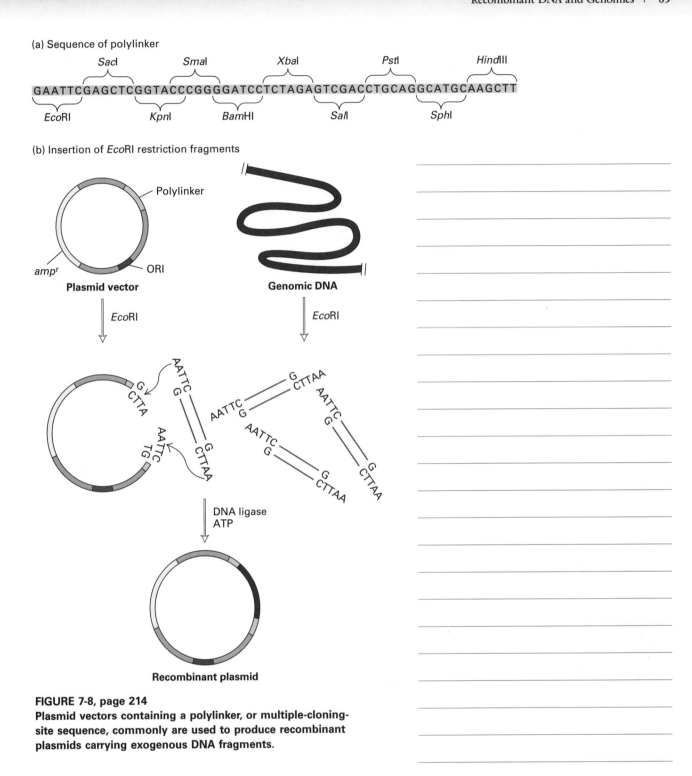

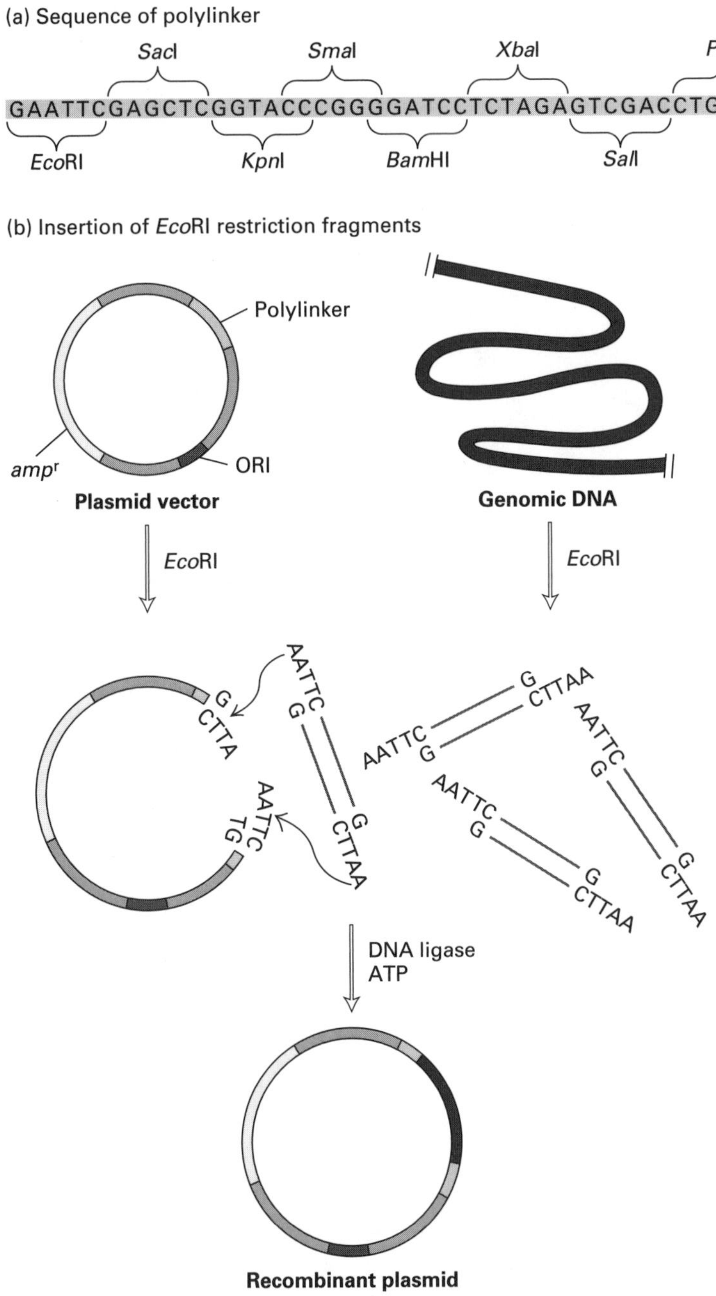

FIGURE 7-8, page 214
Plasmid vectors containing a polylinker, or multiple-cloning-site sequence, commonly are used to produce recombinant plasmids carrying exogenous DNA fragments.

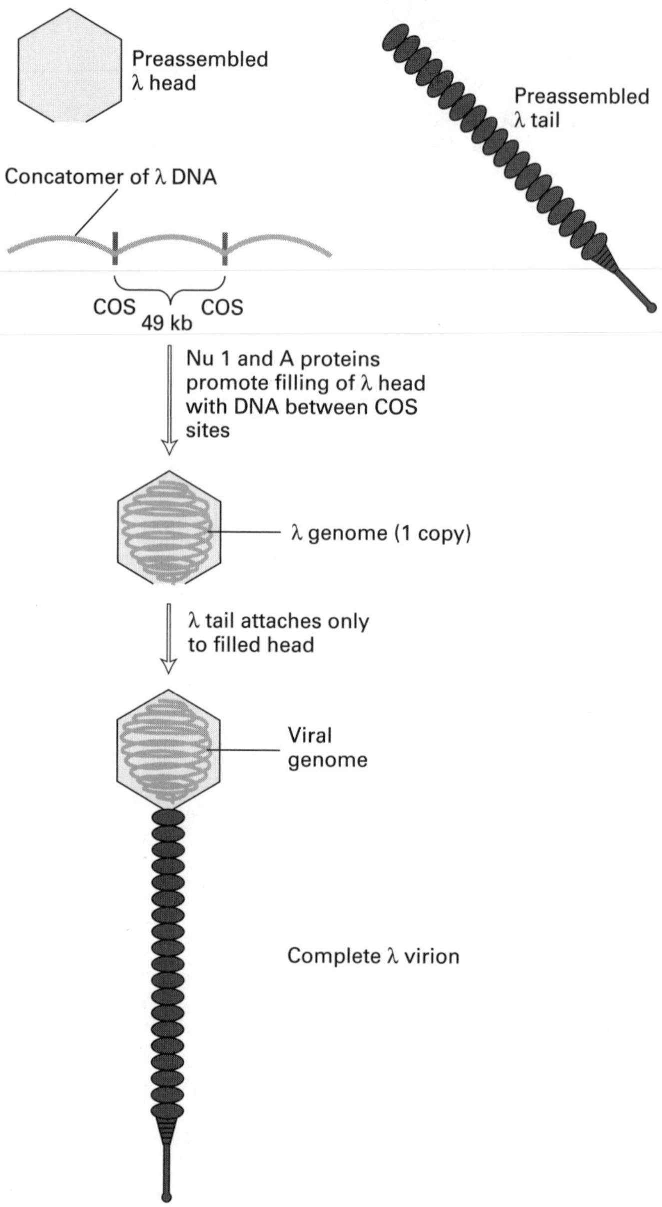

FIGURE 7-11, page 217
Assembly of bacteriophage λ virions.

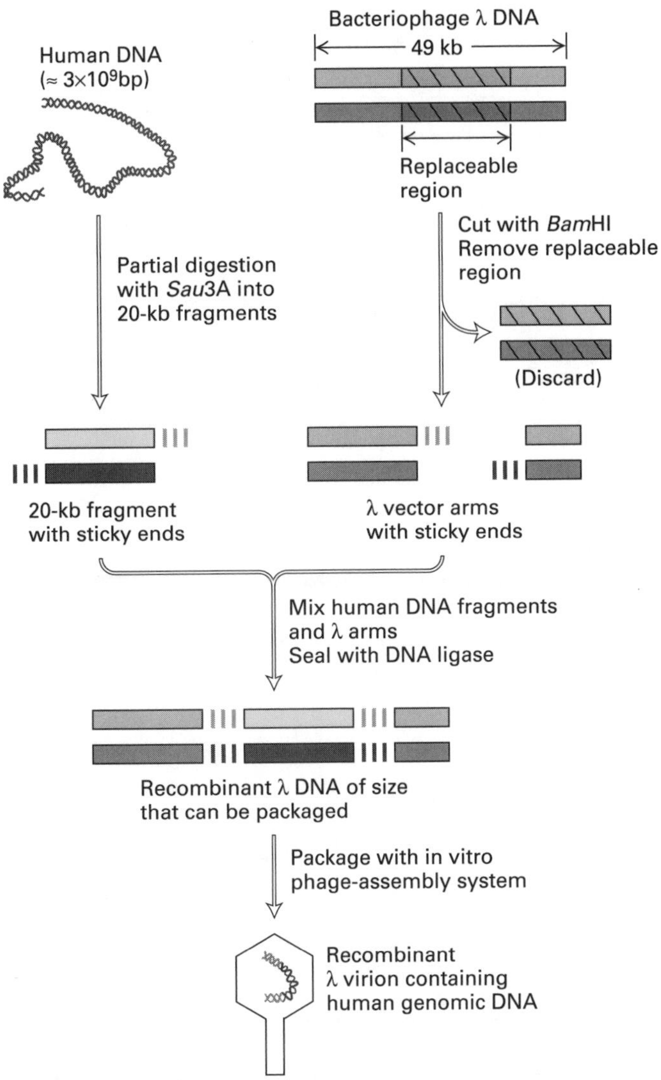

FIGURE 7-12, page 218
Construction of a genomic library of human DNA in a bacteriophage λ vector.

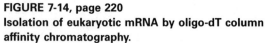

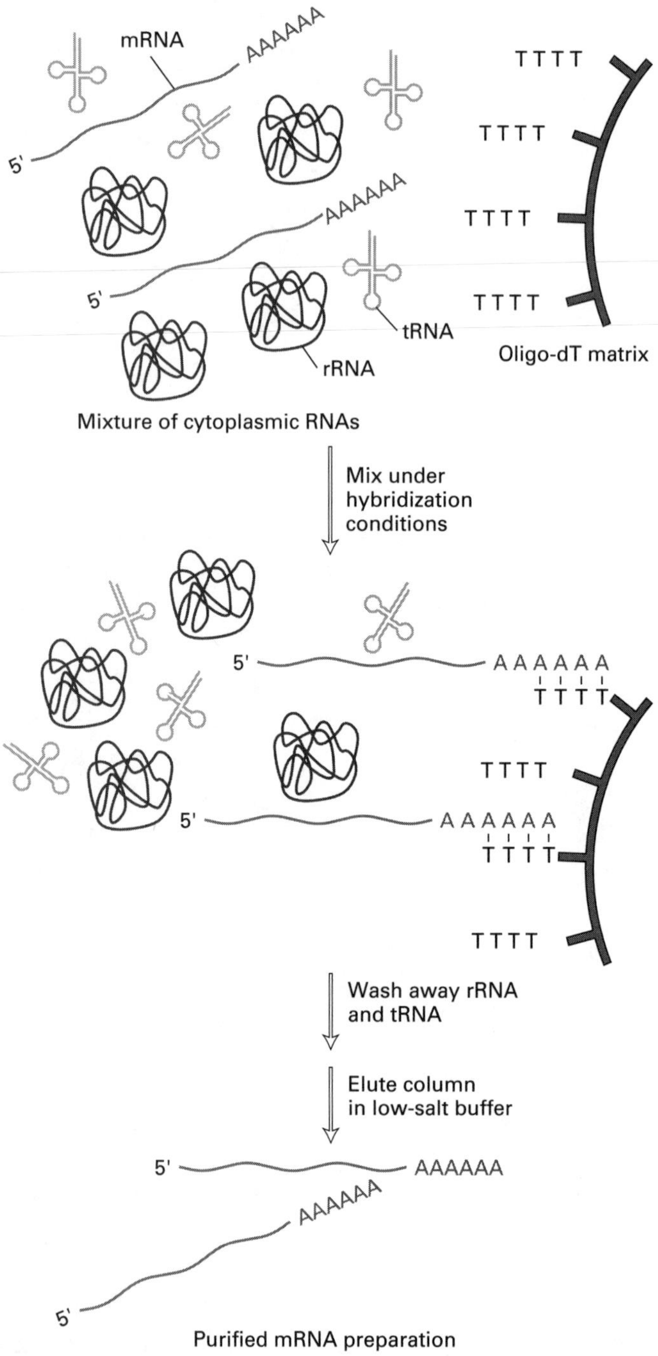

FIGURE 7-14, page 220
Isolation of eukaryotic mRNA by oligo-dT column affinity chromatography.

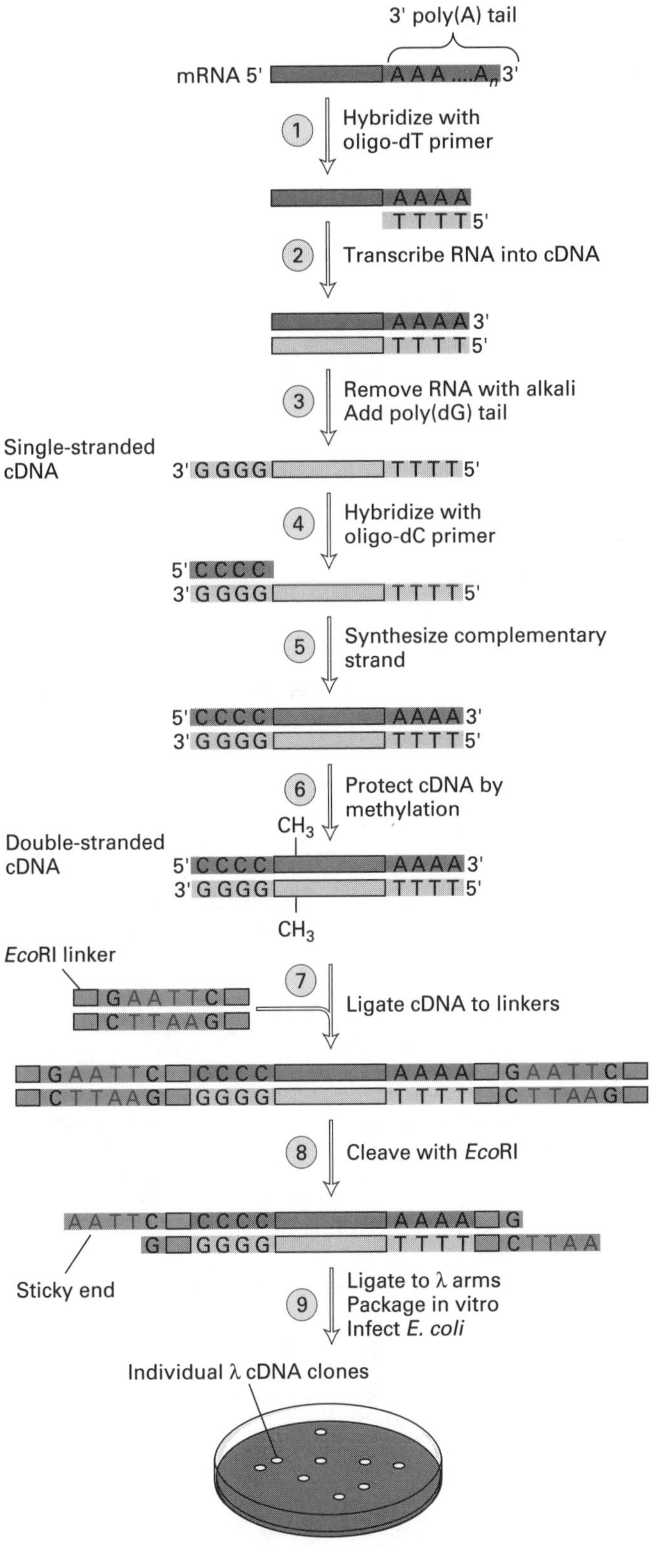

FIGURE 7-15, page 221
Preparation of a bacteriophage λ cDNA library.

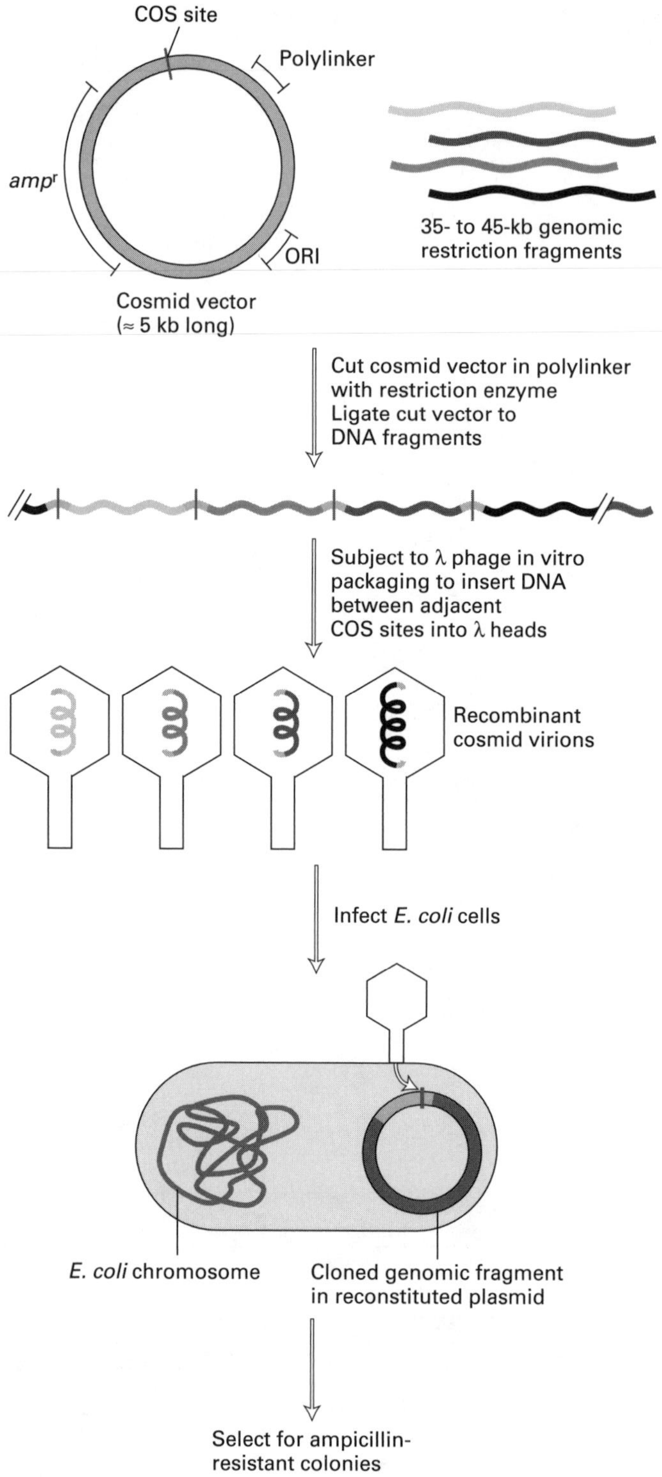
COS site
Polylinker
amp^r
ORI
Cosmid vector
(≈ 5 kb long)
35- to 45-kb genomic
restriction fragments
Cut cosmid vector in polylinker
with restriction enzyme
Ligate cut vector to
DNA fragments
Subject to λ phage in vitro
packaging to insert DNA
between adjacent
COS sites into λ heads
Recombinant
cosmid virions
Infect E. coli cells
E. coli chromosome
Cloned genomic fragment
in reconstituted plasmid
Select for ampicillin-
resistant colonies

FIGURE 7-16, page 222
General procedure for cloning DNA fragments in cosmid vectors.

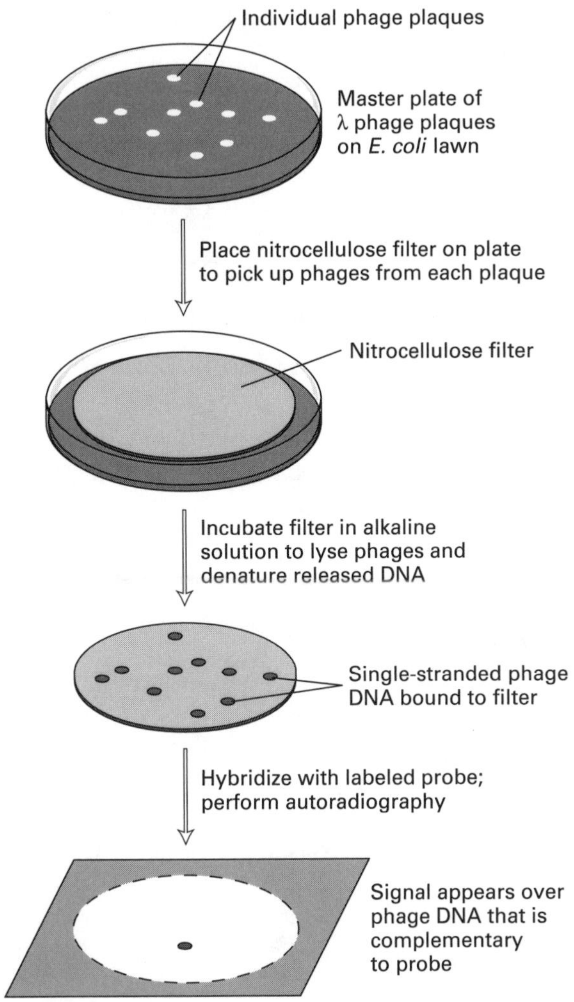
Individual phage plaques
Master plate of λ phage plaques on E. coli lawn
Place nitrocellulose filter on plate to pick up phages from each plaque
Nitrocellulose filter
Incubate filter in alkaline solution to lyse phages and denature released DNA
Single-stranded phage DNA bound to filter
Hybridize with labeled probe; perform autoradiography
Signal appears over phage DNA that is complementary to probe

FIGURE 7-18, page 225
Identification of a specific clone from a λ phage library by membrane hybridization to a radiolabeled probe.

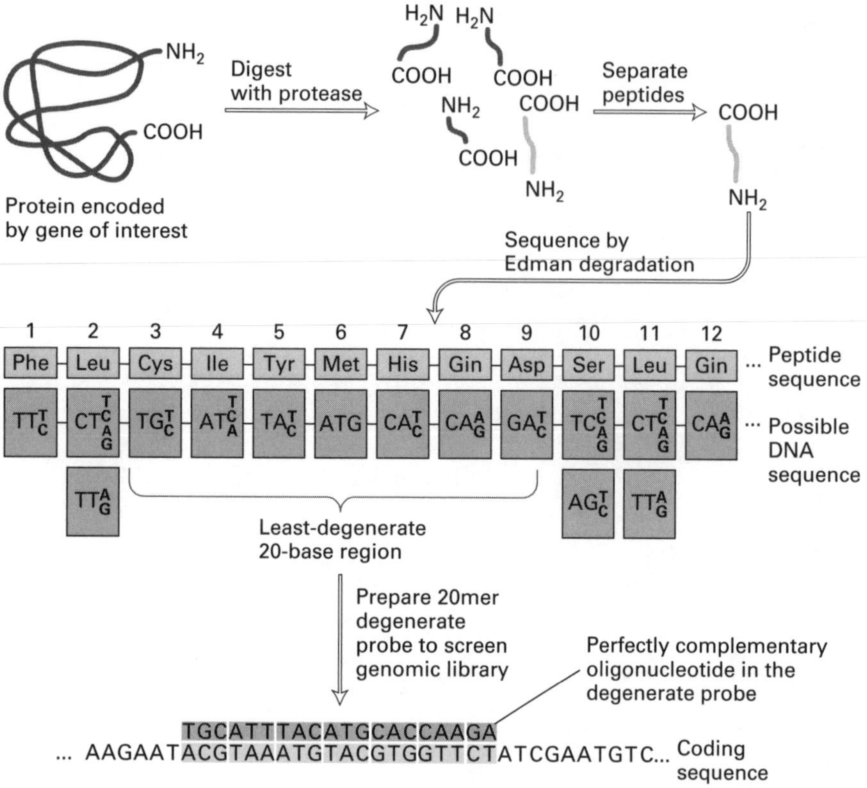

FIGURE 7-19, page 226
Designing oligonucleotide probes based on protein sequence.

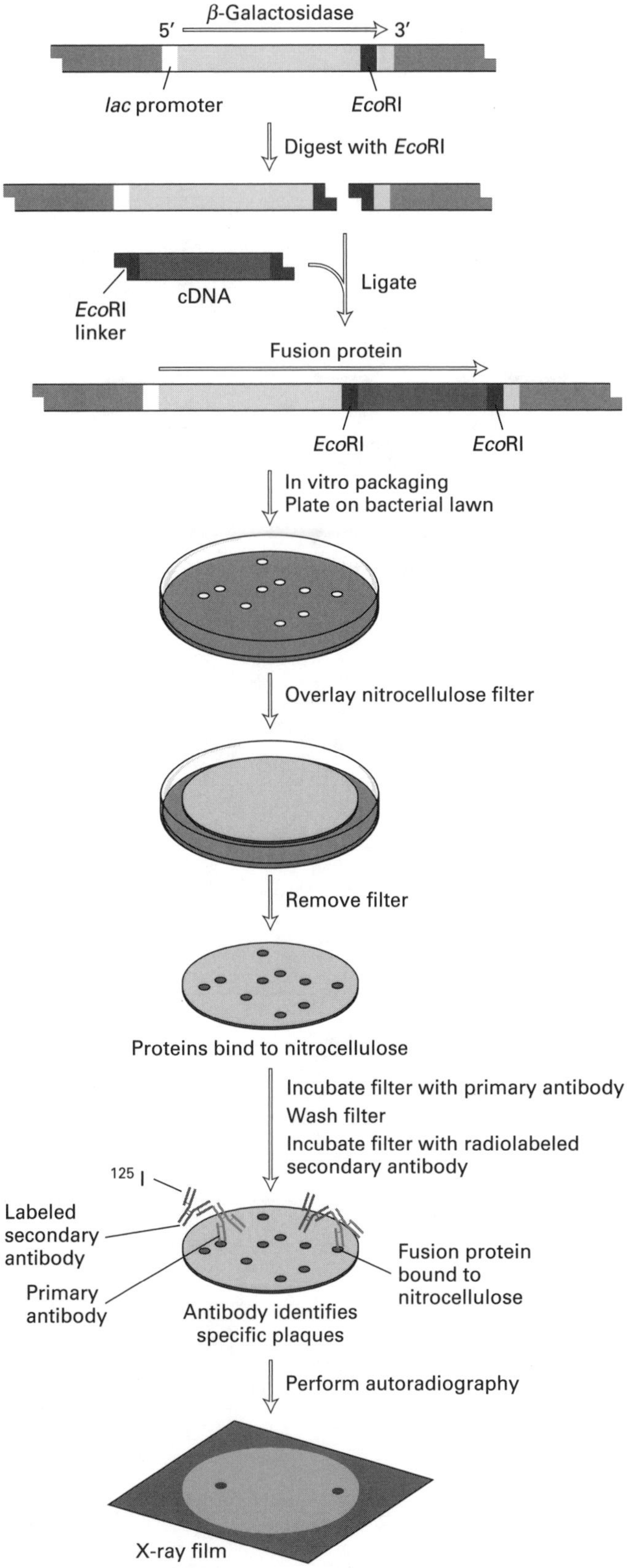

FIGURE 7-21, page 228
Use of λ expression cloning to identify a cloned DNA based on binding of the encoded protein to a specific antibody.

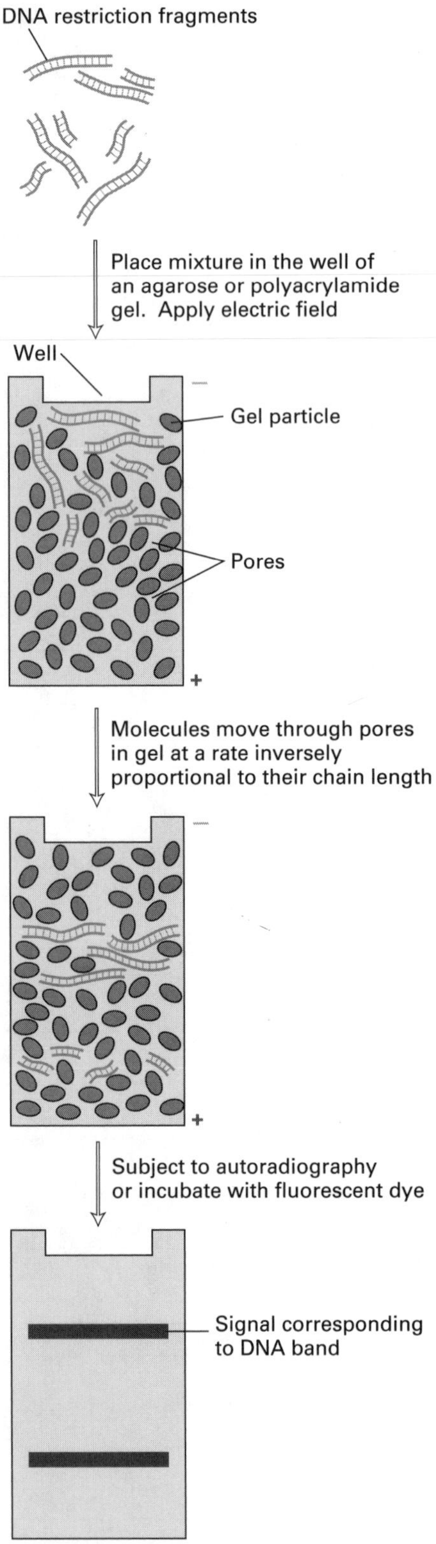

FIGURE 7-22, page 229
Separation of DNA fragments of different lengths by gel electrophoresis.

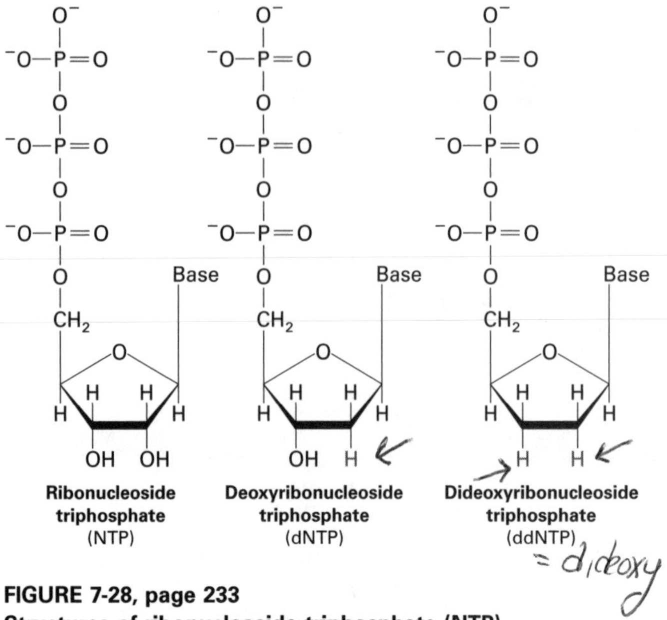

FIGURE 7-28, page 233
Structures of ribonucleoside triphosphate (NTP), deoxyribonucleoside triphosphate (dNTP), and dideoxyribonucleoside triphosphate (ddNTP).

(a)

This allow termination every once in a while

5'
3' ————— 5'

DNA polymerase
+ dNTPs (100 µM)

then

Stop at A   *Stop at G*

+ Dideoxy A   + Dideoxy G   + Dideoxy T   + Dideoxy C
(1 µM)   (1 µM)   (1 µM)   (1 µM)
                                                                   and less [ ]

●——A   ●——G   ●——T   ●——C

●——A   ●——G   ●——T   ●——C

●——A   ●——G   ●——T   ●——C

etc.   etc.   etc.   etc.

⟵——Denature and separate by electrophoresis——⟶

place at different As

(b)

radioactive phosphate to visualize

primer

5' ³²P-TAGCTGACTC 3'
3'   ATCGACTGAGTCAAGAACTATTGGGCTTAA   *Template*

DNA polymerase
+ dATP, dGTP, dCTP, dTTP   *normal nucleotide*
+ ddGTP in low concentration
5' Nucleotide missing 3' OH

5' ³²P-TAGCTGACTCAG 3'
3'   ATCGACTGAGTCAAGAACTATTGGGCTTAA...
              +
5' ³²P-TAGCTGACTCAGTTCTCG 3'
3'   ATCGACTGAGTCAAGAACTATTGGGCTTAA...
              +
5' ³²P-TAGCTGACTCAGTTCTCGATAACCCG 3'
3'   ATCGACTGAGTCAAGAACTATTGGGCTTAA...

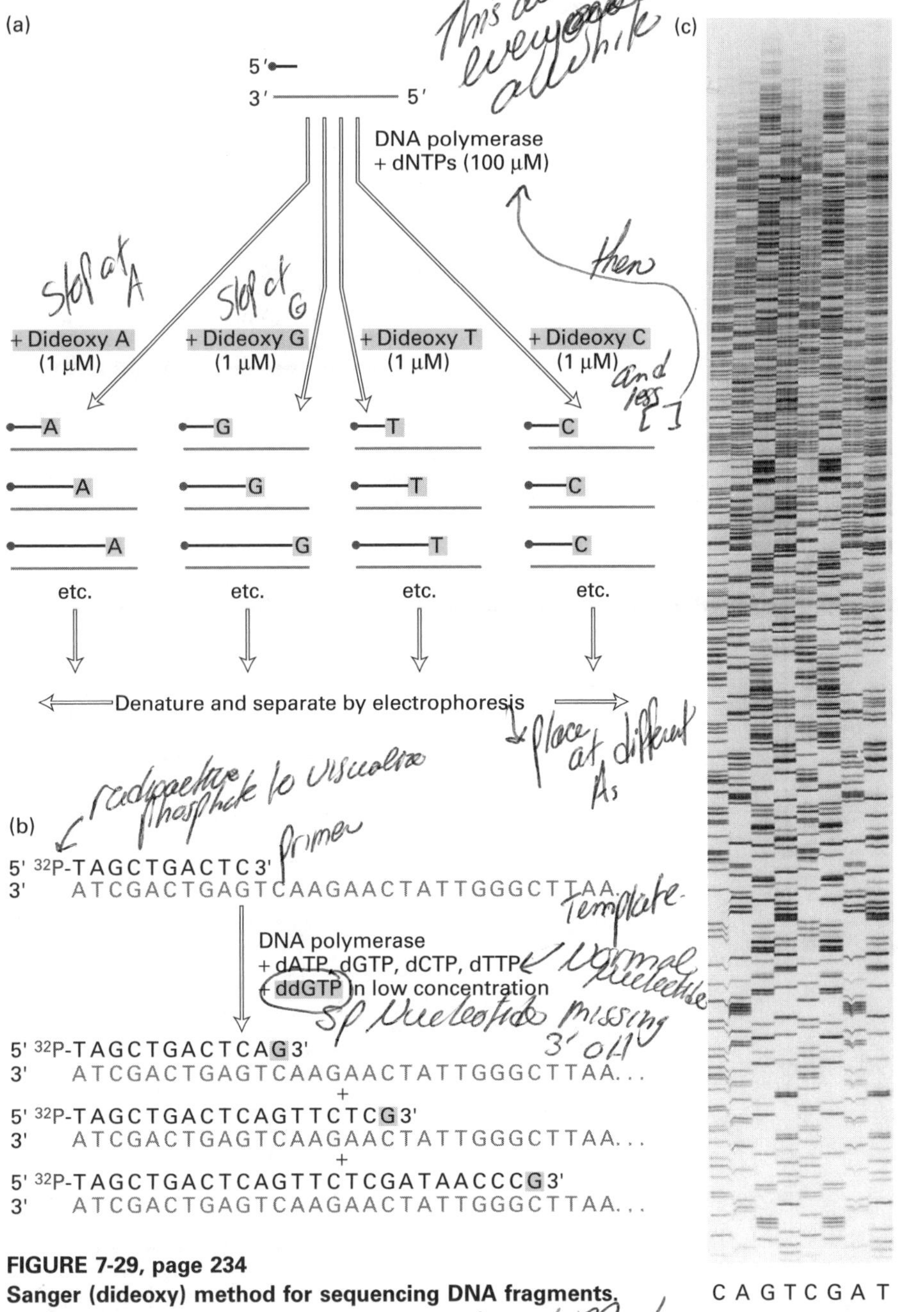

FIGURE 7-29, page 234
Sanger (dideoxy) method for sequencing DNA fragments.

(c)

C A G T C G A T

Polymerase encorparate diff bases until encorparate the dideoxy and thus must terminate. Every G that gets encorparated could be the ddGTP and not the other nucleotides.

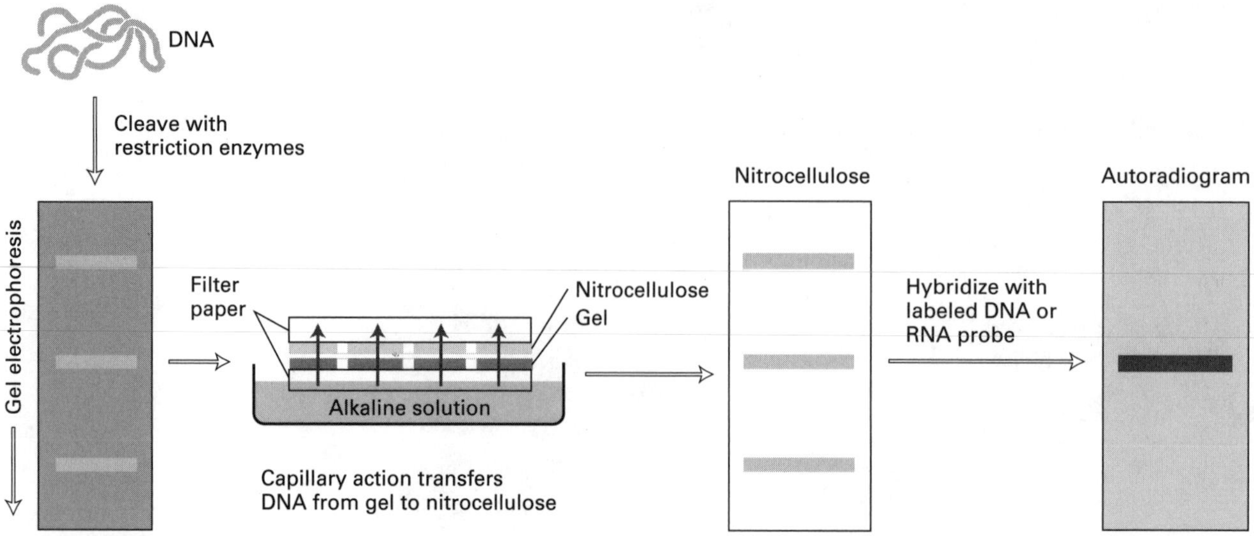

FIGURE 7-32, page 240
The Southern blot technique for detecting the presence of specific DNA sequences following gel electrophoresis of a complex mixture of restriction fragments.

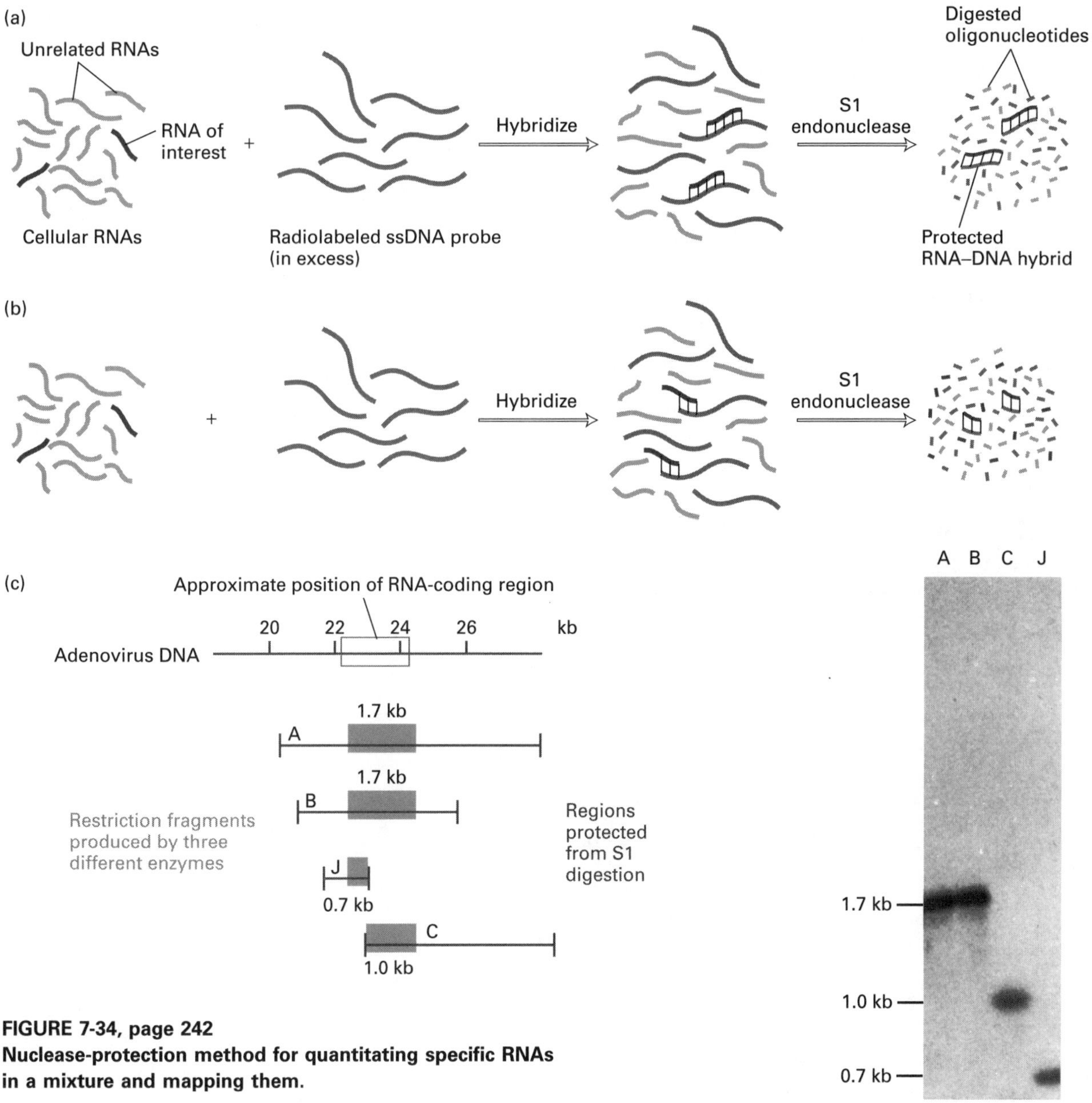

FIGURE 7-34, page 242
Nuclease-protection method for quantitating specific RNAs in a mixture and mapping them.

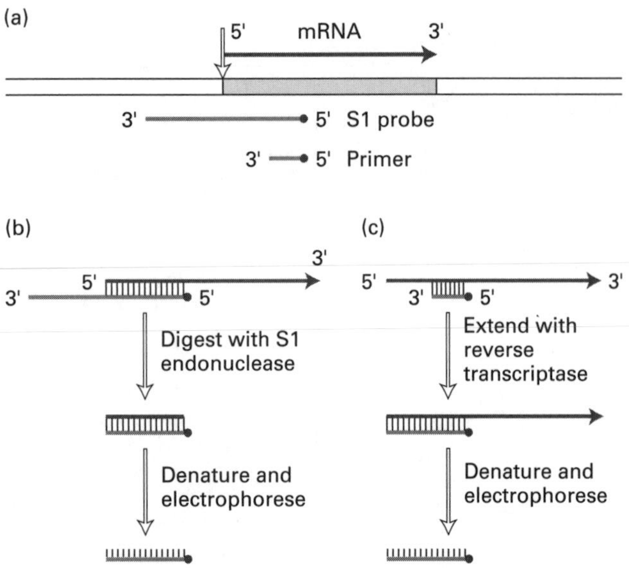

FIGURE 7-35, page 243
Two methods for mapping the start site for transcription of a particular gene in a region of DNA of known sequence.

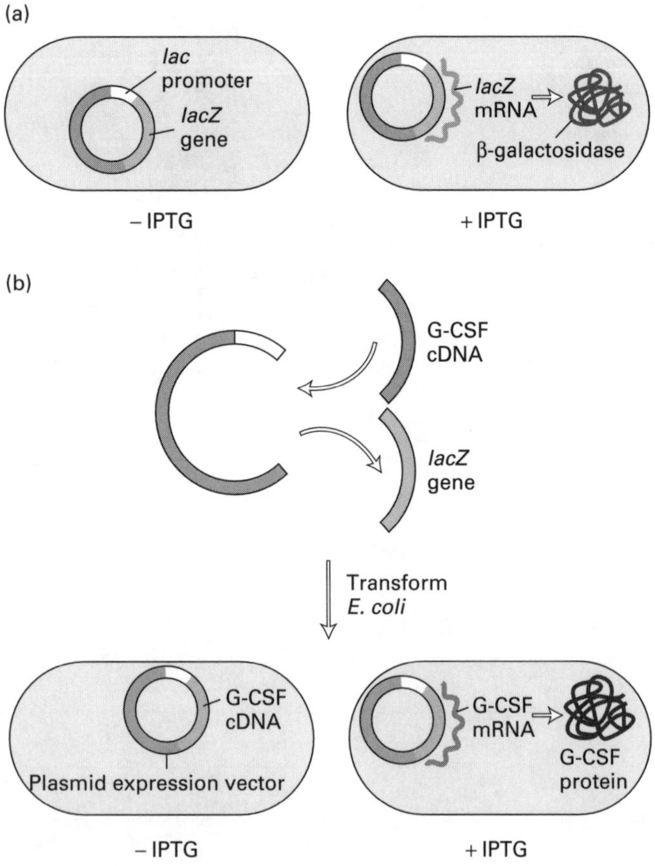

FIGURE 7-36, page 244
A simple *E. coli* expression vector utilizing the *lac* promoter.

MEDIA CONNECTIONS
Polymerase Chain Reaction

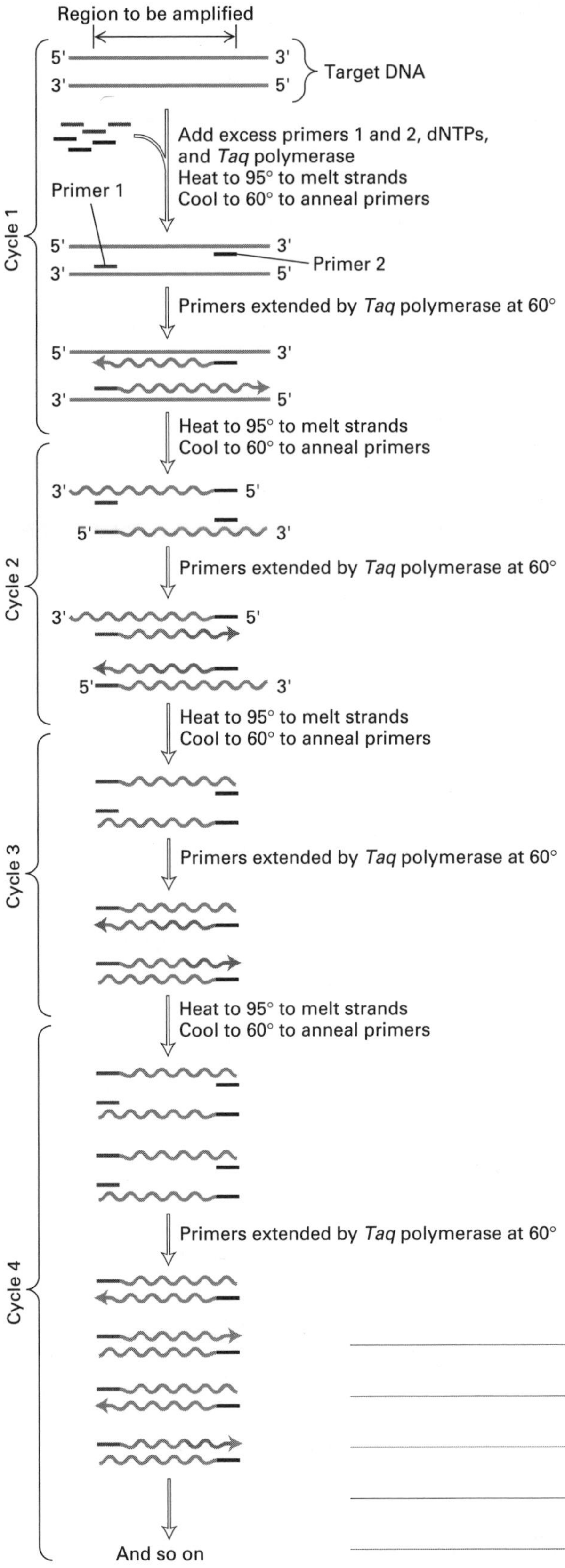

FIGURE 7-38, page 247
The polymerase chain reaction.

Additional Notes

Genetic Analysis in Cell Biology

8

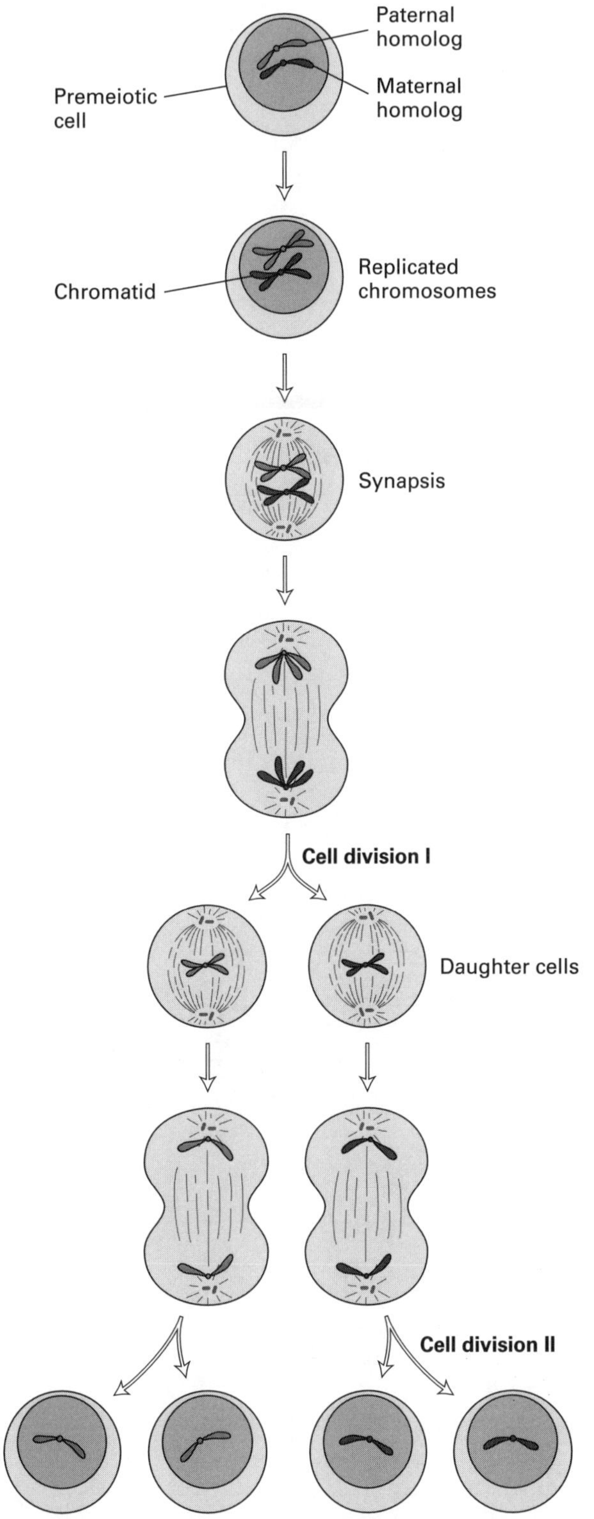

FIGURE 8-2, page 256
Meiosis.

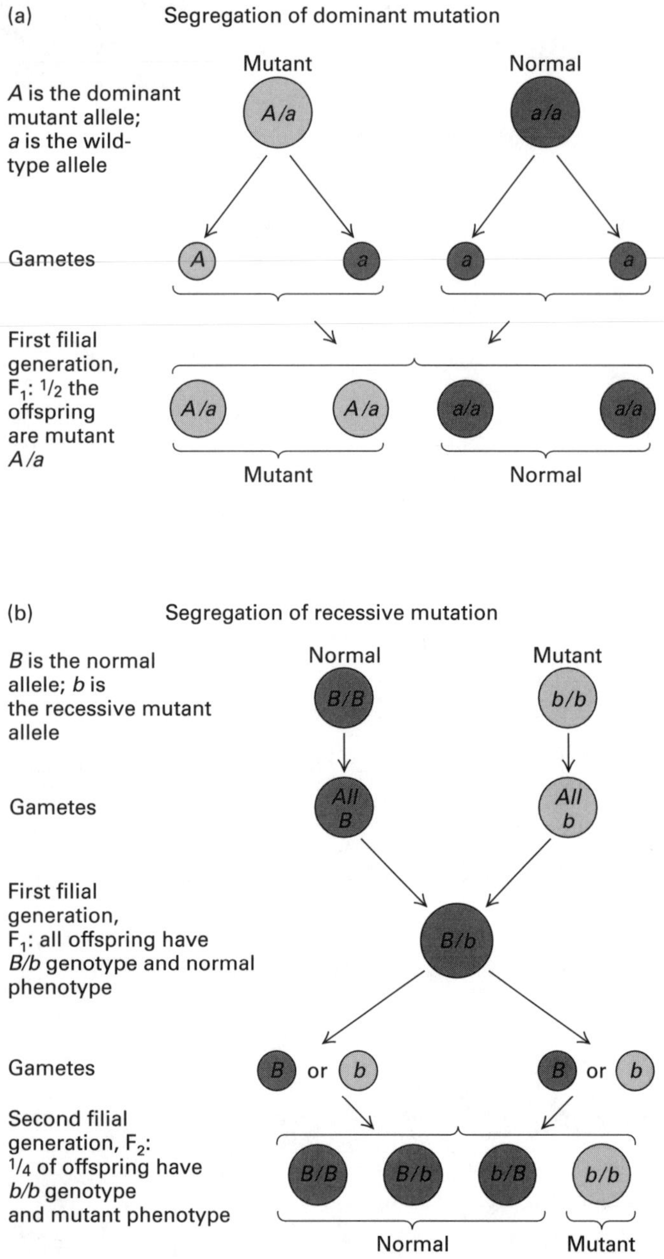

A is the dominant mutant allele; *a* is the wild-type allele

Gametes

First filial generation, F_1: 1/2 the offspring are mutant *A/a*

(b) Segregation of recessive mutation

B is the normal allele; *b* is the recessive mutant allele

Gametes

First filial generation, F_1: all offspring have *B/b* genotype and normal phenotype

Gametes

Second filial generation, F_2: 1/4 of offspring have *b/b* genotype and mutant phenotype

FIGURE 8-3, page 257
Segregation patterns of dominant and recessive mutations.

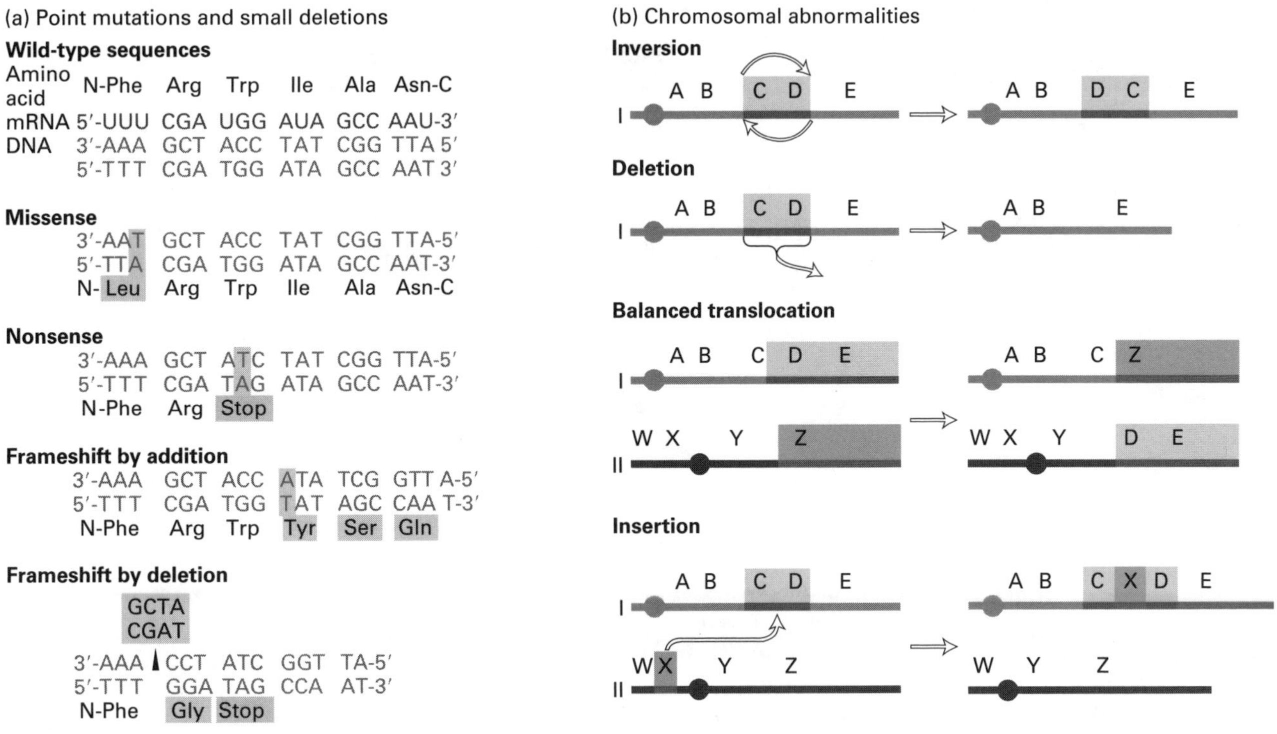
(a) Point mutations and small deletions
Wild-type sequences
Amino acid N-Phe Arg Trp Ile Ala Asn-C
mRNA 5′-UUU CGA UGG AUA GCC AAU-3′
DNA 3′-AAA GCT ACC TAT CGG TTA 5′
5′-TTT CGA TGG ATA GCC AAT 3′
Missense
3′-AAT GCT ACC TAT CGG TTA-5′
5′-TTA CGA TGG ATA GCC AAT-3′
N- Leu Arg Trp Ile Ala Asn-C
Nonsense
3′-AAA GCT ATC TAT CGG TTA-5′
5′-TTT CGA TAG ATA GCC AAT-3′
N-Phe Arg Stop
Frameshift by addition
3′-AAA GCT ACC ATA TCG GTT A-5′
5′-TTT CGA TGG TAT AGC CAA T-3′
N-Phe Arg Trp Tyr Ser Gln
Frameshift by deletion
GCTA
CGAT
3′-AAA CCT ATC GGT TA-5′
5′-TTT GGA TAG CCA AT-3′
N-Phe Gly Stop
(b) Chromosomal abnormalities
Inversion
A B C D E
A B D C E
Deletion
A B C D E
A B E
Balanced translocation
A B C D E
A B C Z
W X Y Z
W X Y D E
Insertion
A B C D E
A B C X D E
W X Y Z
W Y Z
I
II

FIGURE 8-4, page 258
Different types of mutations.

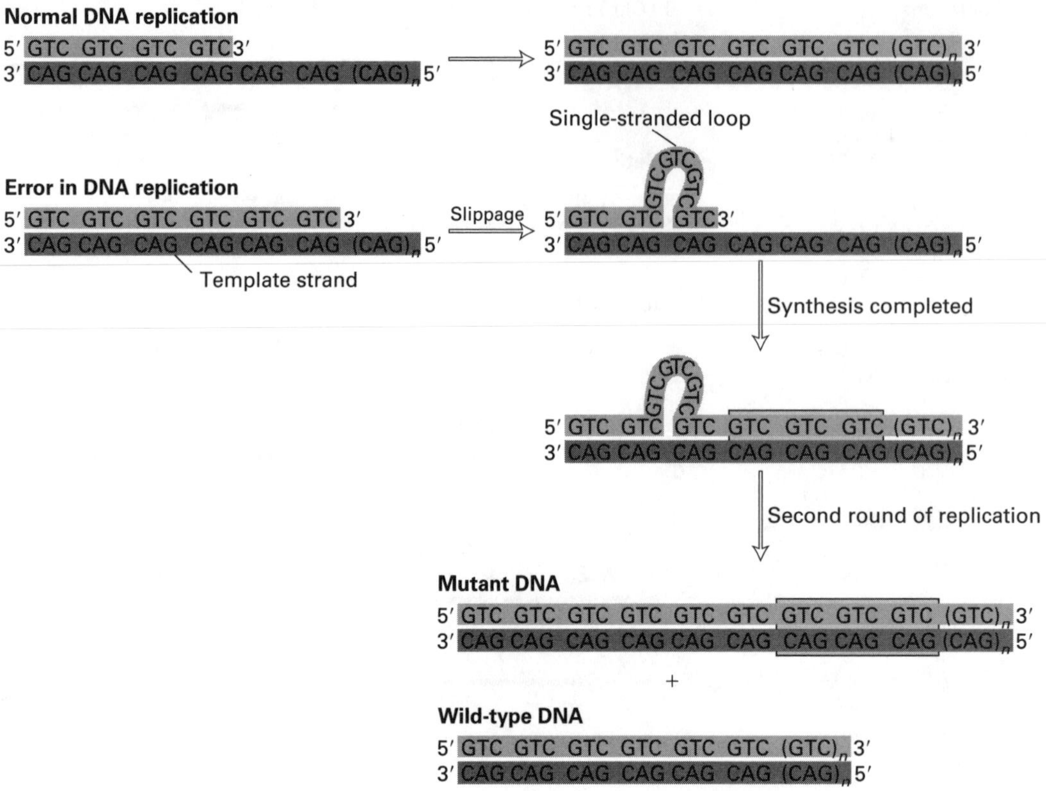

FIGURE 8-5, page 259
One mechanism by which errors in DNA replication produce spontaneous mutations.

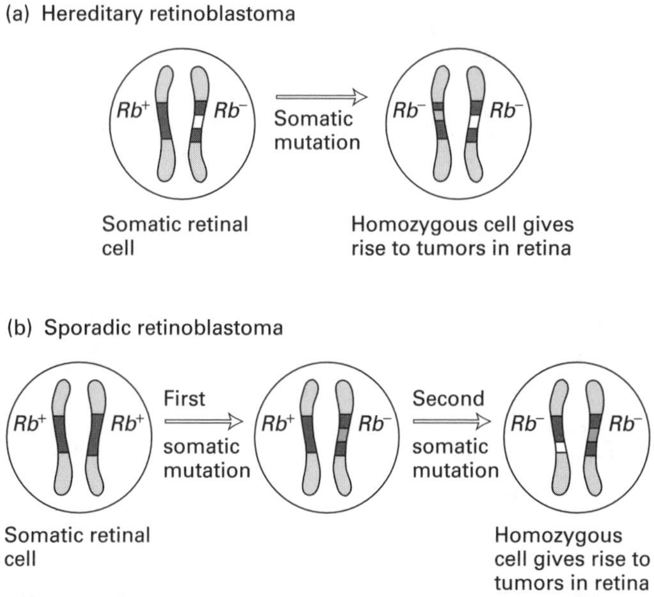

FIGURE 8-7, page 260
Role of spontaneous somatic mutation in retinoblastoma, a childhood disease marked by retinal tumors.

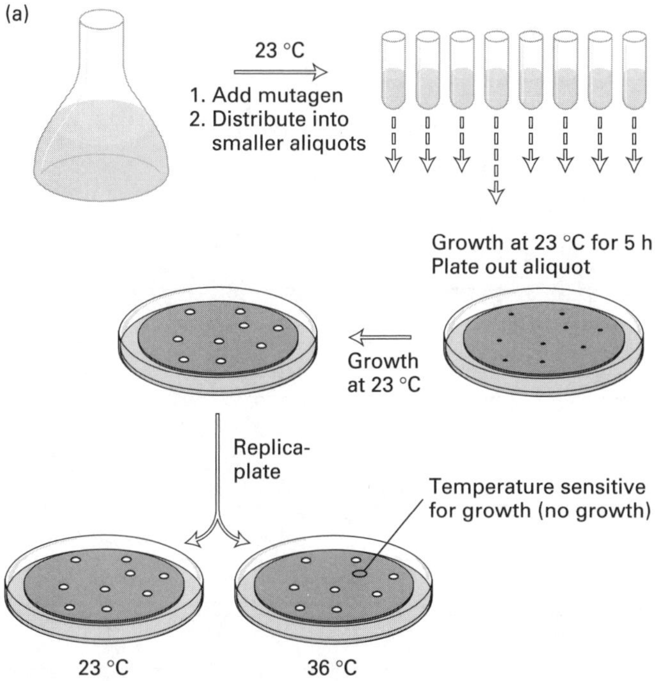

FIGURE 8-9, page 262
Two-step genetic screen used to identify cell-cycle mutants in yeast.

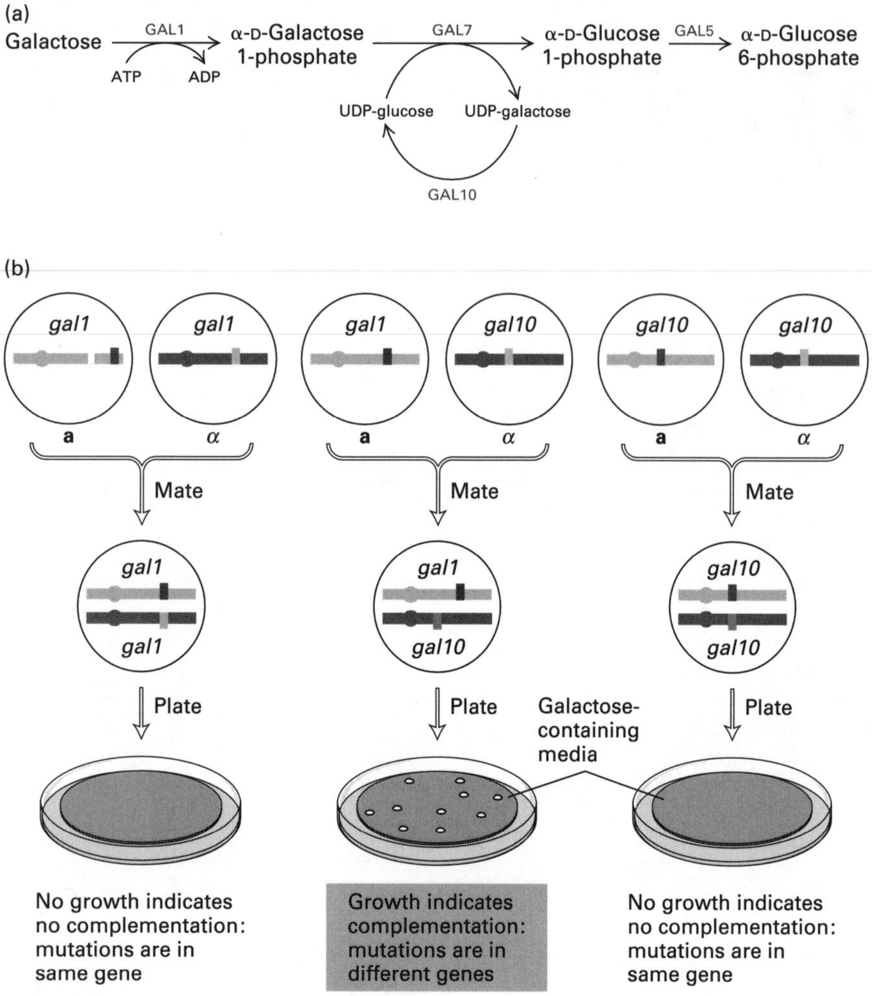

FIGURE 8-11, page 264
Complementation analysis in *S. cerevisiae*.

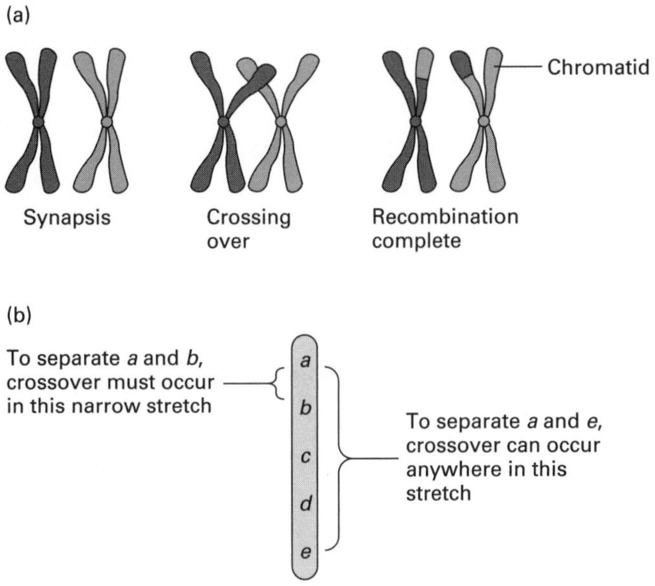

FIGURE 8-18, page 270
Recombination during meiosis.

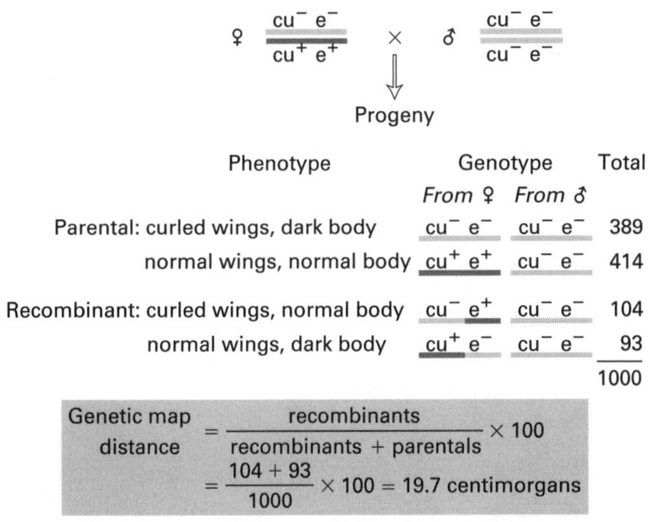

FIGURE 8-19, page 270
Determination of the genetic map distance between two *Drosophila* loci by recombinational analysis.

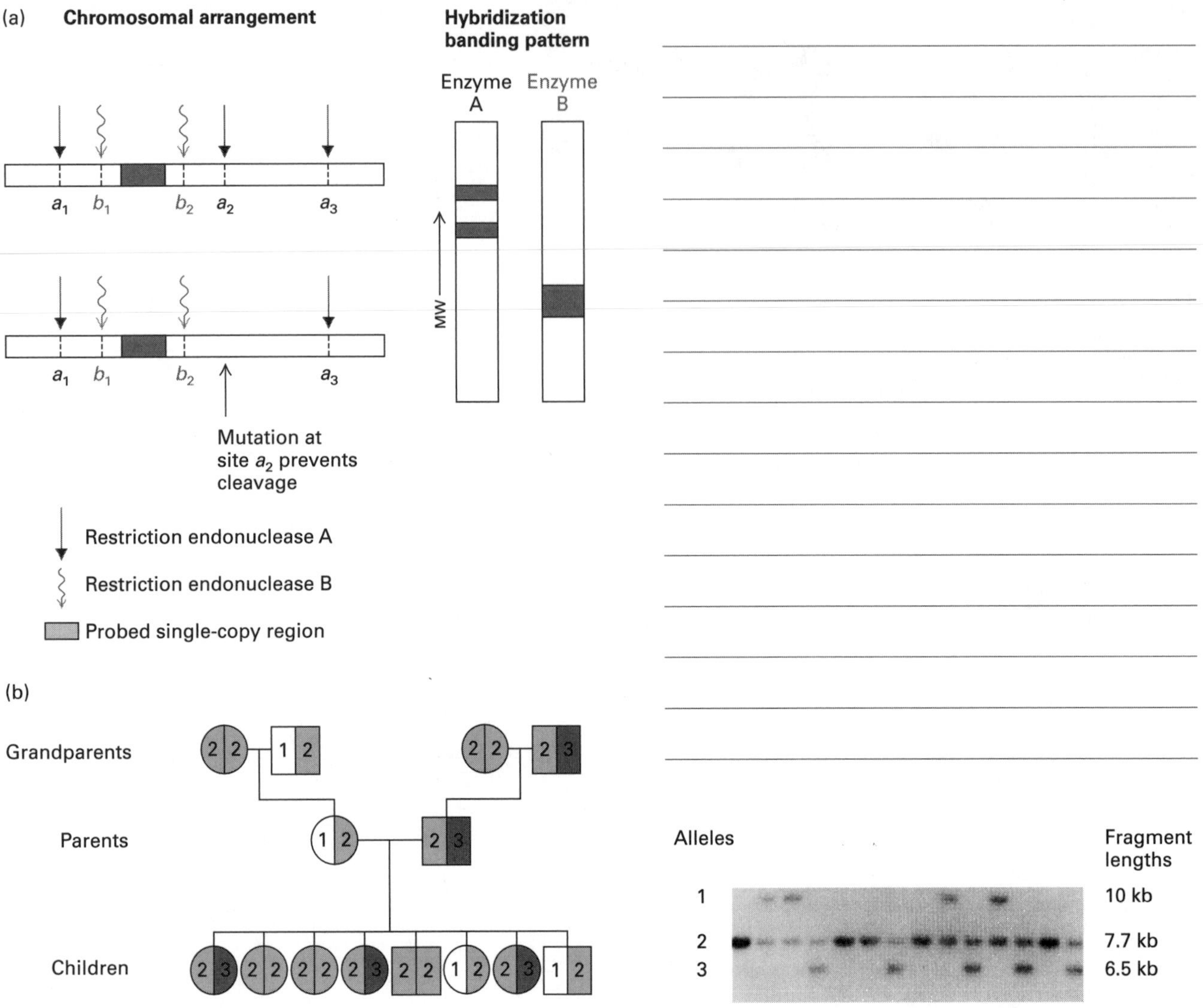

FIGURE 8-20, page 271
Analysis of restriction fragment length polymorphisms (RFLPs).

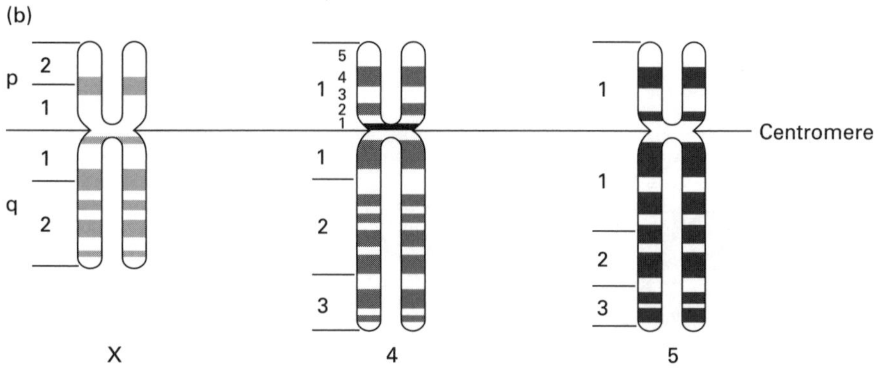

FIGURE 8-21, page 273
G-banding of human chromosomes.

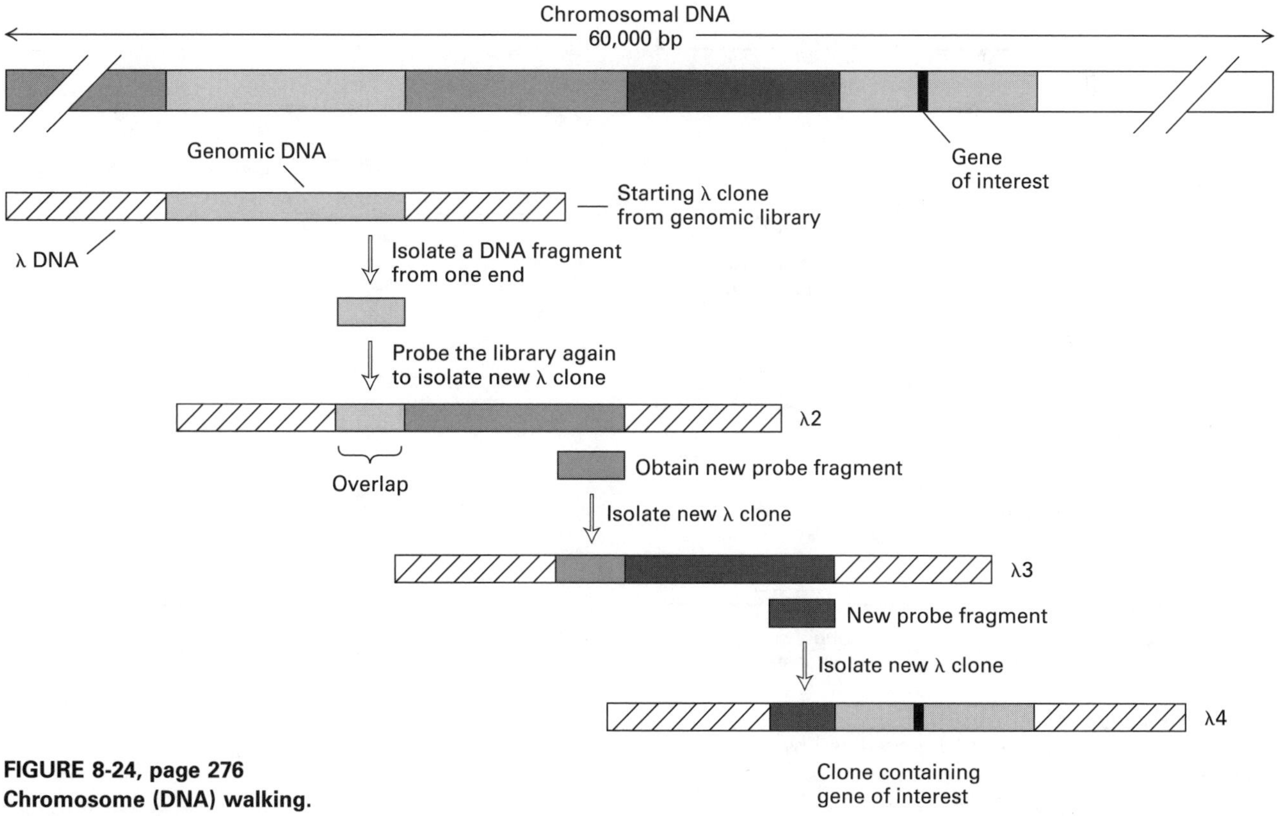

FIGURE 8-24, page 276
Chromosome (DNA) walking.

MEDIA CONNECTIONS
In Vitro Mutagenesis of Cloned Genes

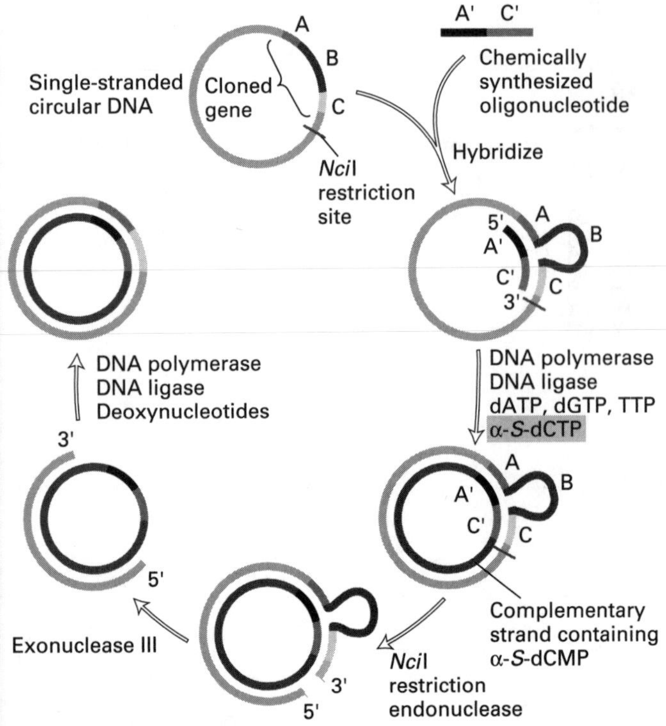

FIGURE 8-29, page 282
In vitro mutagenesis of cloned genes with chemically synthesized oligonucleotides.

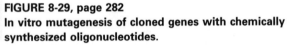

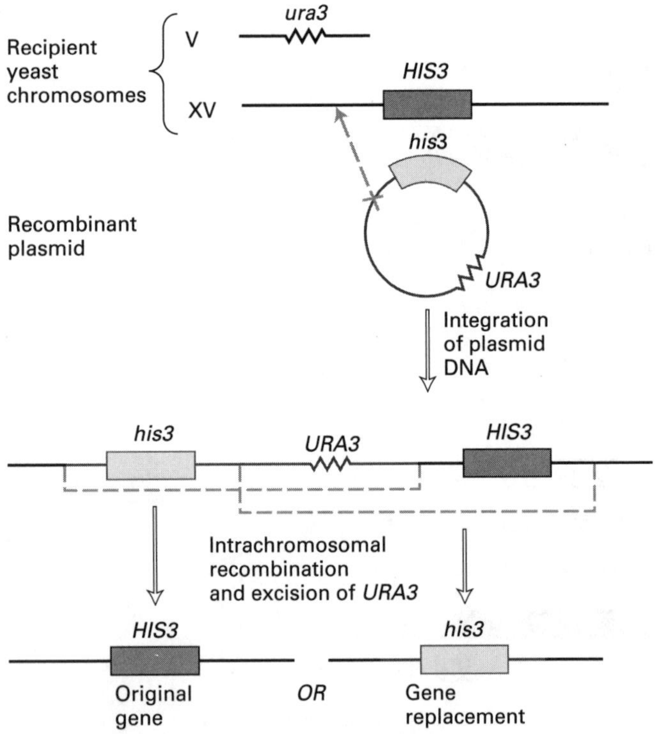

FIGURE 8-30, page 283
Replacement of the normal *HIS3* gene (blue) by homologous recombination with a mutant *his3* gene (yellow) in the yeast *S. cerevisiae*.

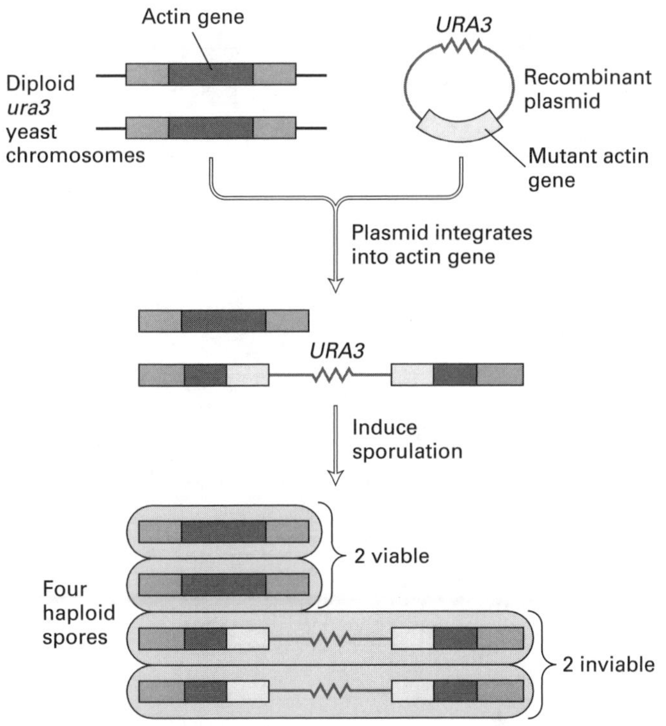

FIGURE 8-31, page 283
Demonstration that actin gene is required for yeast viability by gene-targeted knockout.

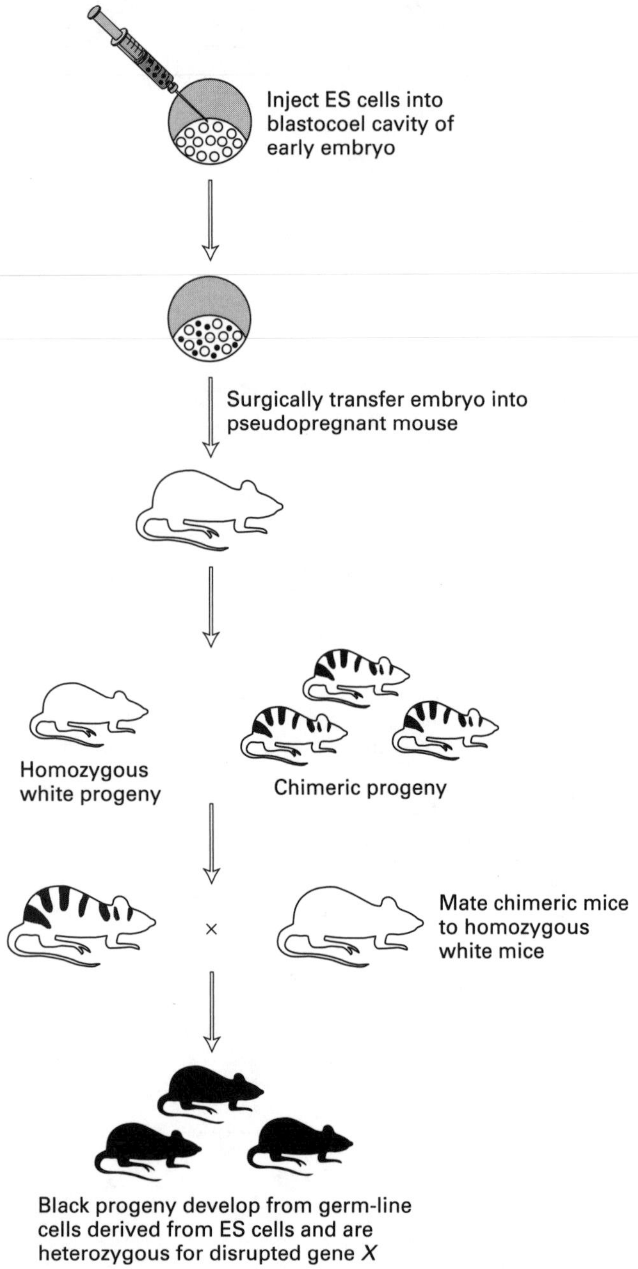
Inject ES cells into blastocoel cavity of early embryo
Surgically transfer embryo into pseudopregnant mouse
Homozygous white progeny
Chimeric progeny
×
Mate chimeric mice to homozygous white mice
Black progeny develop from germ-line cells derived from ES cells and are heterozygous for disrupted gene *X*

FIGURE 8-34, page 285
General procedure for producing gene-targeted knockout mice.

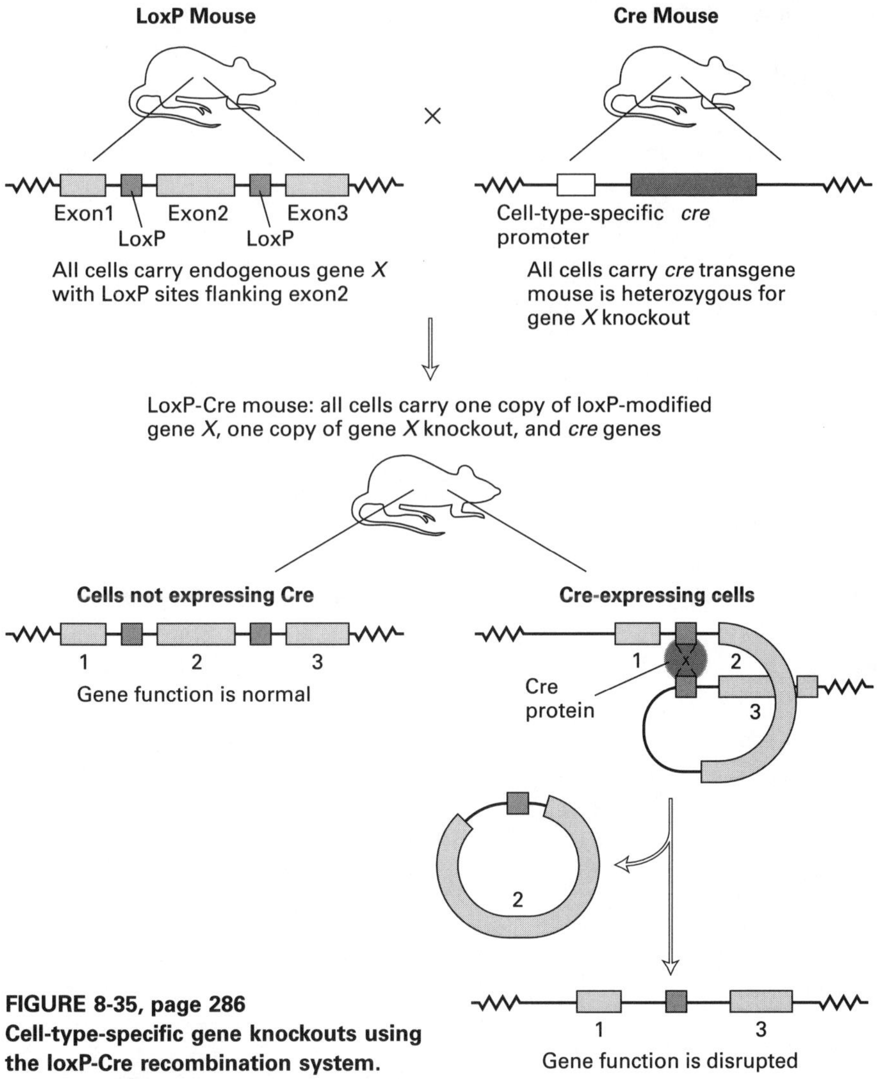
LoxP Mouse
Cre Mouse
×
Exon1
Exon2
Exon3
LoxP
LoxP
All cells carry endogenous gene X with LoxP sites flanking exon2
Cell-type-specific promoter
cre
All cells carry cre transgene mouse is heterozygous for gene X knockout
LoxP-Cre mouse: all cells carry one copy of loxP-modified gene X, one copy of gene X knockout, and cre genes
Cells not expressing Cre
1
2
3
Gene function is normal
Cre-expressing cells
1
2
Cre protein
3
2
1
3
Gene function is disrupted

FIGURE 8-35, page 286
Cell-type-specific gene knockouts using the loxP-Cre recombination system.

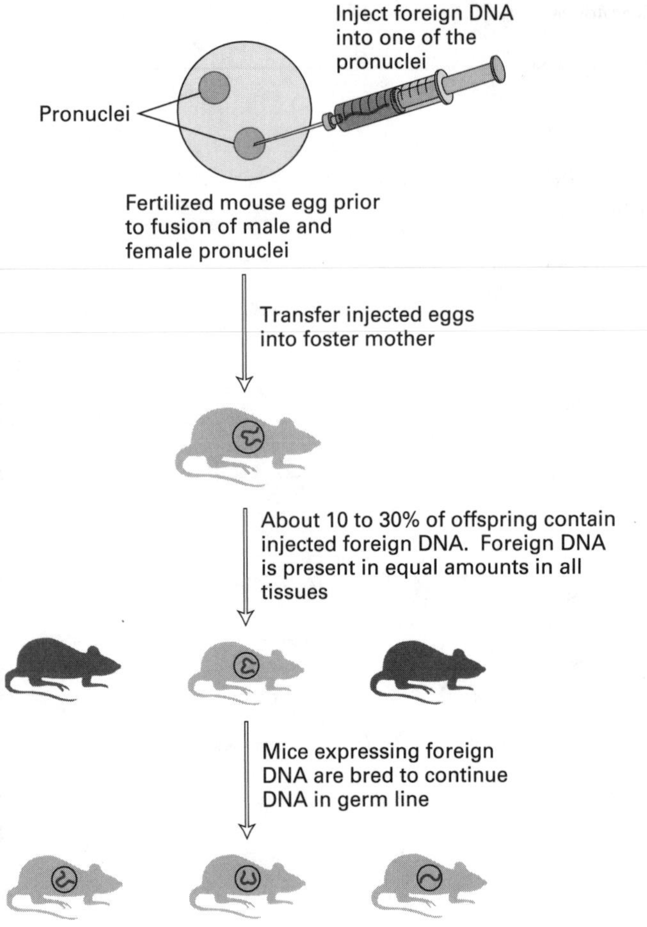
Inject foreign DNA into one of the pronuclei
Pronuclei
Fertilized mouse egg prior to fusion of male and female pronuclei
Transfer injected eggs into foster mother
About 10 to 30% of offspring contain injected foreign DNA. Foreign DNA is present in equal amounts in all tissues
Mice expressing foreign DNA are bred to continue DNA in germ line

FIGURE 8-36, page 287
General procedure for producing transgenic mice.

Additional Notes

Molecular Structure of Genes and Chromosomes

9

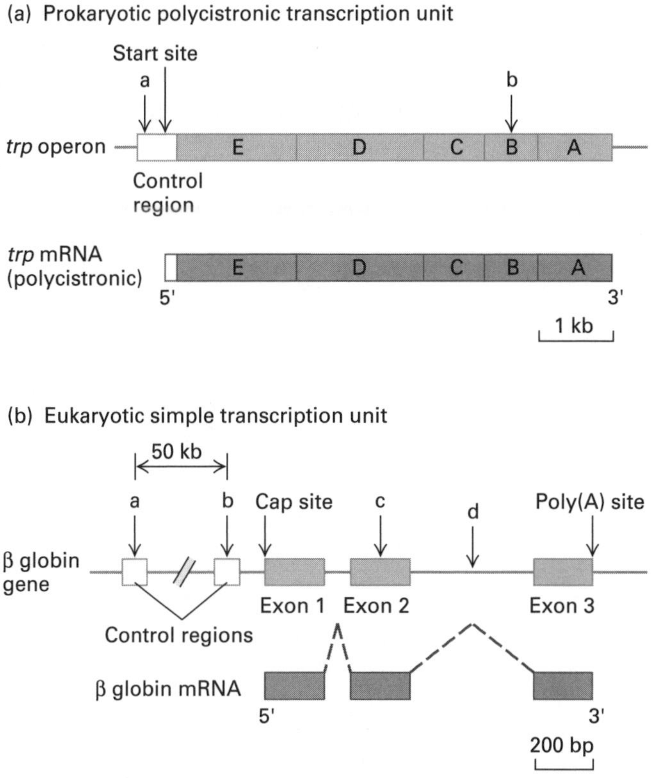

FIGURE 9-1, page 296
Comparison of bacterial operons and simple eukaryotic transcription units.

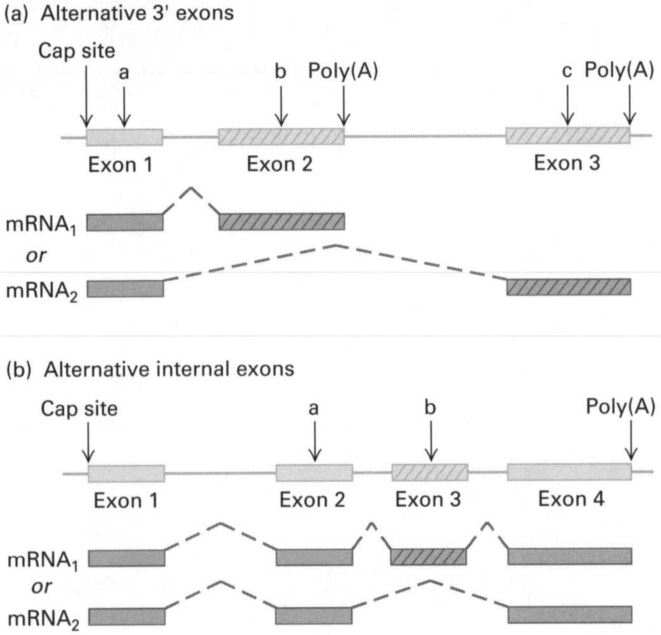

FIGURE 9-2, page 297
Two examples of complex eukaryotic transcription units and the effect of mutations on expression of the encoded proteins.

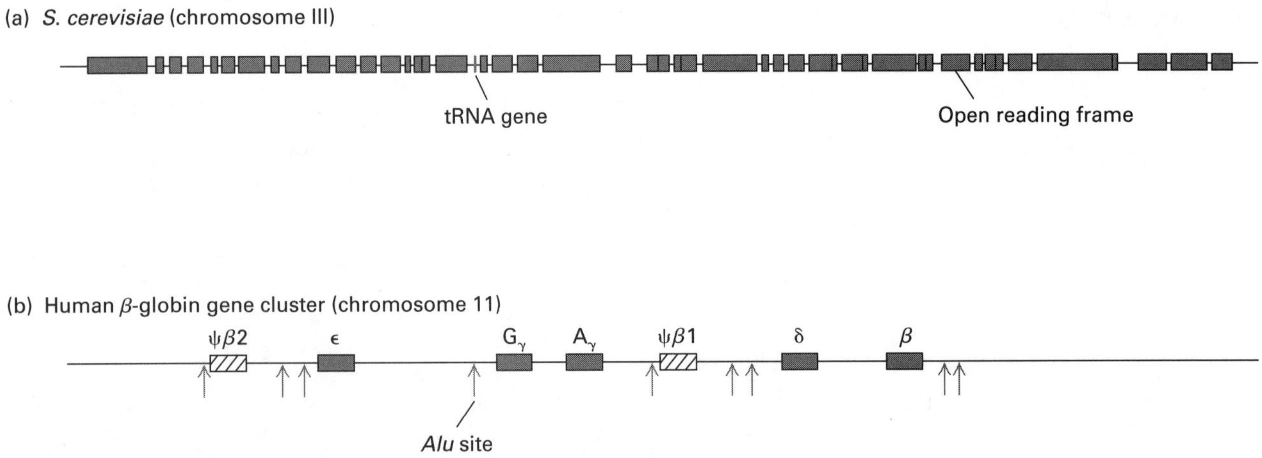

FIGURE 9-3, page 298
Diagrams of ≈80-kb region from chromosome III of the yeast *S. cerevisiae* and the β-globin gene cluster on human chromosome 11.

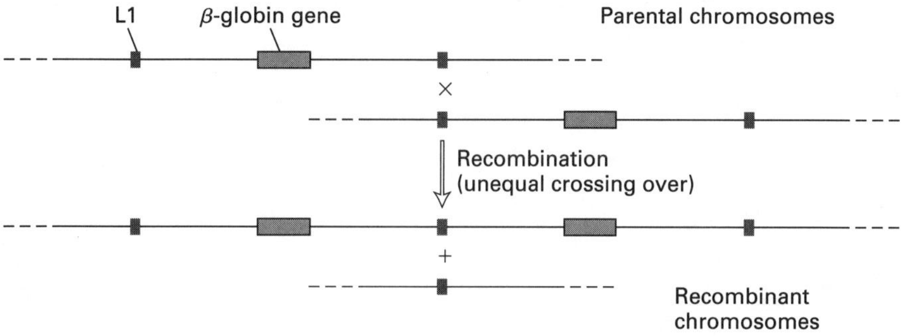

FIGURE 9-5, page 300
Gene duplication resulting from unequal crossing over.

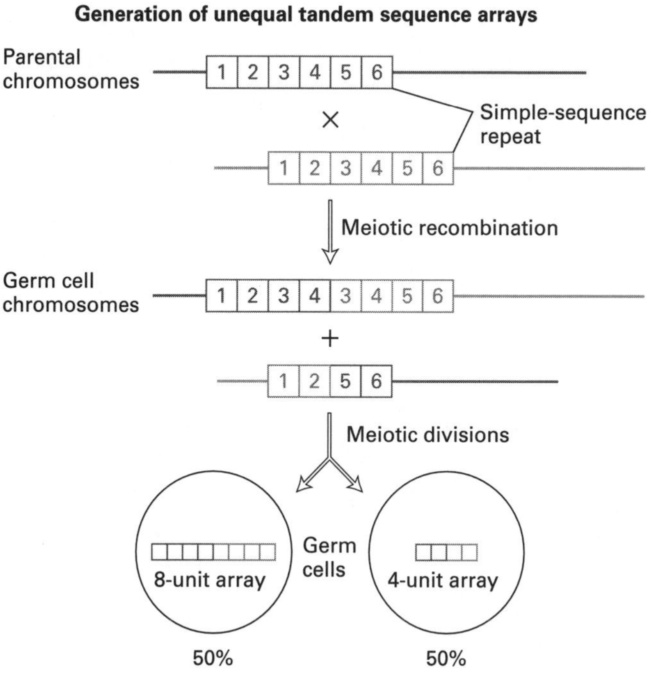

FIGURE 9-7, page 302
Unequal crossing over during meiosis can generate differences in lengths of simple-sequence DNA tandem arrays.

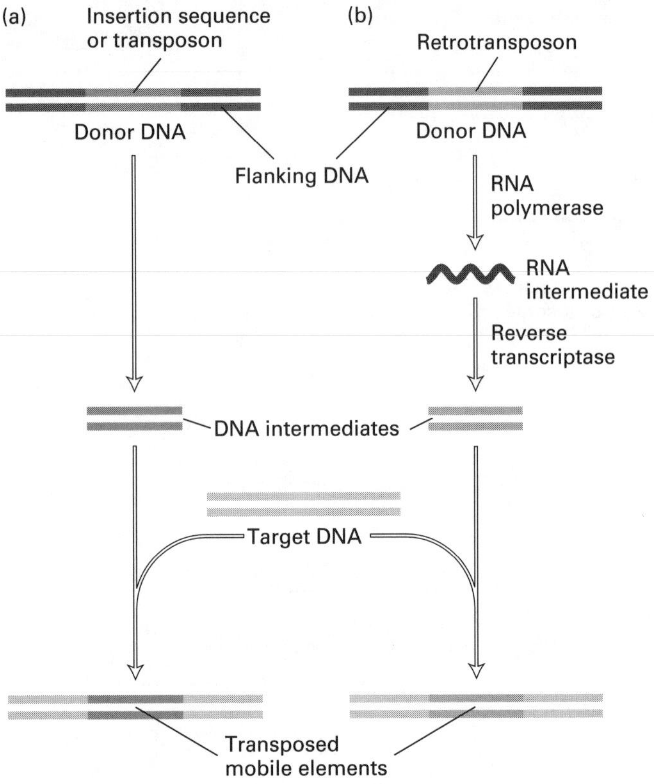

FIGURE 9-10, page 304
Classification of mobile elements into two major classes.

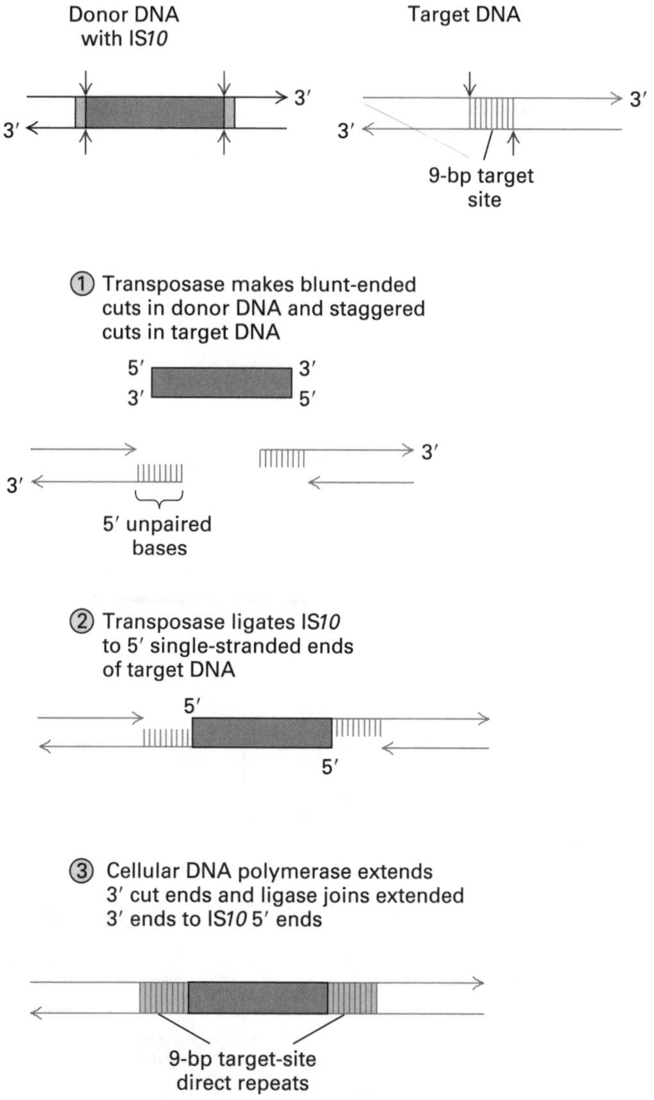
Donor DNA
with IS*10*
Target DNA
3′
3′
3′
3′
9-bp target
site
① Transposase makes blunt-ended
cuts in donor DNA and staggered
cuts in target DNA
5′
3′
3′
5′
3′
3′
5′ unpaired
bases
② Transposase ligates IS*10*
to 5′ single-stranded ends
of target DNA
5′
5′
③ Cellular DNA polymerase extends
3′ cut ends and ligase joins extended
3′ ends to IS*10* 5′ ends
9-bp target-site
direct repeats

FIGURE 9-12, page 306
Model for nonreplicative transposition of bacterial insertion sequences.

MEDIA CONNECTIONS
Retroviral Reverse Transcription

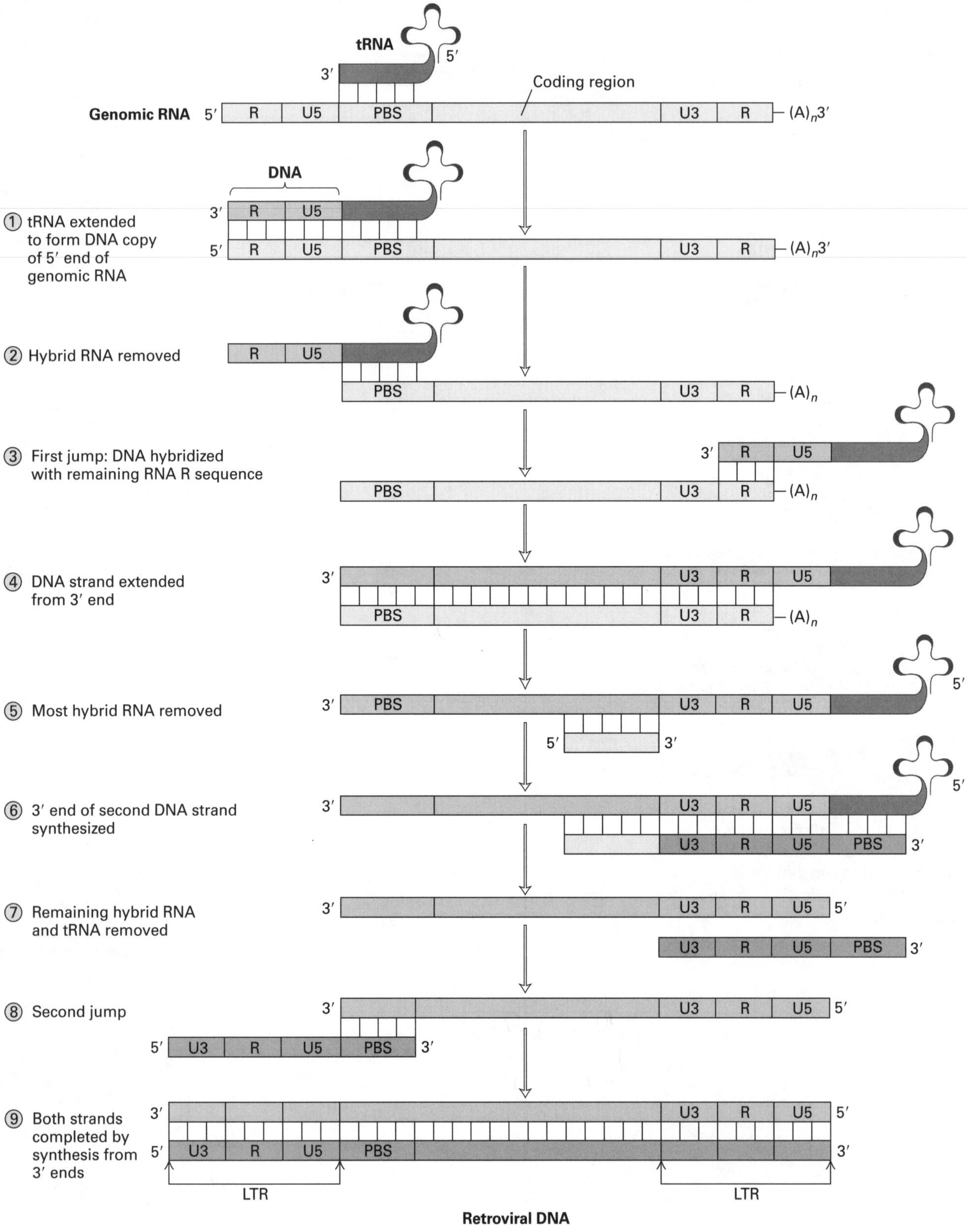
tRNA
5′
3′
Coding region
Genomic RNA 5′ R U5 PBS U3 R (A)n3′
DNA
① tRNA extended to form DNA copy of 5′ end of genomic RNA
② Hybrid RNA removed
③ First jump: DNA hybridized with remaining RNA R sequence
④ DNA strand extended from 3′ end
⑤ Most hybrid RNA removed
⑥ 3′ end of second DNA strand synthesized
⑦ Remaining hybrid RNA and tRNA removed
⑧ Second jump
⑨ Both strands completed by synthesis from 3′ ends
LTR
LTR
Retroviral DNA

FIGURE 9-16, page 309
Generation of LTRs during reverse transcription of retroviral genomic RNA.

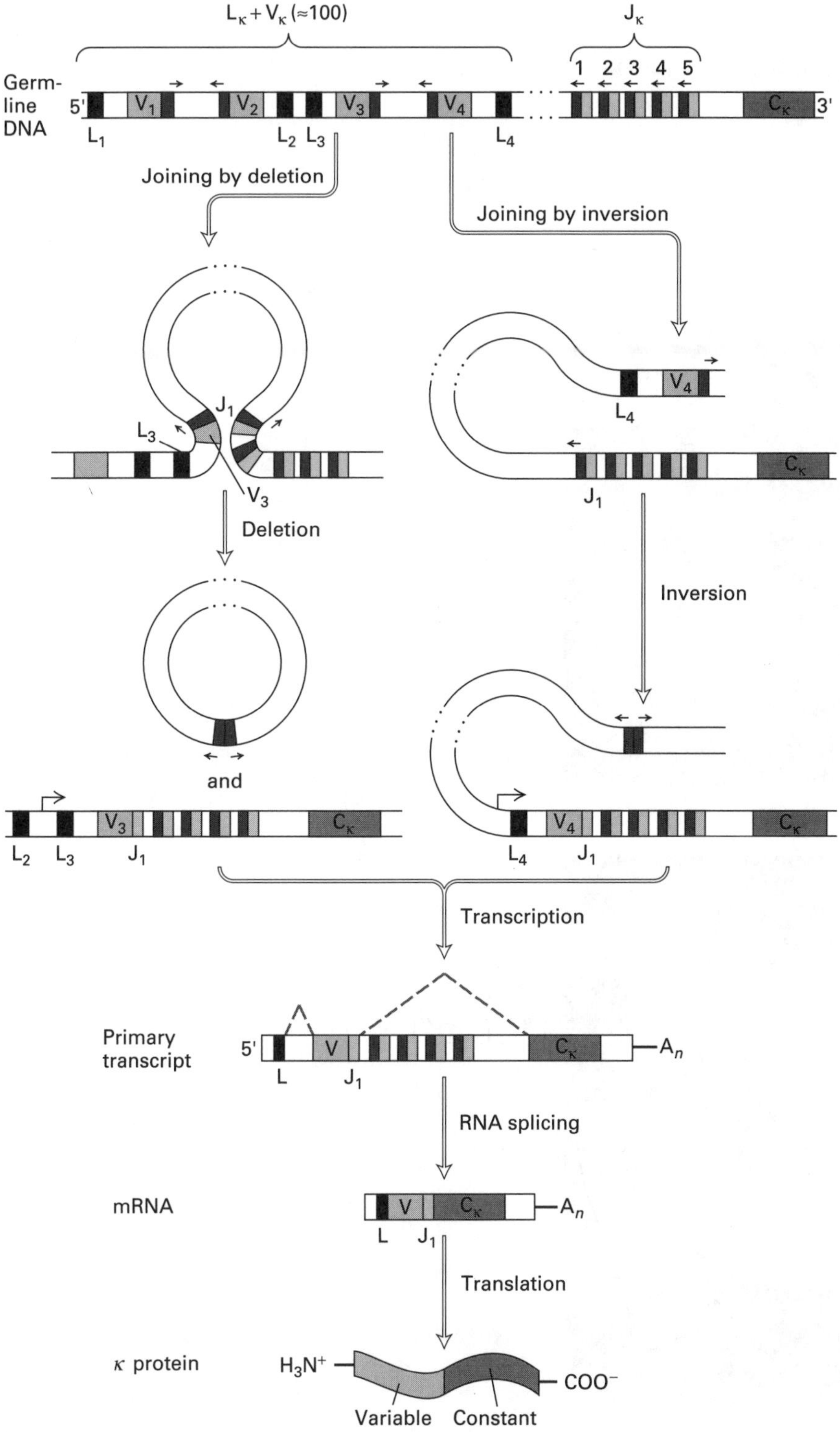

FIGURE 9-24, page 317
Joining of V_κ to J_κ in human germ-line DNA and formation of a κ light chain.

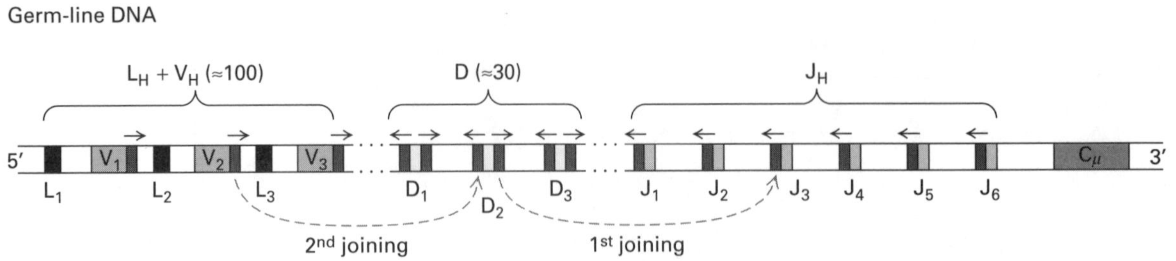

FIGURE 9-25, page 318
Organization of heavy-chain gene segments in human germ-line DNA.

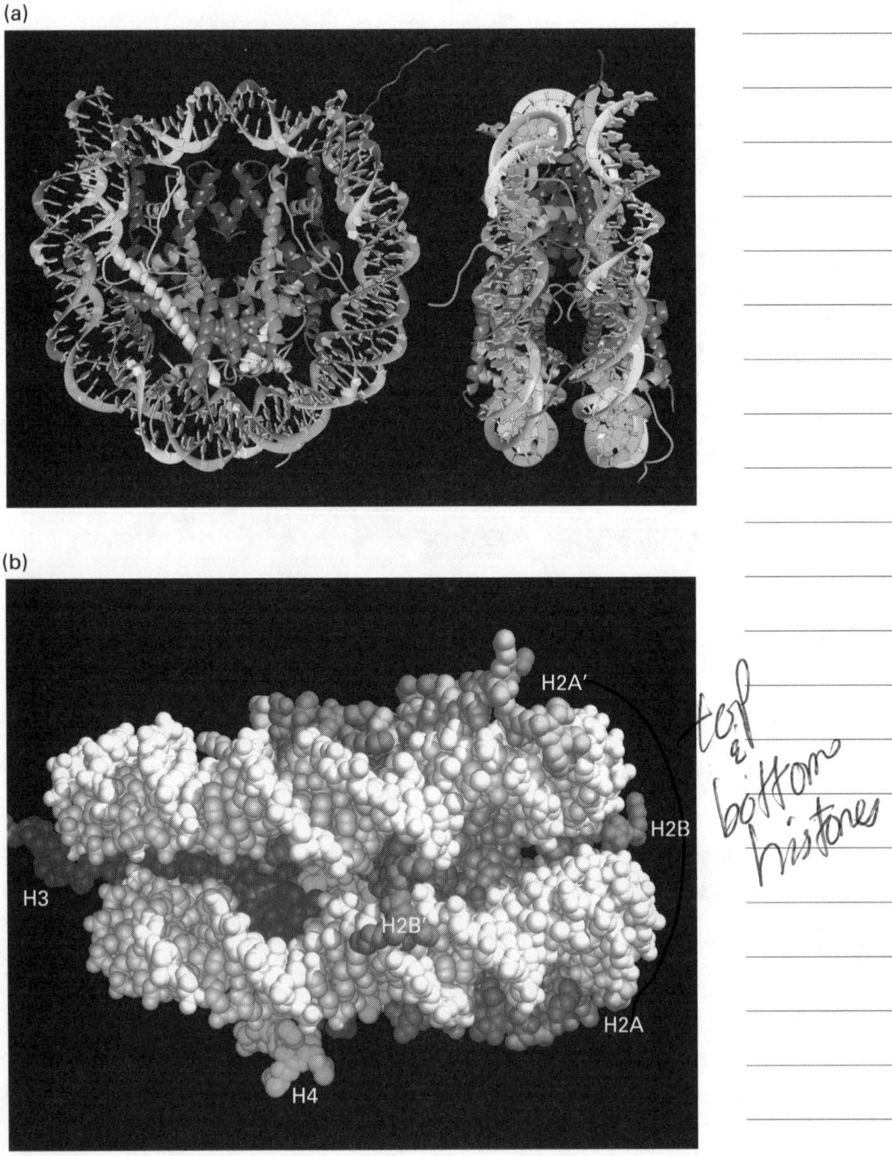

FIGURE 9-30, page 322
Structure of the nucleosome.

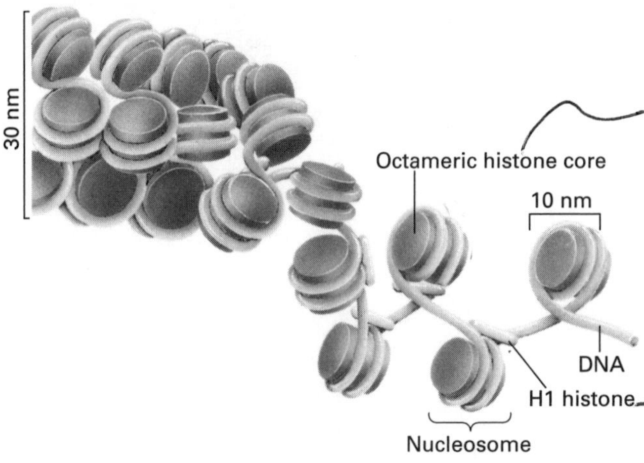

FIGURE 9-31, page 323
Solenoid model of the 30-nm condensed chromatin fiber in a side view.

Octomer around which DNA is wrapped.
made up of 2 copies of H2A, H2B, H3, H4
associated w/ each nucleosome

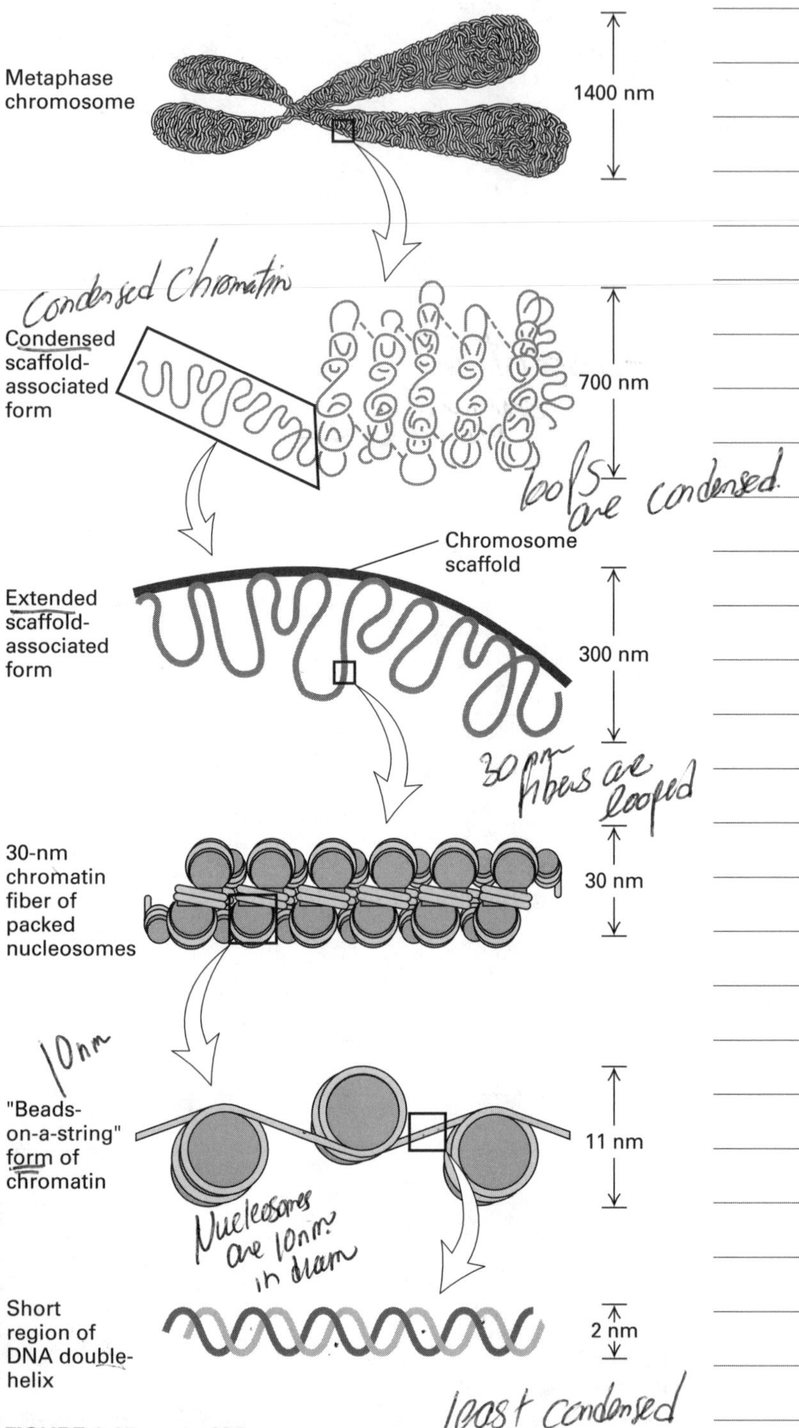

FIGURE 9-35, page 326
Model for the packing of chromatin and the chromosome scaffold in metaphase chromosomes.

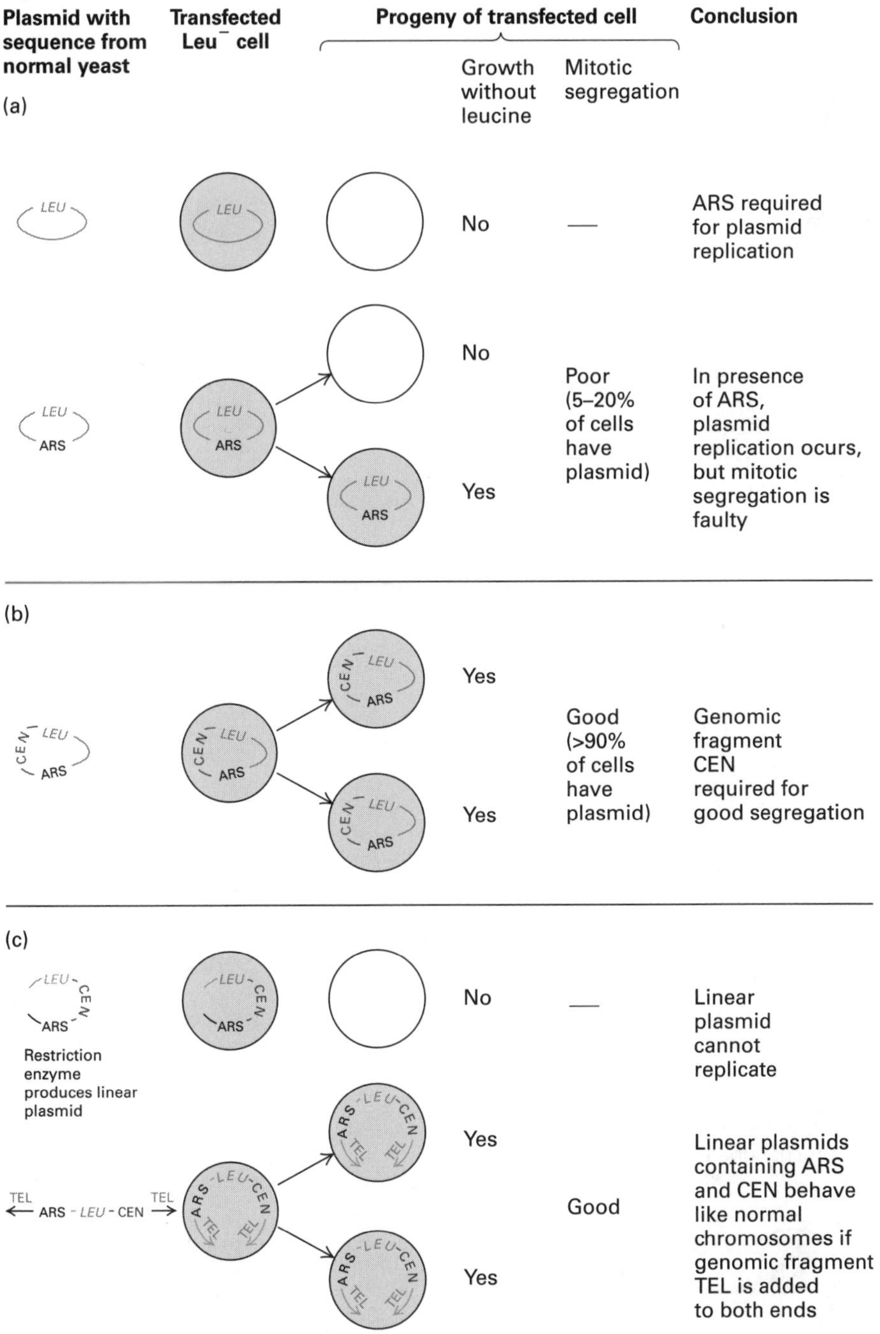

FIGURE 9-40, page 330
Experimental demonstration of functional chromosomal elements in experiments with yeast cells that lack an enzyme necessary for leucine synthesis (*leu*⁻ cells).

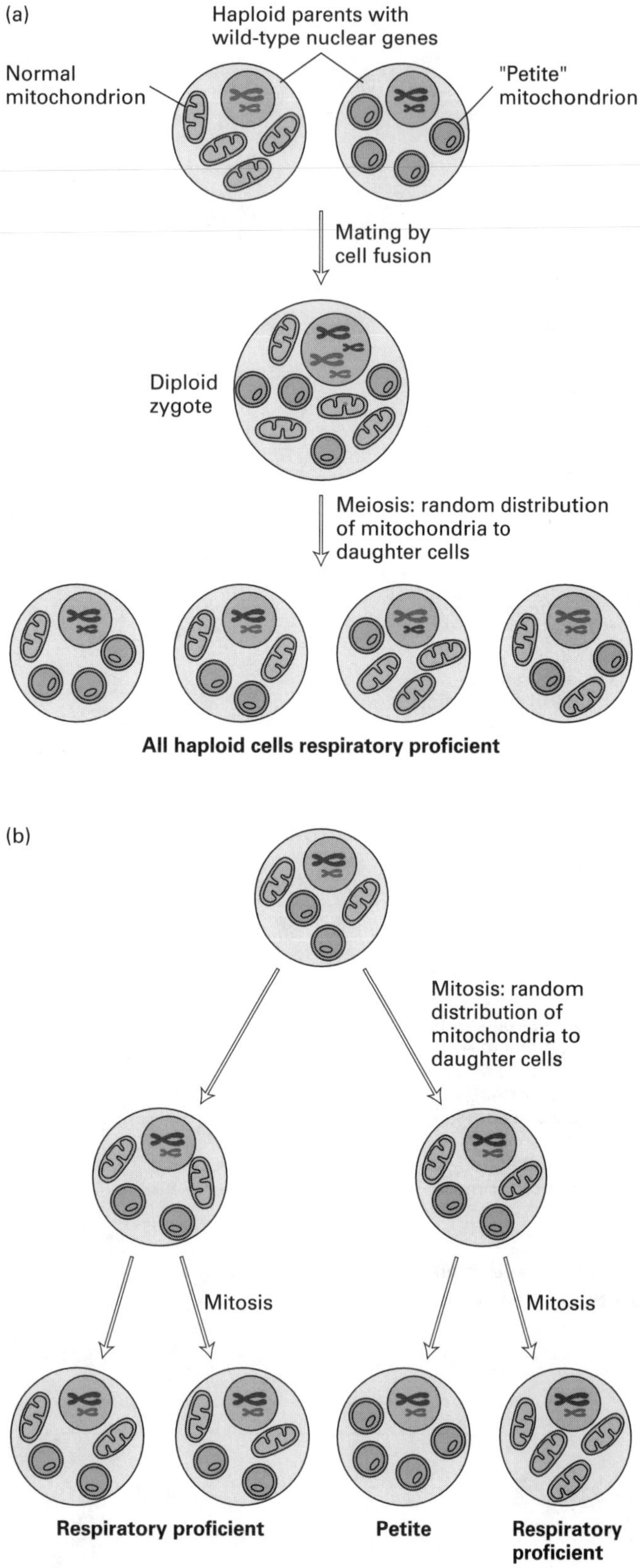

FIGURE 9-43, page 333
Cytoplasmic inheritance of the *petite* mutation in yeast.

Additional Notes

Regulation of Transcription Initiation

10

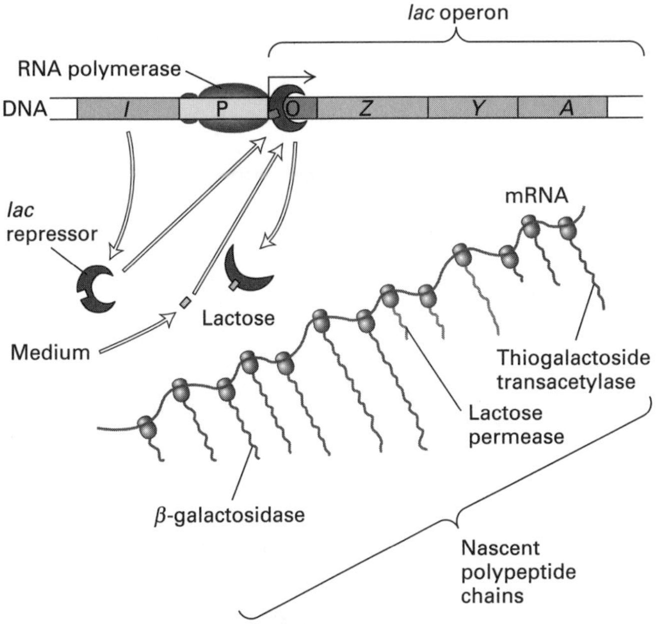

FIGURE 10-2, page 343
Jacob and Monod model of transcriptional regulation of the *lac* operon by *lac* repressor.

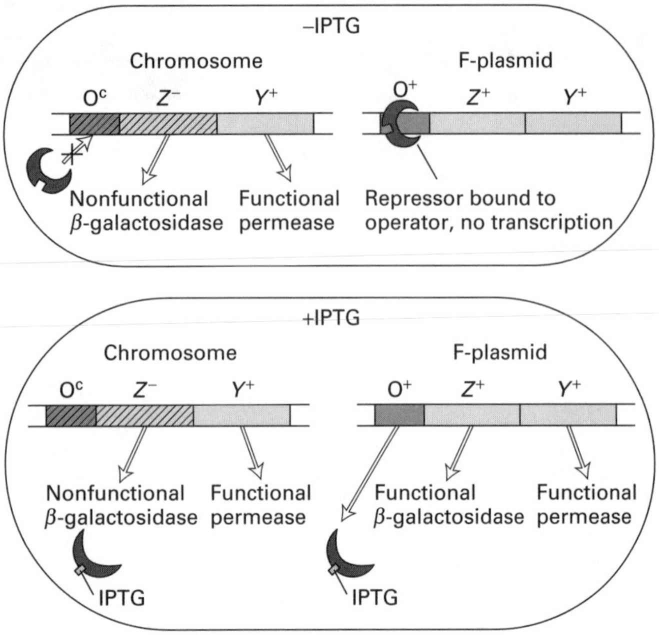

FIGURE 10-3, page 344
Experimental demonstration that O^c mutations are cis-acting.

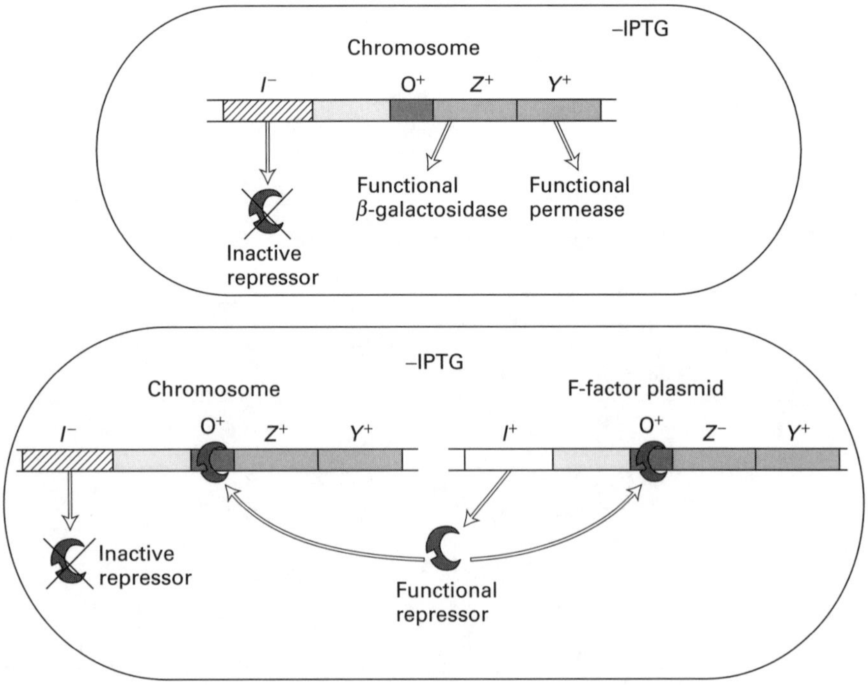

FIGURE 10-4, page 345
Experimental demonstration that the *lacI*$^+$ gene is trans-acting.

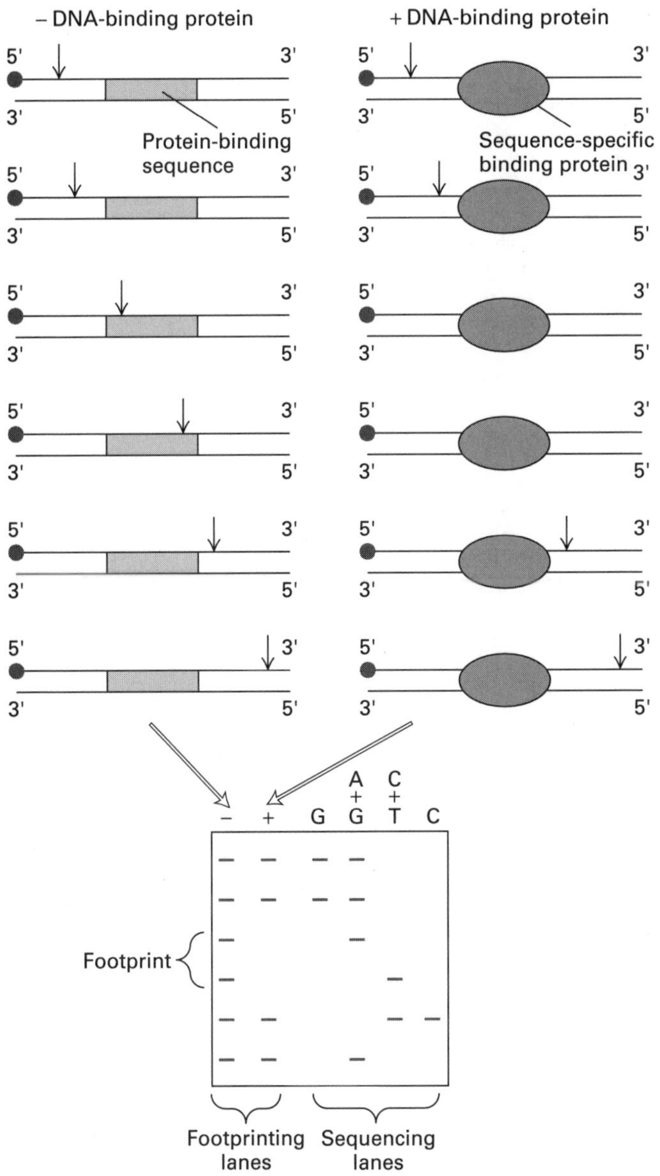

FIGURE 10-6, page 346
DNase I footprinting, a common technique for identifying protein-binding sites in DNA.

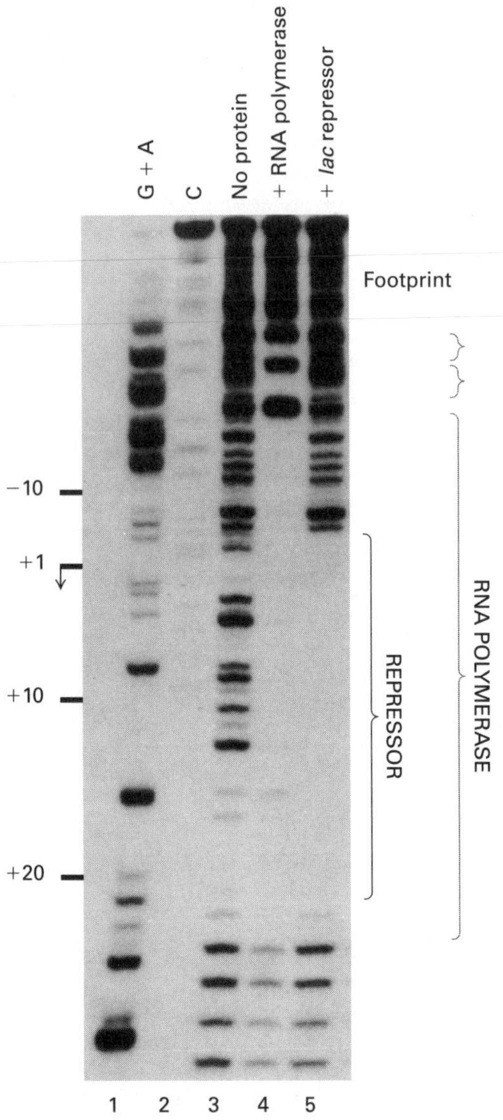

FIGURE 10-8, page 348
Footprints of RNA polymerase and *lac* repressor on *lac* control-region DNA.

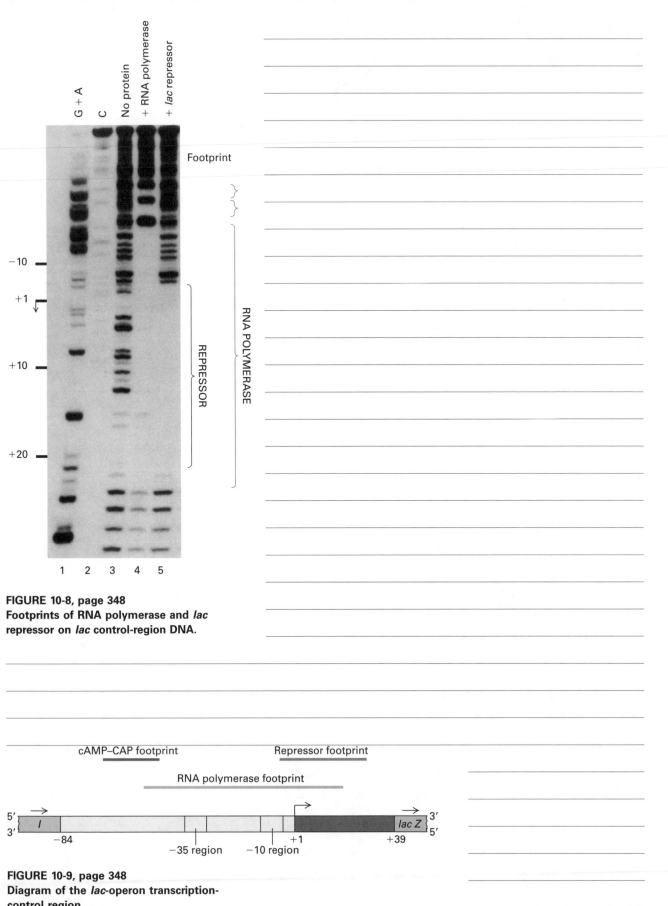

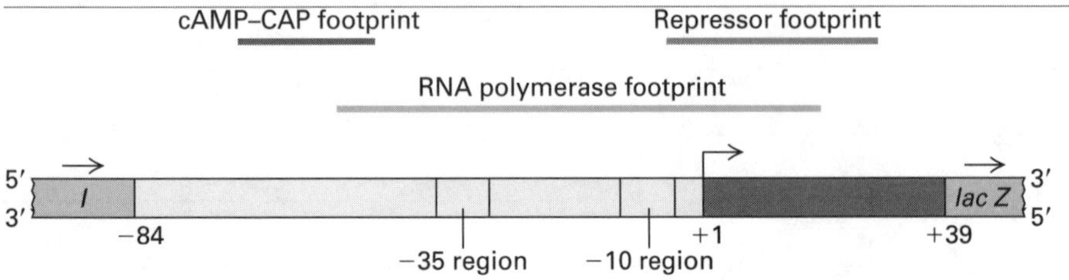

FIGURE 10-9, page 348
Diagram of the *lac*-operon transcription-control region.

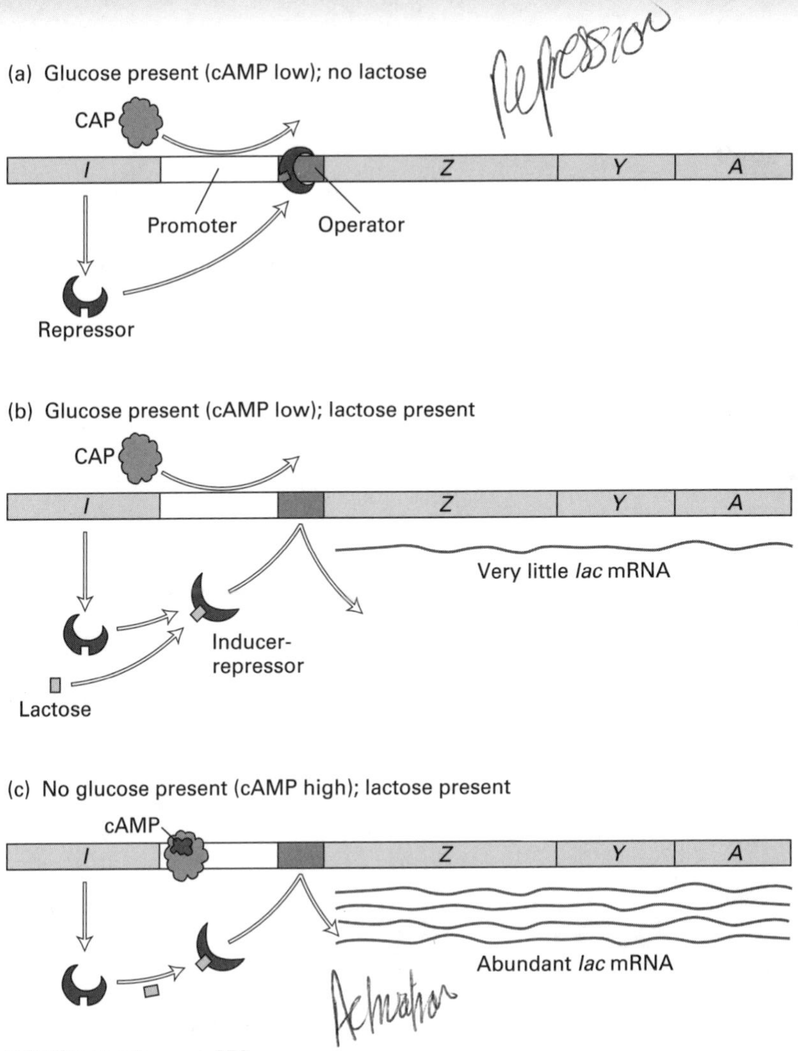

FIGURE 10-16, page 353
Negative and positive transcriptional control of the *lac* operon by the *lac* repressor and cAMP-CAP, respectively.

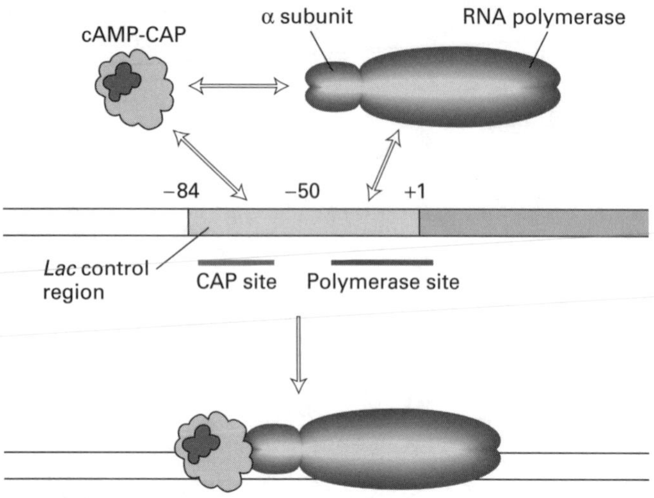

FIGURE 10-17, page 354
Cooperative binding of *E. coli* RNA polymerase and cAMP-CAP to the *lac* promoter.

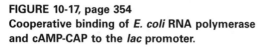

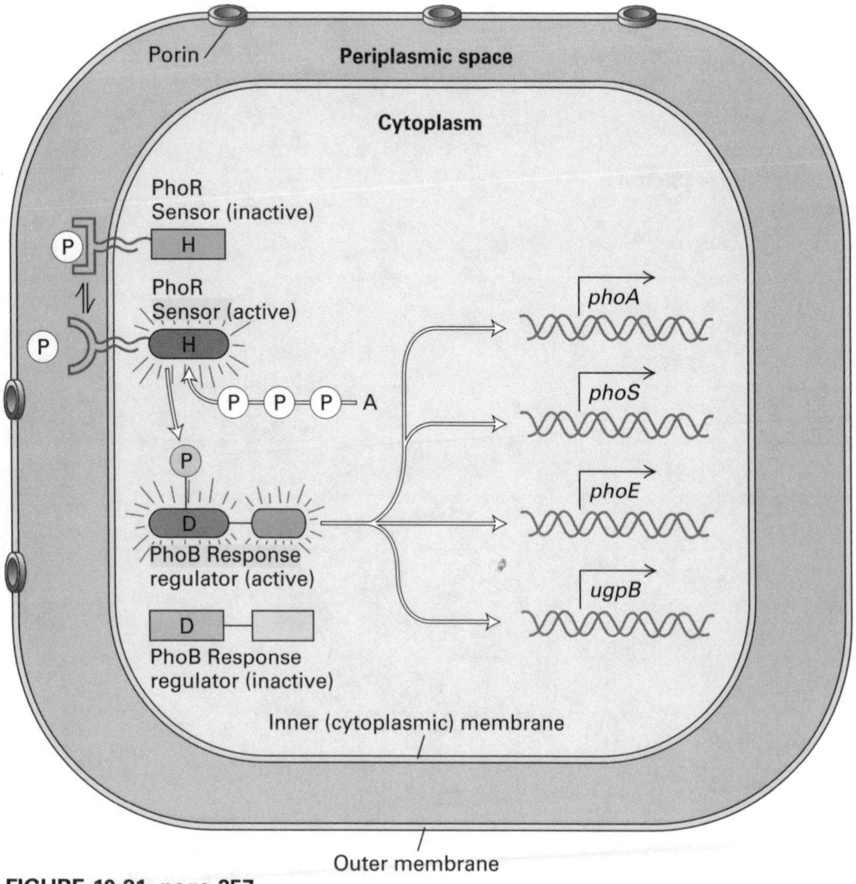

FIGURE 10-21, page 357
The PhoR/PhoB two-component regulatory system in *E. coli.*

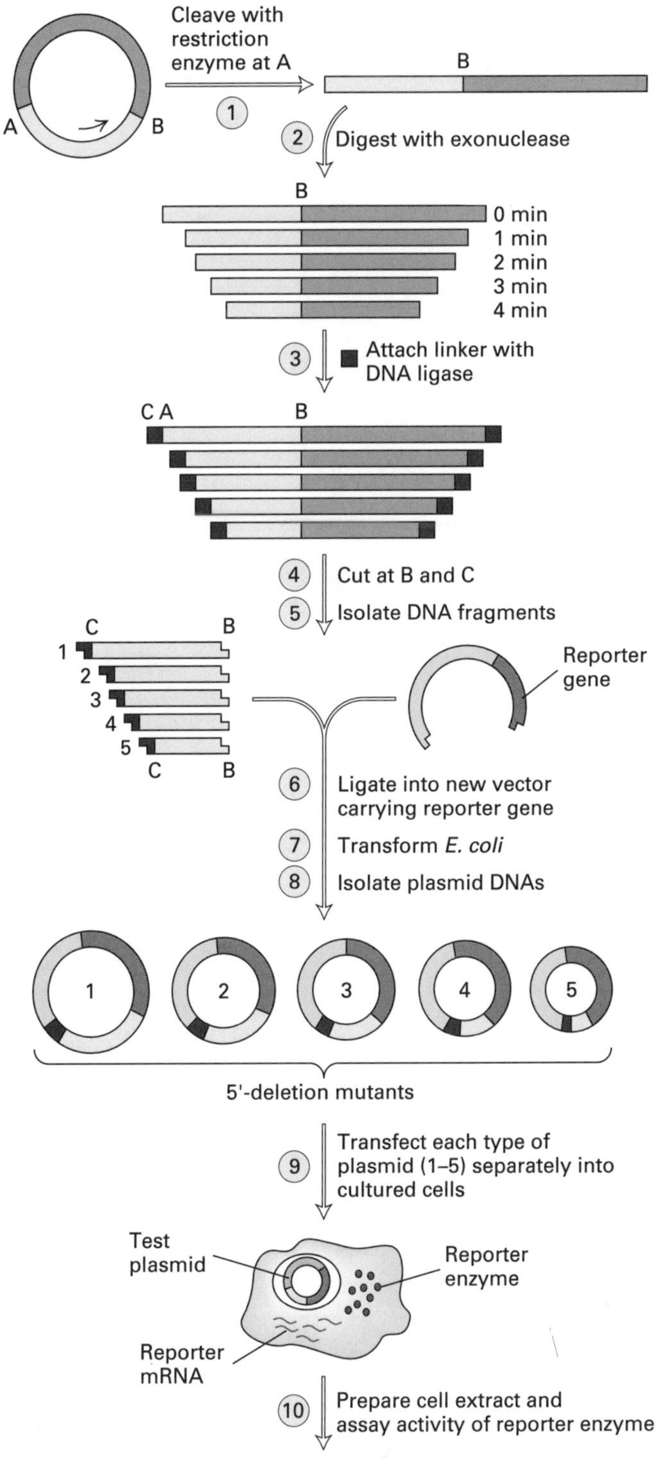

Plasmid no.	Reporter-gene expression
1	+++
2	+++
3	+
4	+
5	−

FIGURE 10-24, page 360
Construction and analysis of a 5′-deletion series to locate transcription-control sequences in DNA upstream of a eukaryotic gene.

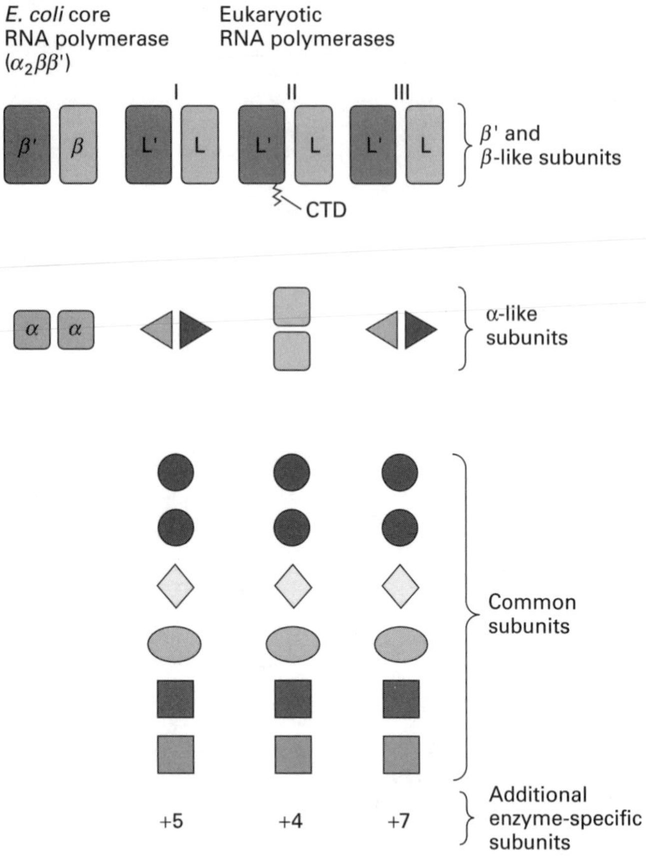

FIGURE 10-26, page 362
Schematic representation of the subunit structure of yeast nuclear RNA polymerases and comparison with *E. coli* RNA core polymerase.

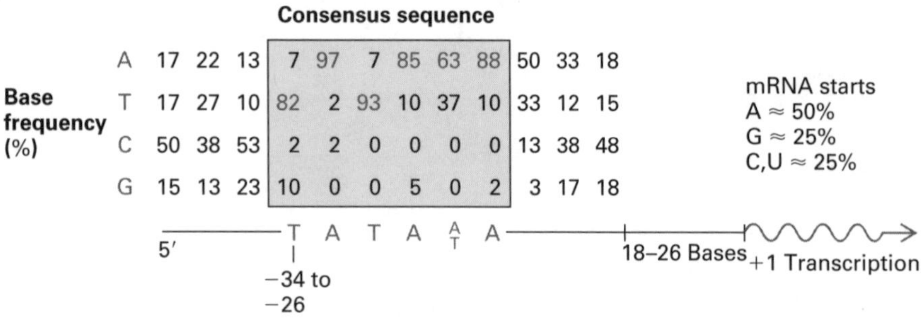

A	17	22	13	7	97	7	85	63	88	50	33	18
T	17	27	10	82	2	93	10	37	10	33	12	15
C	50	38	53	2	2	0	0	0	0	13	38	48
G	15	13	23	10	0	0	5	0	2	3	17	18

FIGURE 10-30, page 365
Comparison of nucleotide sequences upstream of the start site in 60 different vertebrate protein-coding genes.

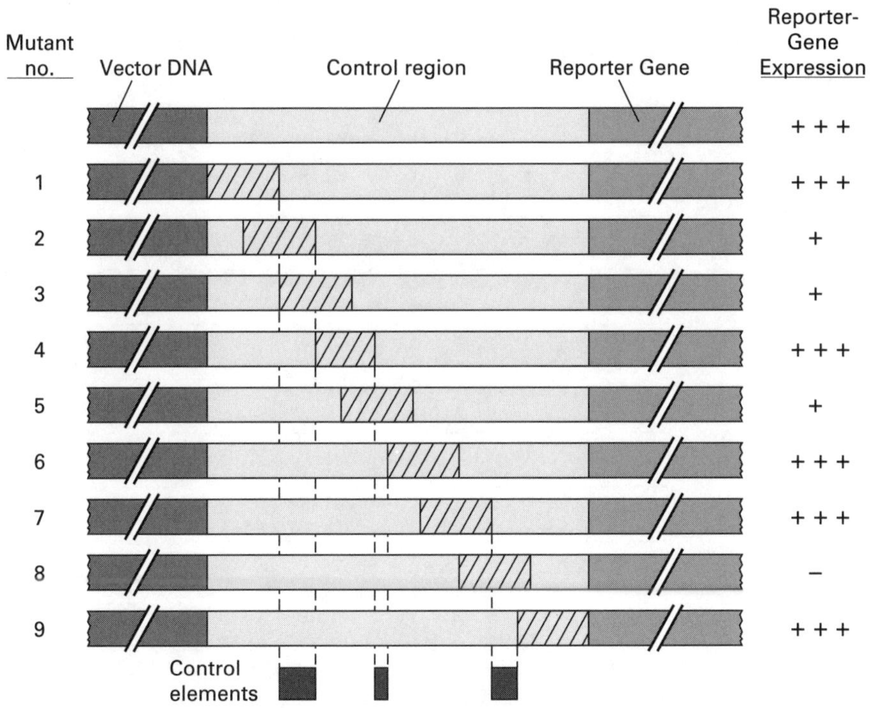

FIGURE 10-31, page 366
Analysis of linker scanning mutations to identify transcription-control elements.

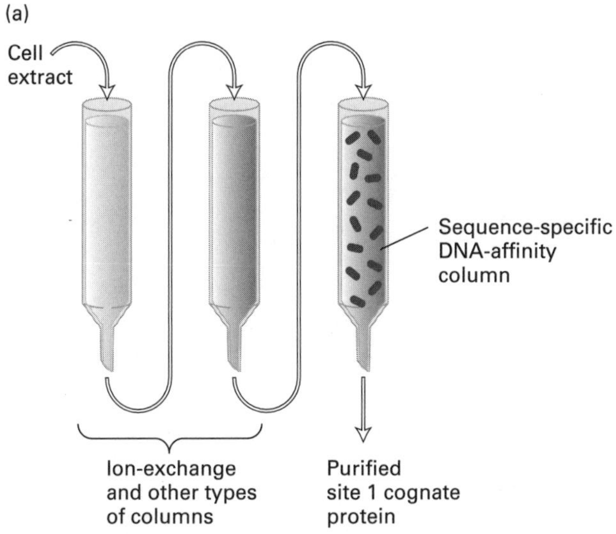

FIGURE 10-35, page 370
Transcription-factor purification.

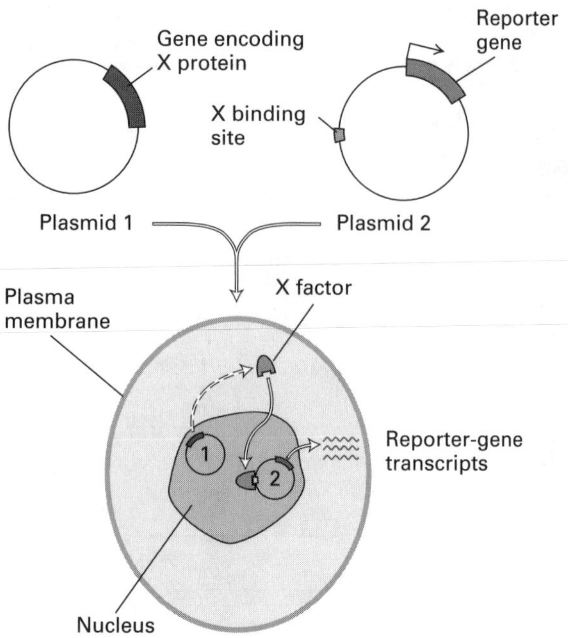

FIGURE 10-37, page 371
In vivo assay for transcription factor activity.

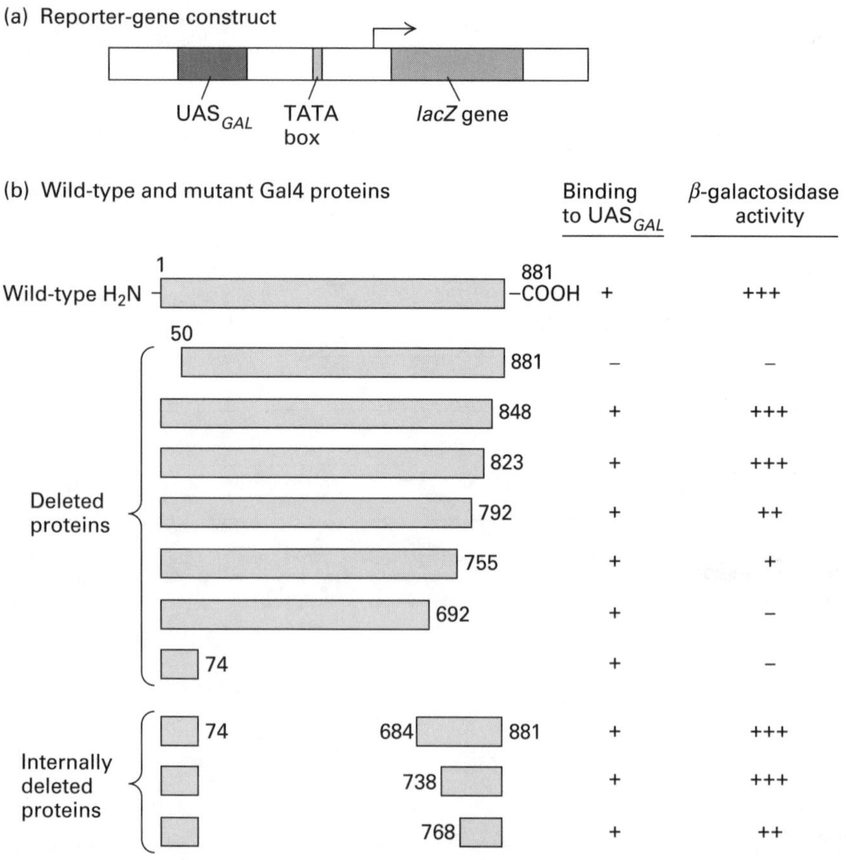

FIGURE 10-38, page 372
Experimental demonstration of separate functional domains in yeast Gal4 protein.

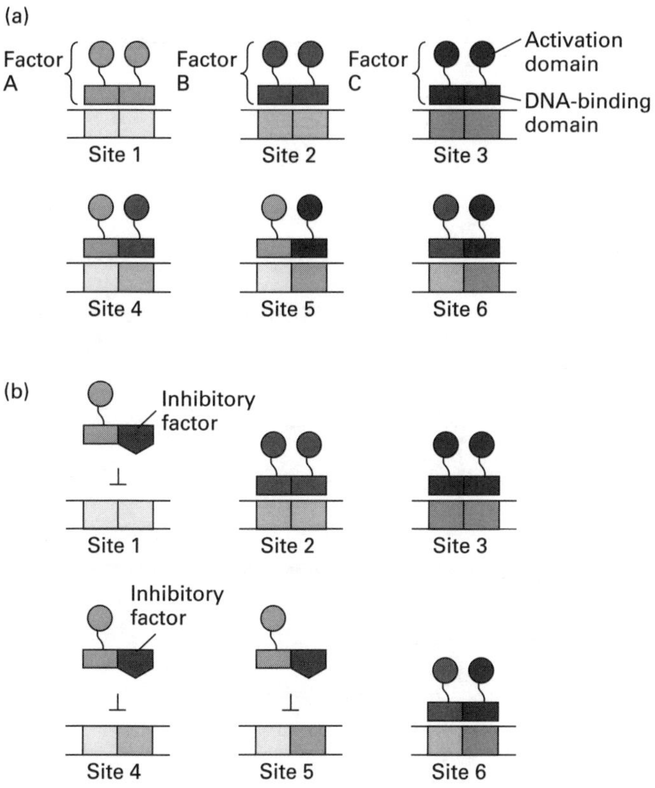

FIGURE 10-45, page 377
Combinatorial possibilities due to formation of heterodimeric transcription factors.

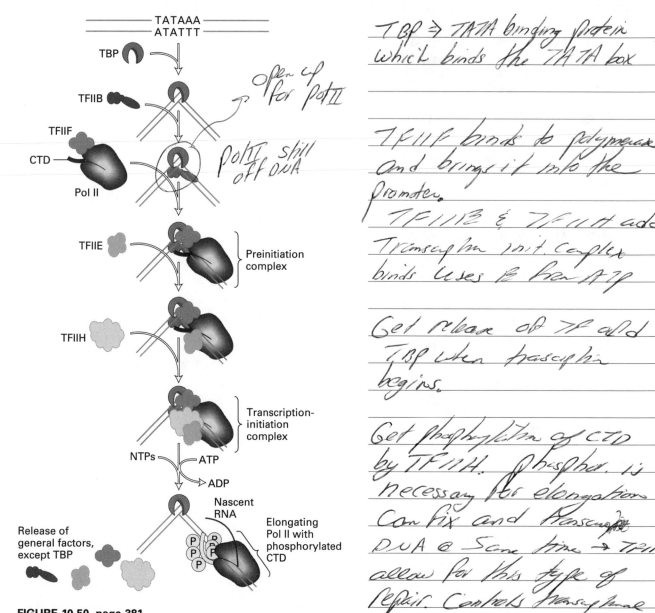

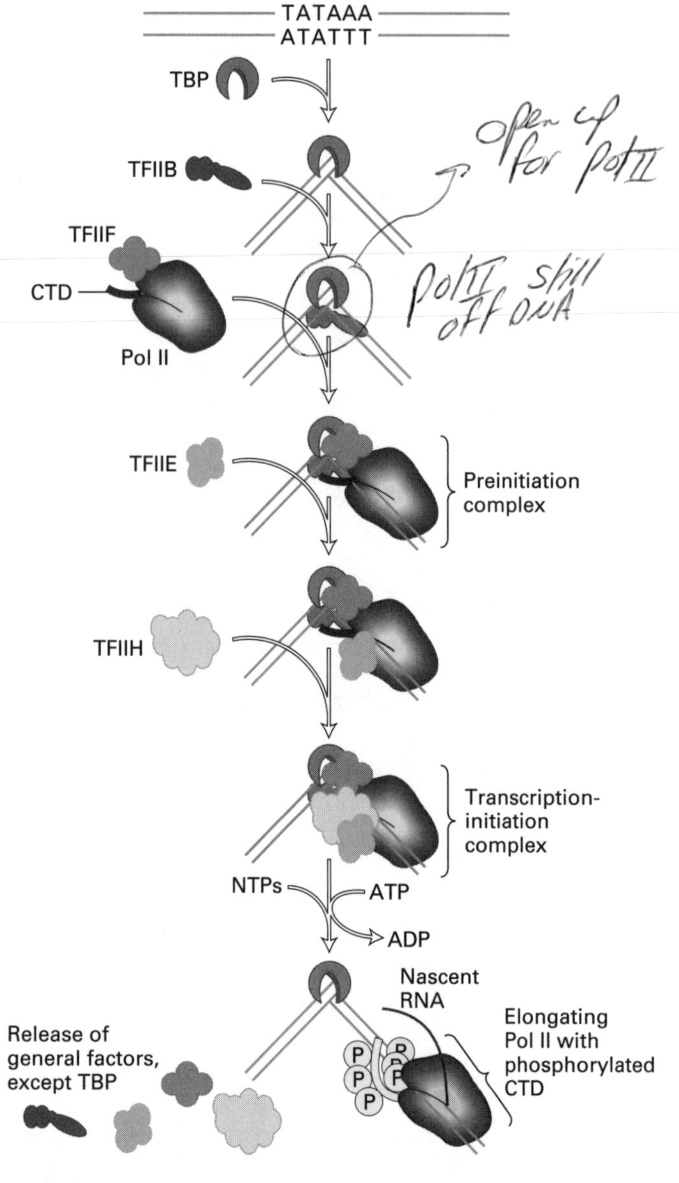

FIGURE 10-50, page 381
Stepwise assembly of a transcription-initiation complex from isolated RNA polymerase II (Pol II) and general transcription factors.

TBP ⇒ TATA binding protein which binds the TATA box

TFIIF binds to polymerase and brings it into the promoter.
TFIIB & TFIIH add
Transcription init. complex binds uses B from ATP

Get release of TF and TBP when transcription begins.

Get phosphorylation of CTD by TFIIH. Phosphor. is necessary for elongation complex and transcribe DNA @ same time → TFIIH allow for this type of repair. Controls transcriptional elongation and does repair

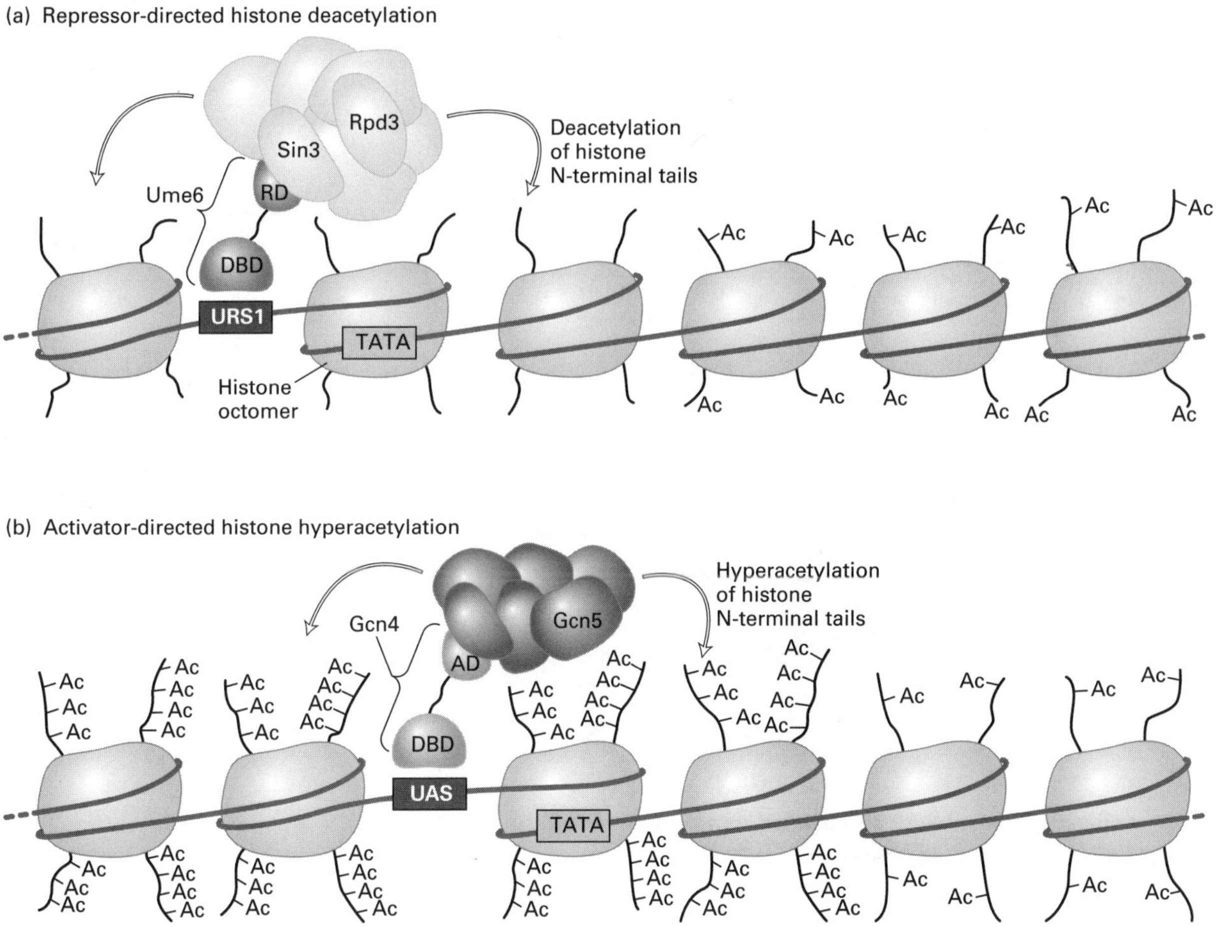

FIGURE 10-58, page 388
Role of deacetylation and hyperacetylation of histone N-terminal tails in yeast transcription control.

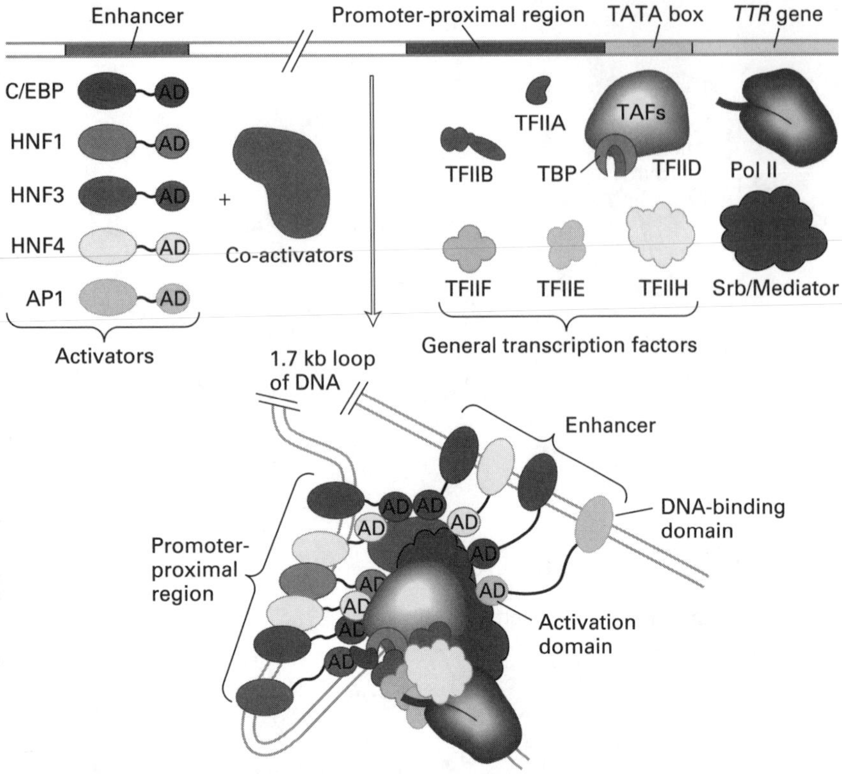

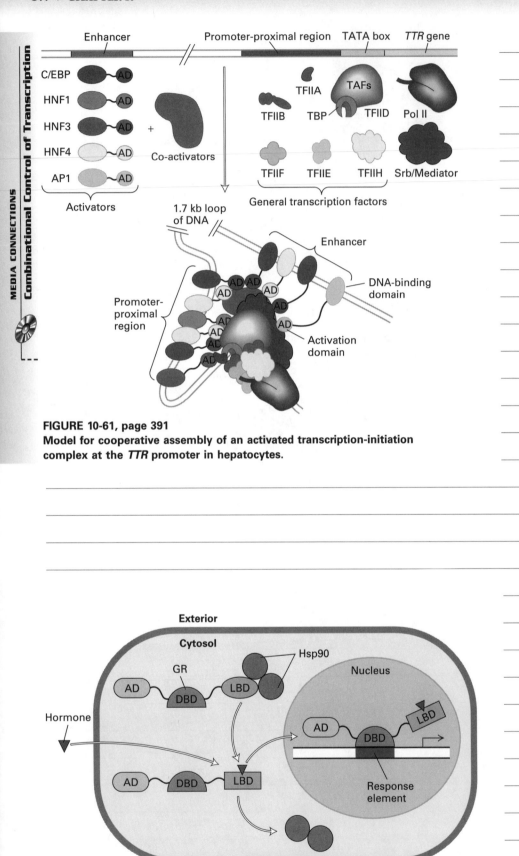

FIGURE 10-61, page 391
Model for cooperative assembly of an activated transcription-initiation complex at the *TTR* promoter in hepatocytes.

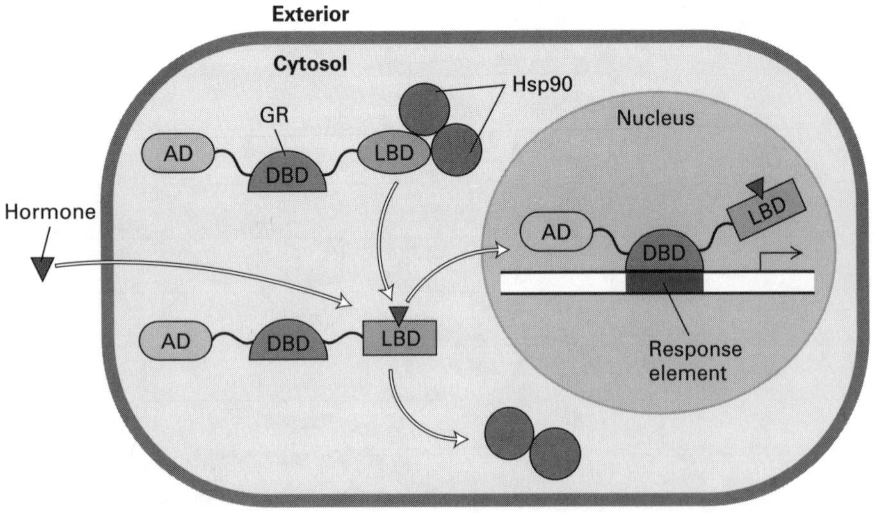

FIGURE 10-67, page 395
Model of hormone-dependent gene activation by the glucocorticoid receptor (GR).

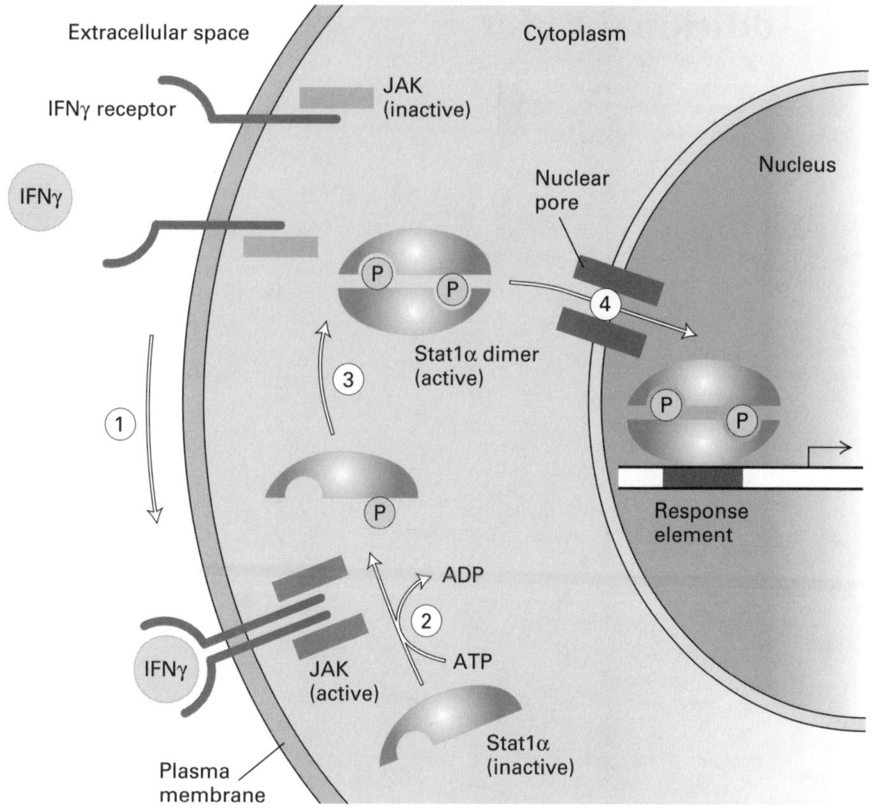

FIGURE 10-68, page 396
Model of IFNγ-mediated gene activation by phosphorylation and dimerization of Stat1α.

Additional Notes

RNA Processing, Nuclear Transport, and Post-Transcriptional Control

11

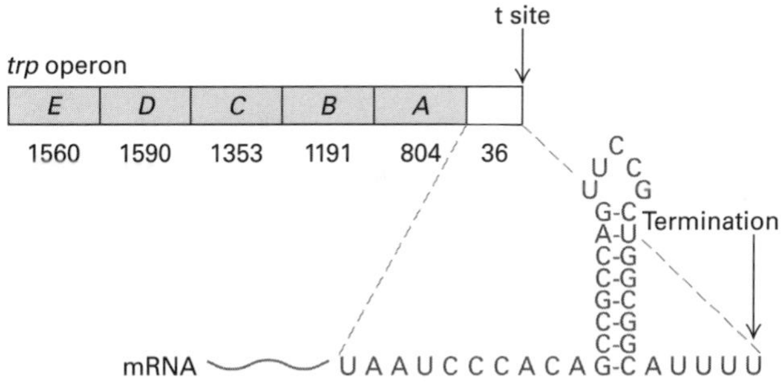

FIGURE 11-1, page 405
Sequence of *trp* termination (t) site, a Rho-independent site.

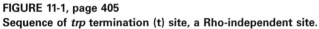

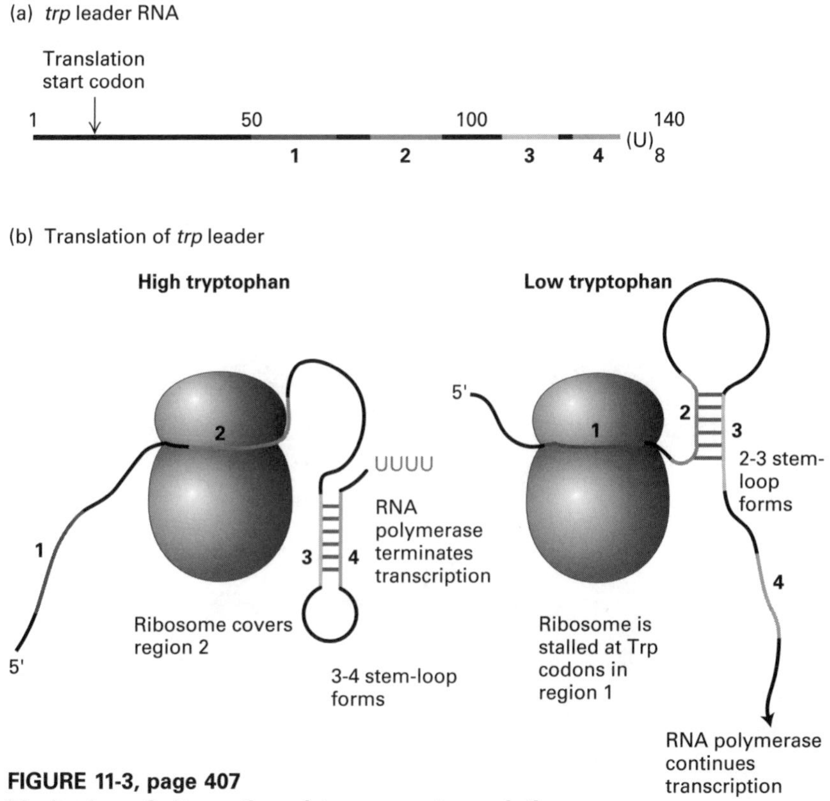

FIGURE 11-3, page 407
Mechanism of attenuation of *trp*-operon transcription.

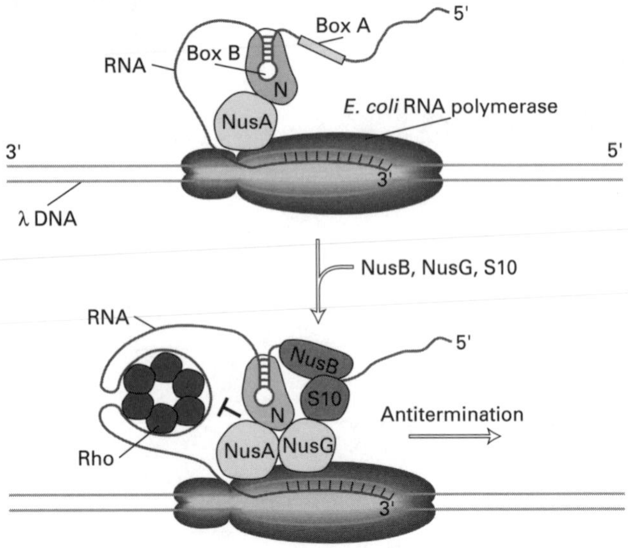

FIGURE 11-5, page 408
Antitermination by λ-phage N protein and *E. coli* cellular proteins.

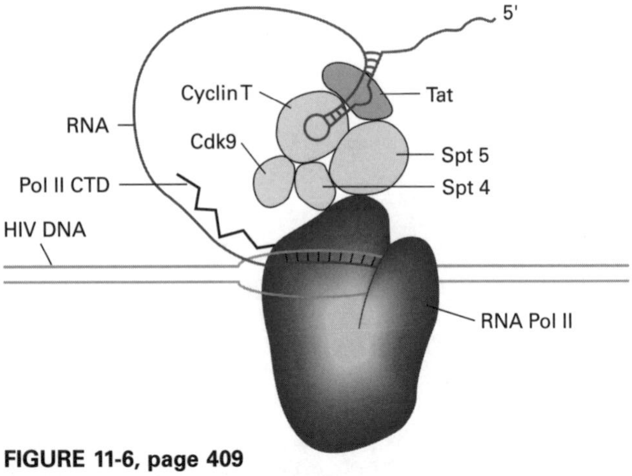

FIGURE 11-6, page 409
Antitermination complex composed of HIV Tat protein and several eukaryotic cellular proteins.

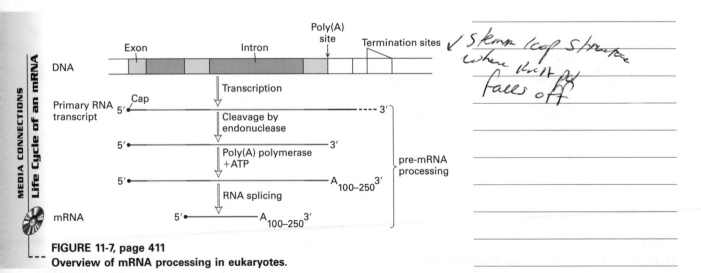

MEDIA CONNECTIONS
Life Cycle of an mRNA

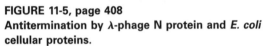

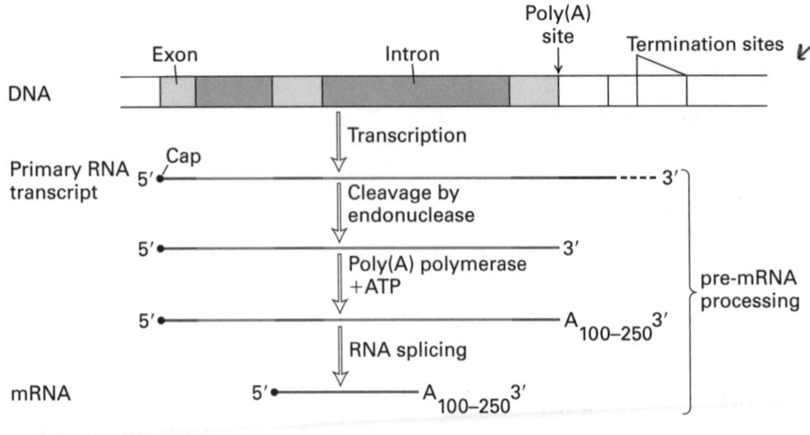

✓ Stem loop structure when RNA pol falls off

FIGURE 11-7, page 411
Overview of mRNA processing in eukaryotes.

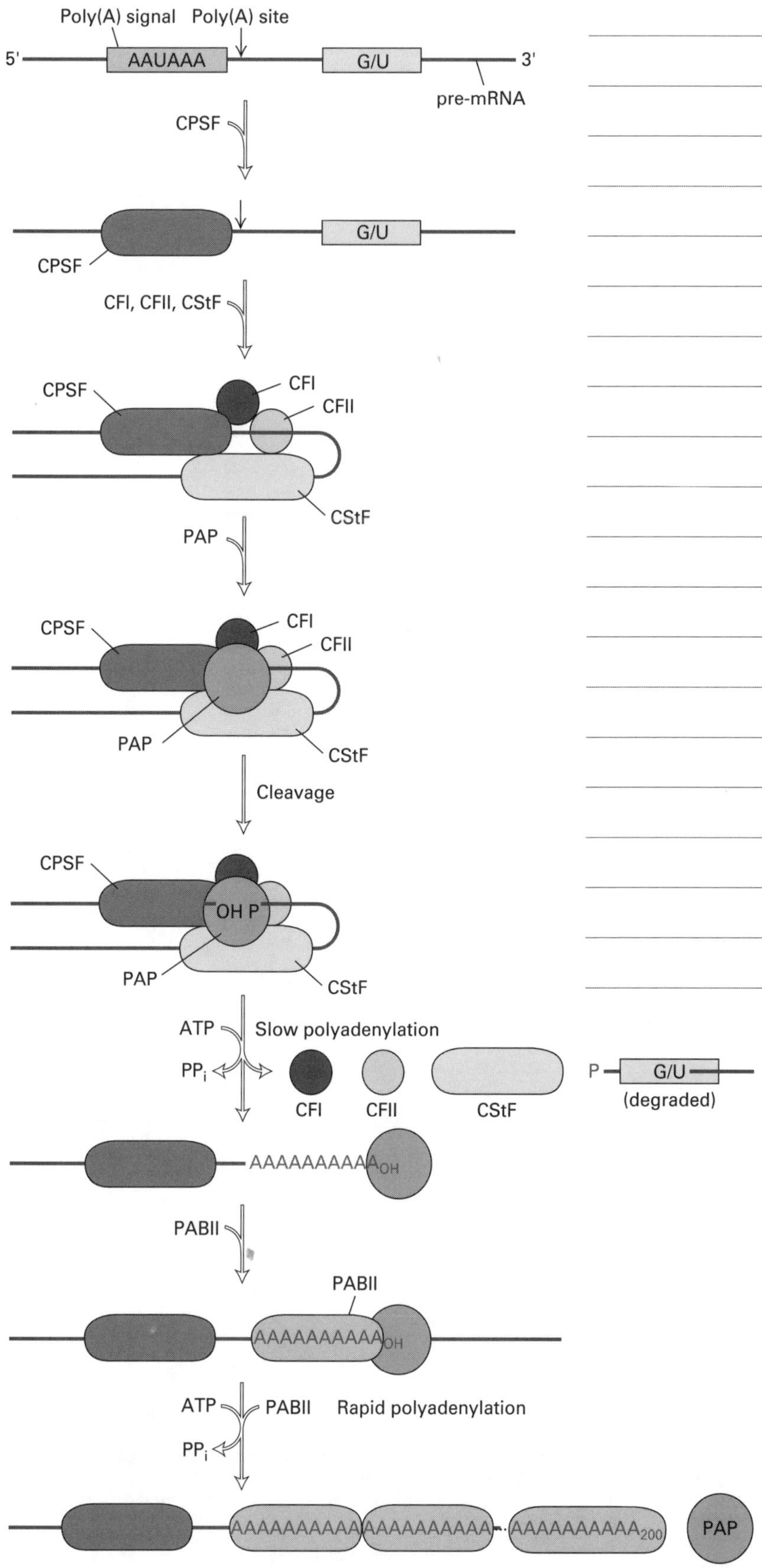

FIGURE 11-12, page 414
Model for cleavage and polyadenylation of pre-mRNAs in mammalian cells.

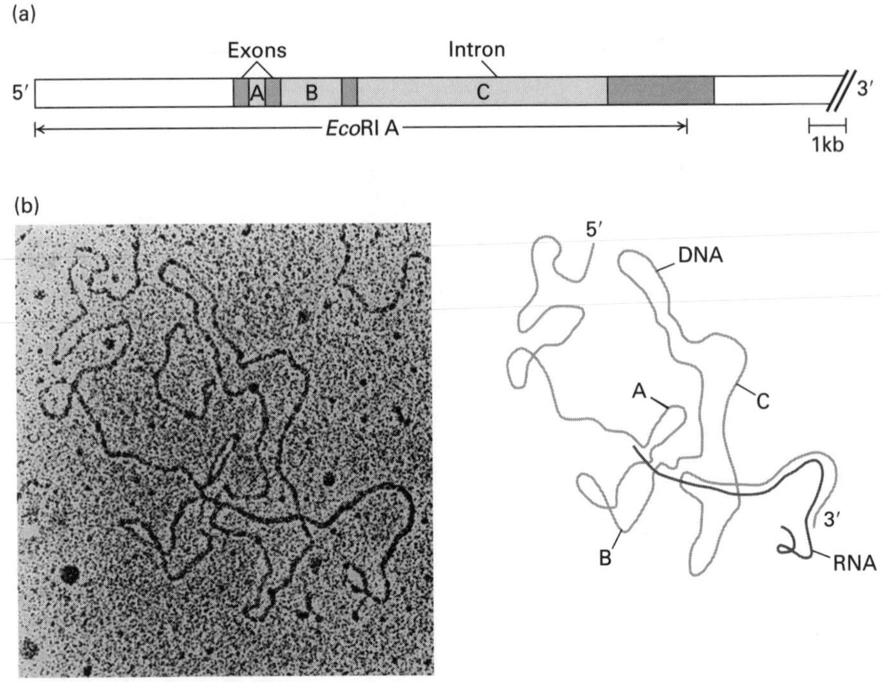

FIGURE 11-13, page 415
Demonstration that introns are spliced out by electron microscopy of RNA-DNA hybrid between adenovirus DNA and the mRNA encoding hexon, a major viral protein.

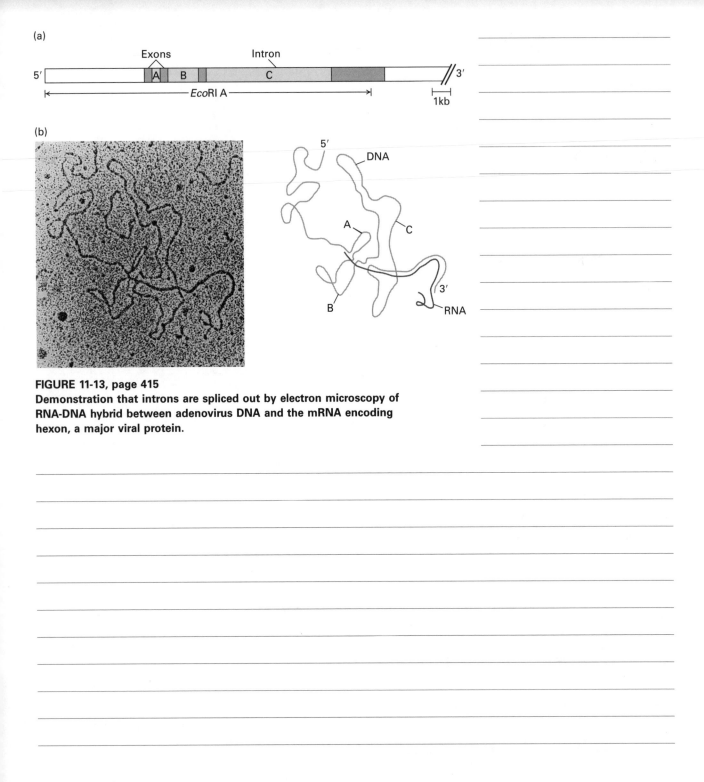

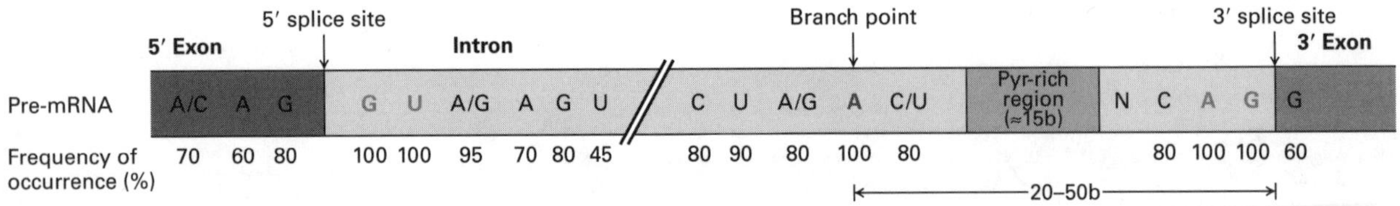

FIGURE 11-14, page 416
Consensus sequences around 5′ and 3′ splice sites in vertebrate pre-mRNAs.

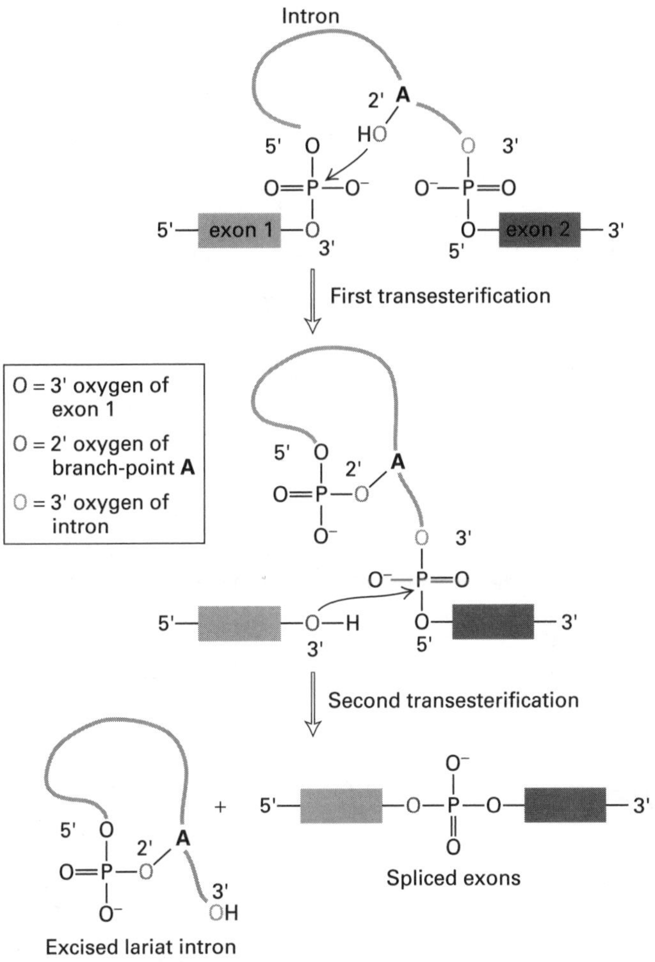

FIGURE 11-16, page 417
Splicing of exons in pre-mRNA occurs via two transesterification reactions.

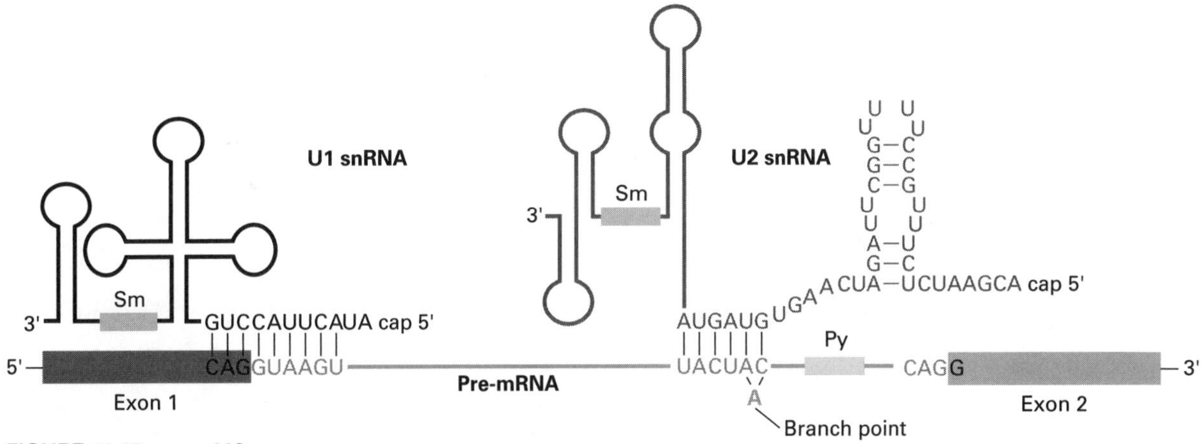

FIGURE 11-17, page 418
Diagram of interactions between pre-mRNA, U1 snRNA, and U2 snRNA early in the splicing process.

MEDIA CONNECTIONS
mRNA Splicing

MEDIA CONNECTIONS
mRNA Splicing

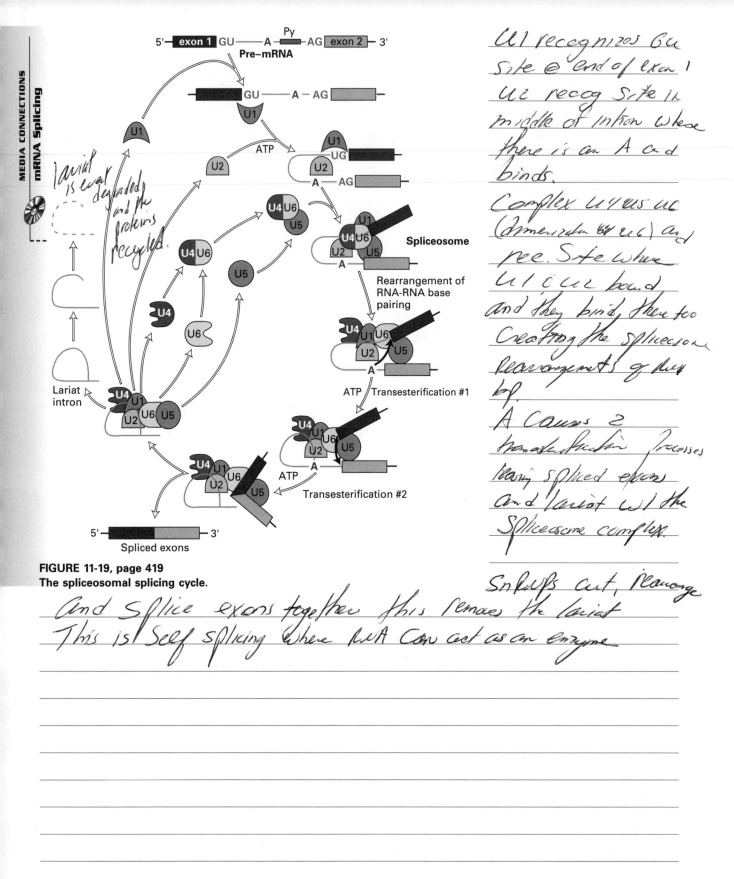

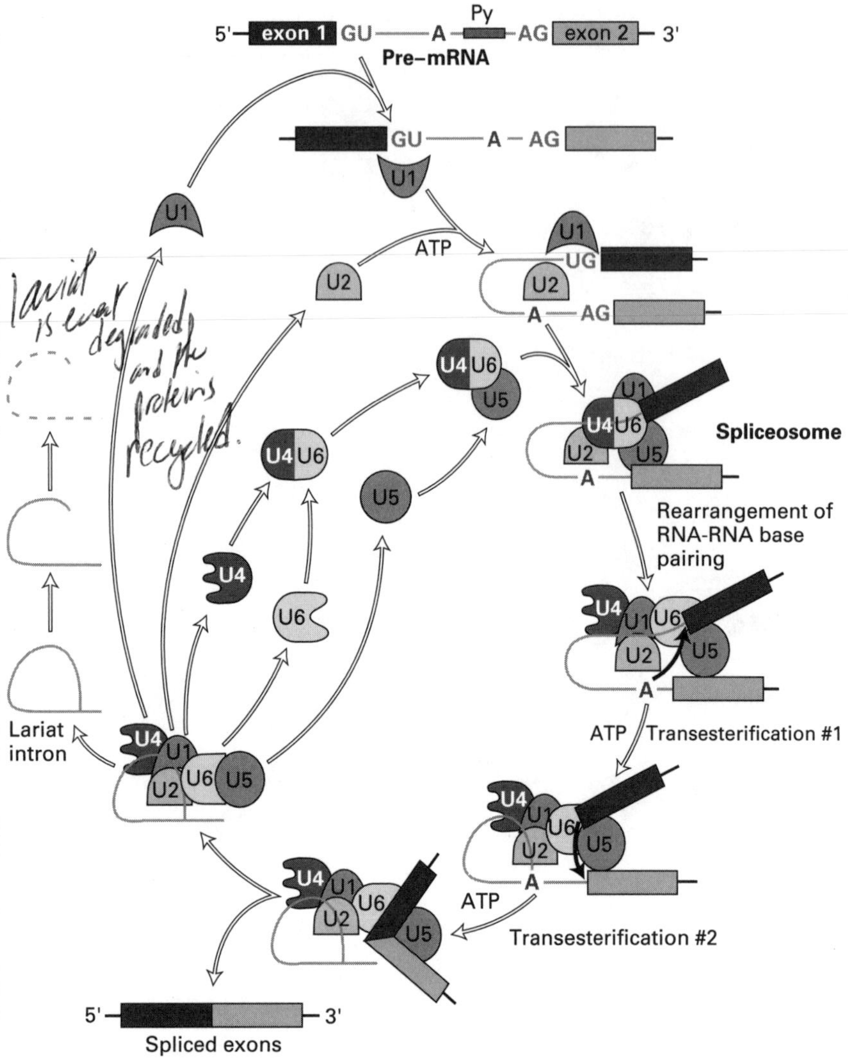

FIGURE 11-19, page 419
The spliceosomal splicing cycle.

U1 recognizes GU site @ end of exon 1
U2 recog site in middle of intron where there is an A and binds.
Complex U4 U5 U6 (dimerization of U4 U6) and rec. site where U1 & U2 bound and they bind there too creating the spliceosome
Rearrangements of the bp.
A causes 2 transesterification processes leaving spliced exons and lariat w/ the spliceosome complex.

SnRNPs cut, rearrange and splice exons together this removes the lariat
This is self splicing where RNA can act as an enzyme

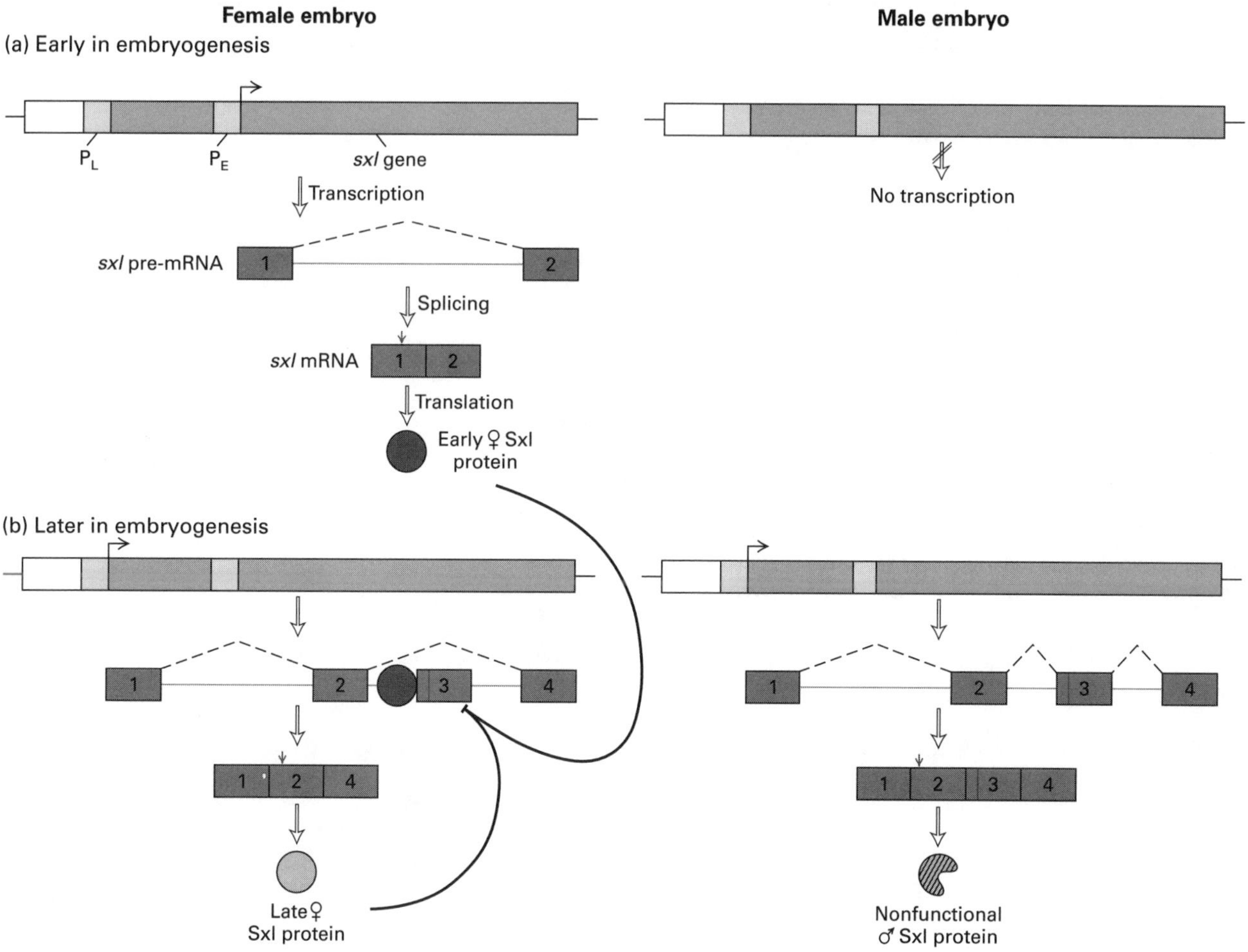

FIGURE 11-25, page 424
Expression of Sex-lethal (Sxl) protein during *Drosophila* embryogenesis.

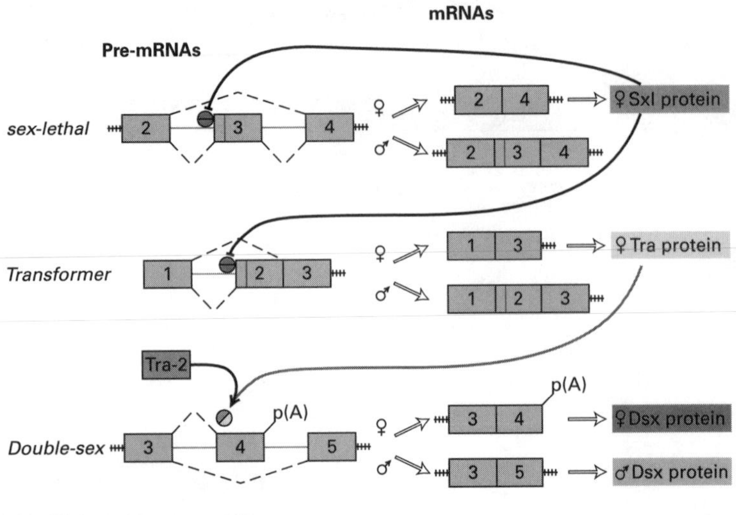

FIGURE 11-26, page 425
Cascade of regulated splicing that controls expression of *sex-lethal (sxl)*, *transformer (tra)*, and *double-sex (dsx)* genes in *Drosophila* embryos.

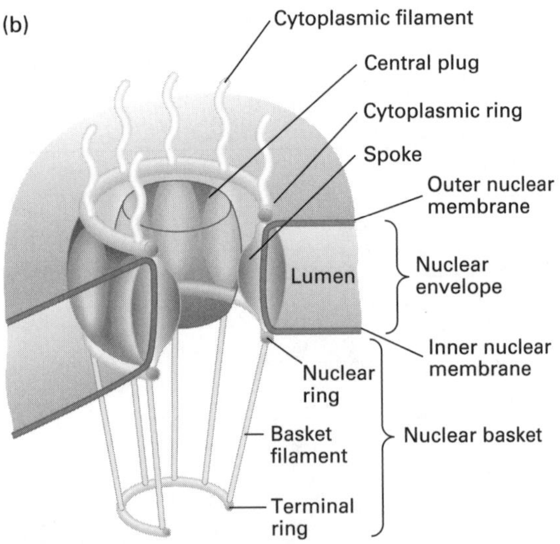

FIGURE 11-28 (b), page 427
Nuclear pore complex. (b) Cut-away model of the NPC. (Adapted from M. Ohno et al., 1998, *Cell* **92**:327.)

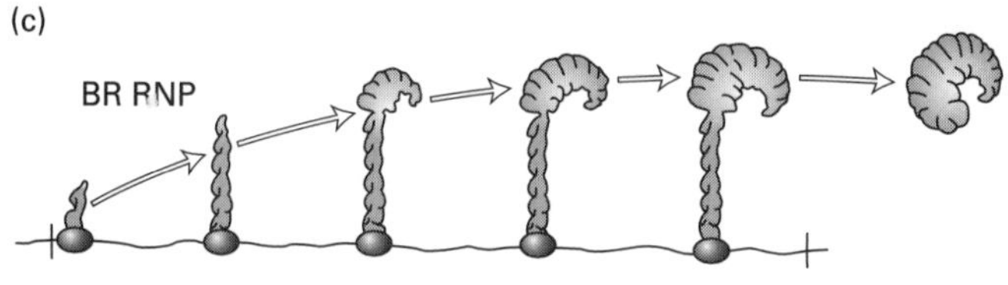

FIGURE 11-29 (c), page 428
Formation of coiled heterogeneous ribonucleoprotein (hn-RNP) during synthesis of the *Chironomous tentans* Balbiani ring (BR) mRNA. (c) A model for the structure and biogenesis of BR hnRNP. (Adated from B. Daneholt, 1997, *Cell* **88**:585.)

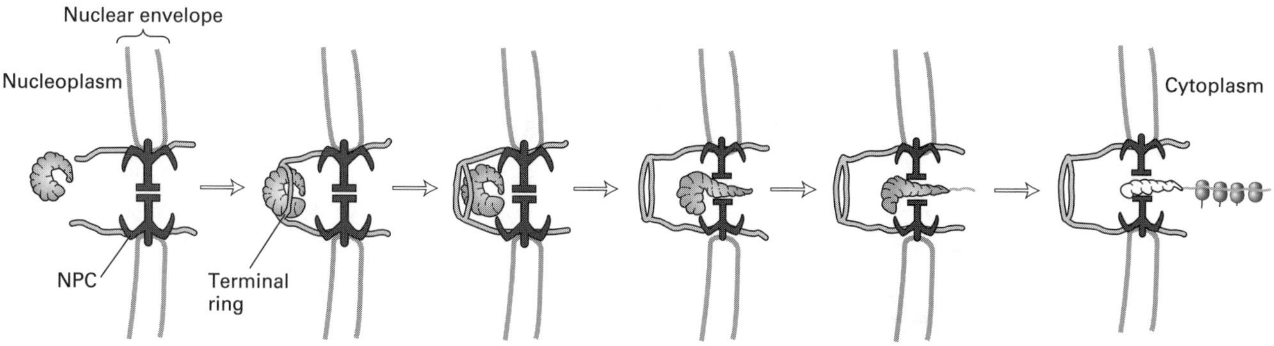

FIGURE 11-31, page 429
Model for passage of mRNPs through the nuclear pore complex (NPC) based on electron microscopic studies of Balbiani ring (BR) mRNP transport in *Chironomous tentans*.

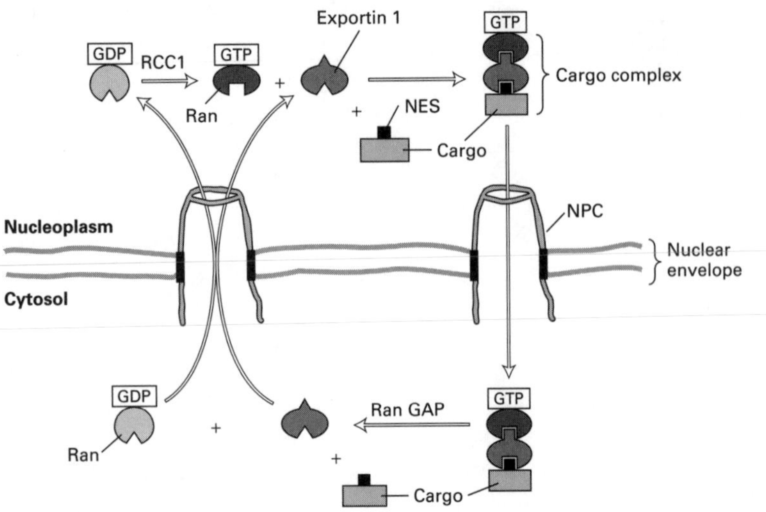

FIGURE 11-33, page 431
Proposed mechanism for the transport of "cargo" proteins containing a leucine-rich nuclear-export signal (NES) from the nucleus to the cytosol.

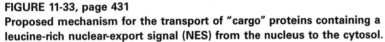

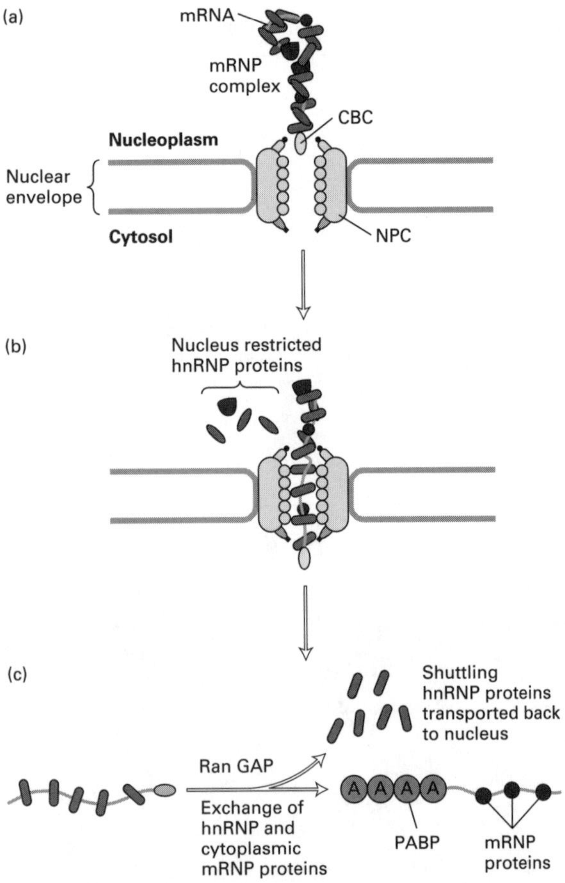

FIGURE 11-34, page 431
Proposed mechanism for hnRNP protein–mediated export of mRNA from the nucleus.

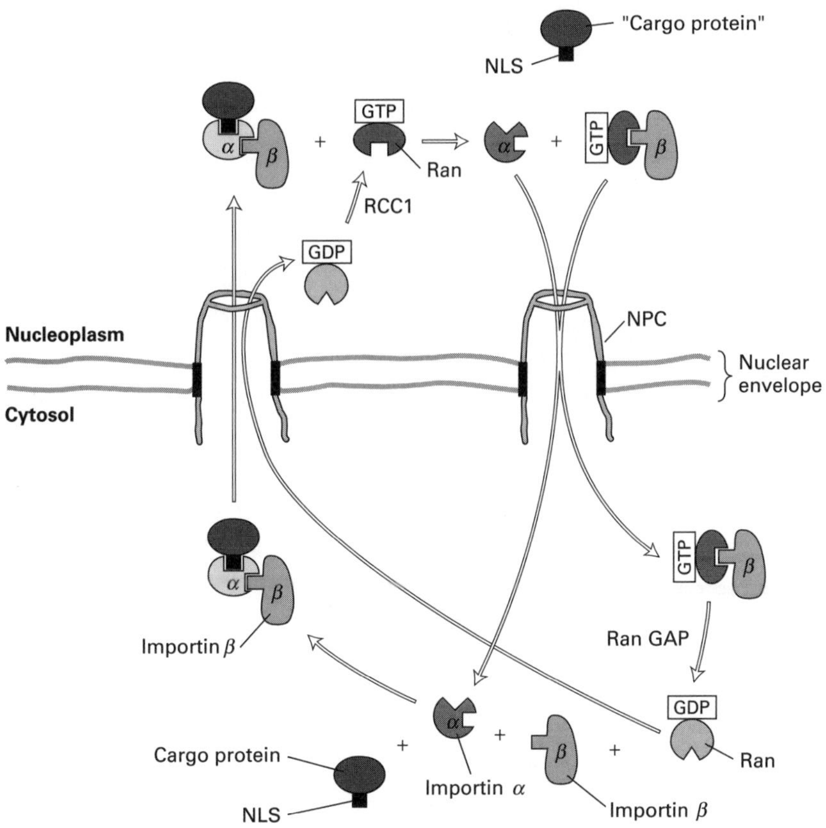

FIGURE 11-37, page 434
Proposed mechanism for the transport of "cargo" proteins containing a basic nuclear-localization signal (NLS) from the cytoplasm to the nucleus.

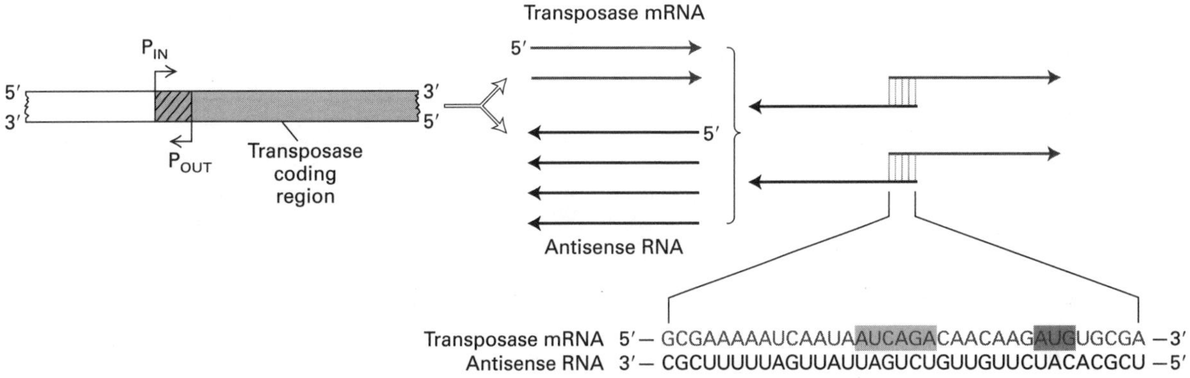

FIGURE 11-46, page 443
Antisense control of translation of transposase mRNA encoded by IS*10*, a bacterial mobile element.

Additional Notes

DNA Replication, Repair, and Recombination

12

MEDIA CONNECTIONS **Bidirectional Replication of DNA**

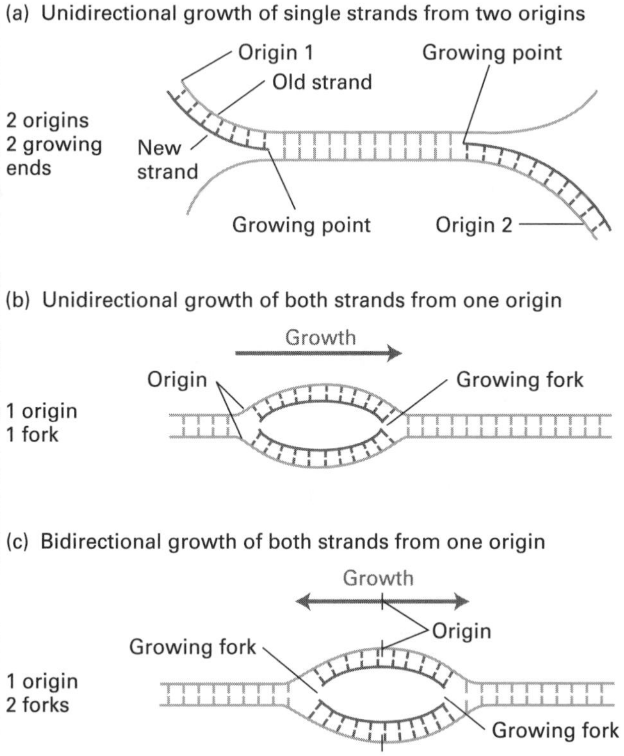

FIGURE 12-2, page 455
Three mechanisms of DNA strand growth that are consistent with semiconservative replication.

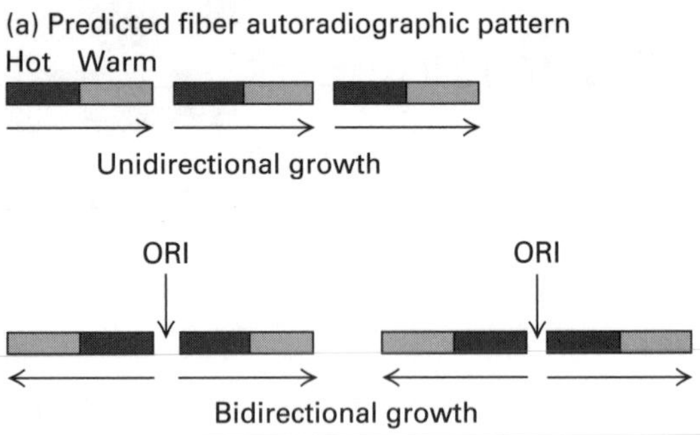

FIGURE 12-3, page 455
Demonstration of bidirectional growth of cellular DNA chains by fiber autoradiography.

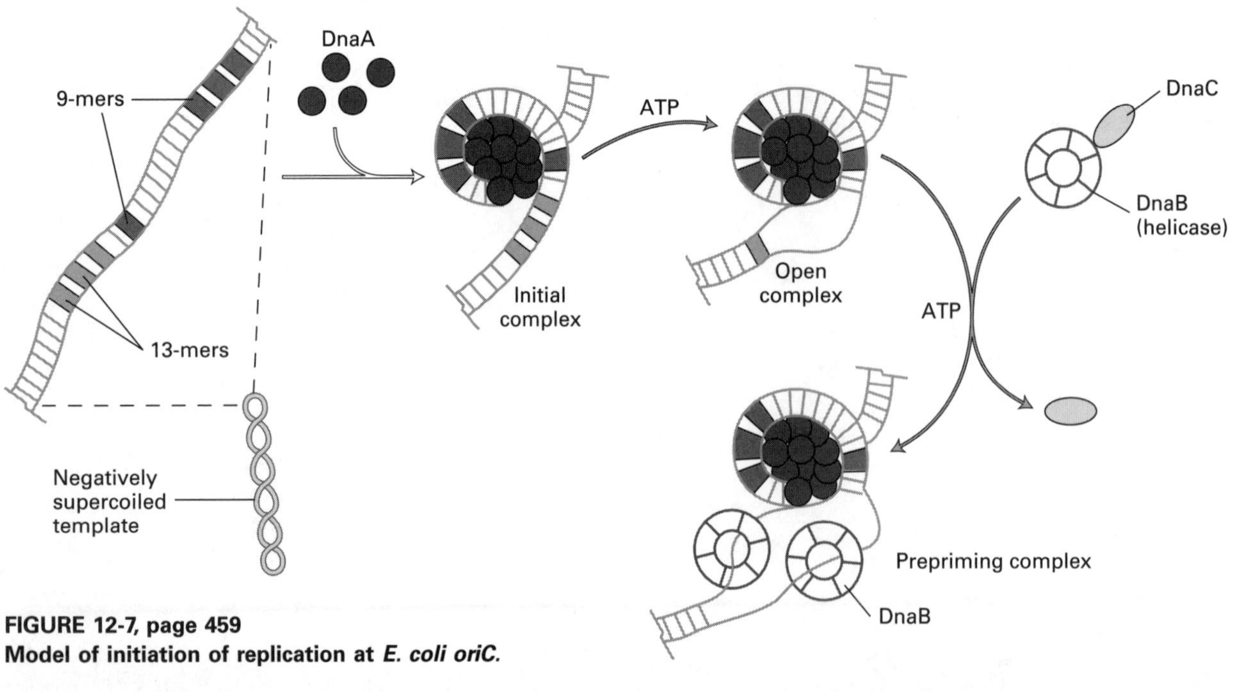

FIGURE 12-7, page 459
Model of initiation of replication at *E. coli oriC.*

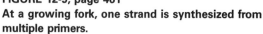

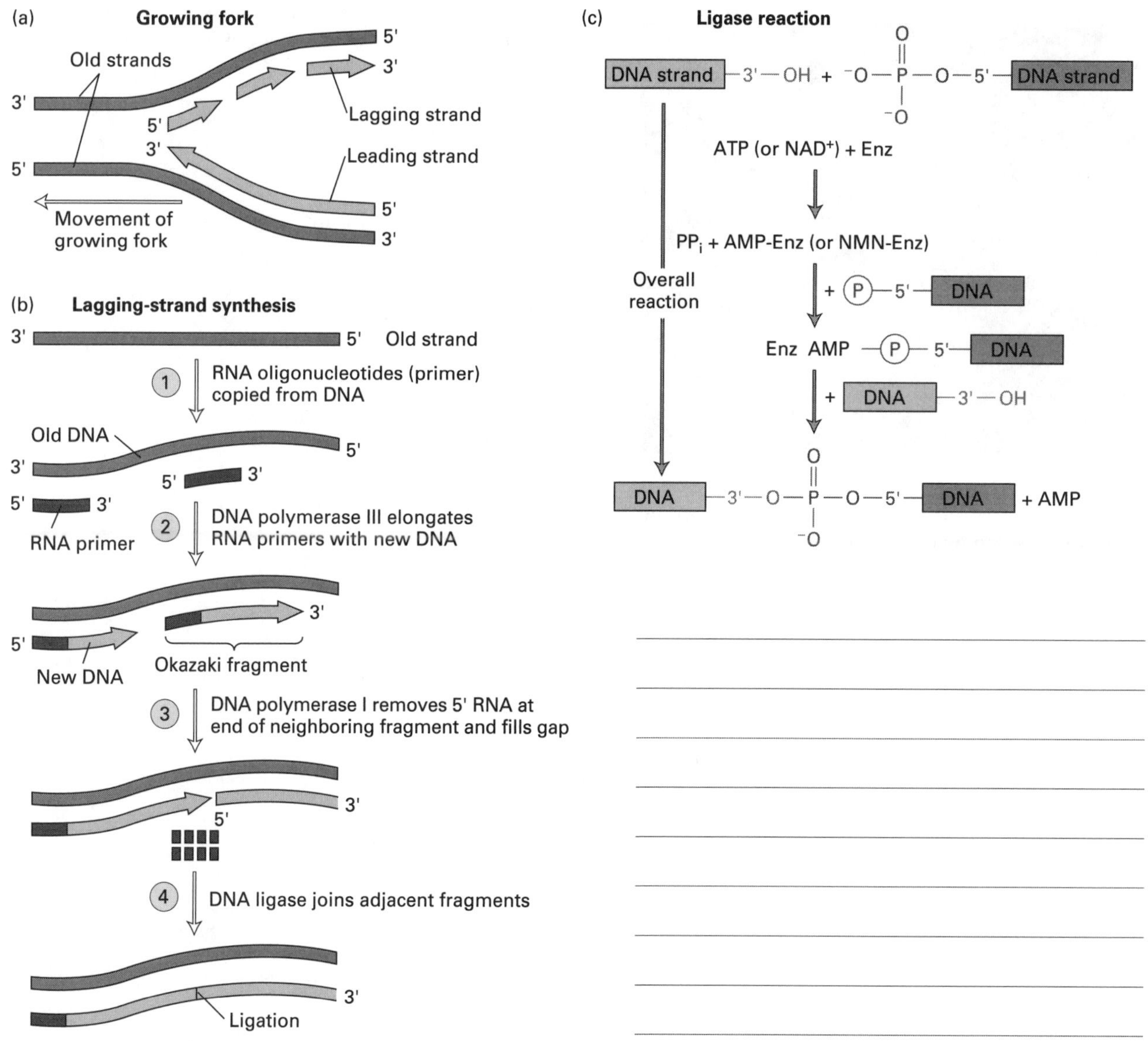

FIGURE 12-9, page 461
At a growing fork, one strand is synthesized from multiple primers.

MEDIA CONNECTIONS Nucleotide Polymerization by DNA Polymerase

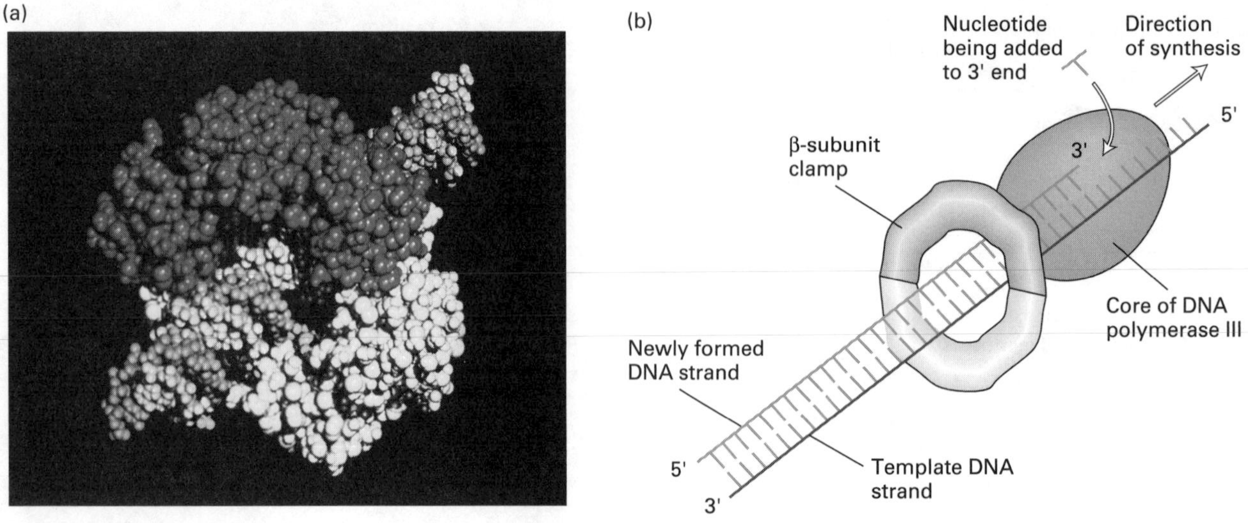

FIGURE 12-10, page 463
A β-subunit dimer tethers the core of *E. coli* DNA polymerase III to DNA, thereby increasing its processivity.

MEDIA CONNECTIONS Coordination of Leading and Lagging Strand Synthesis

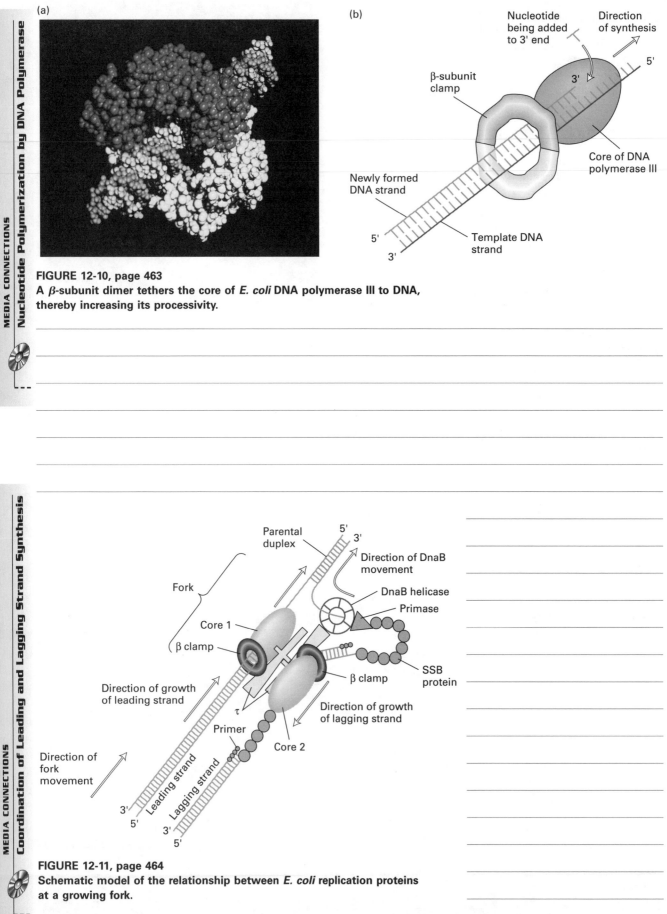

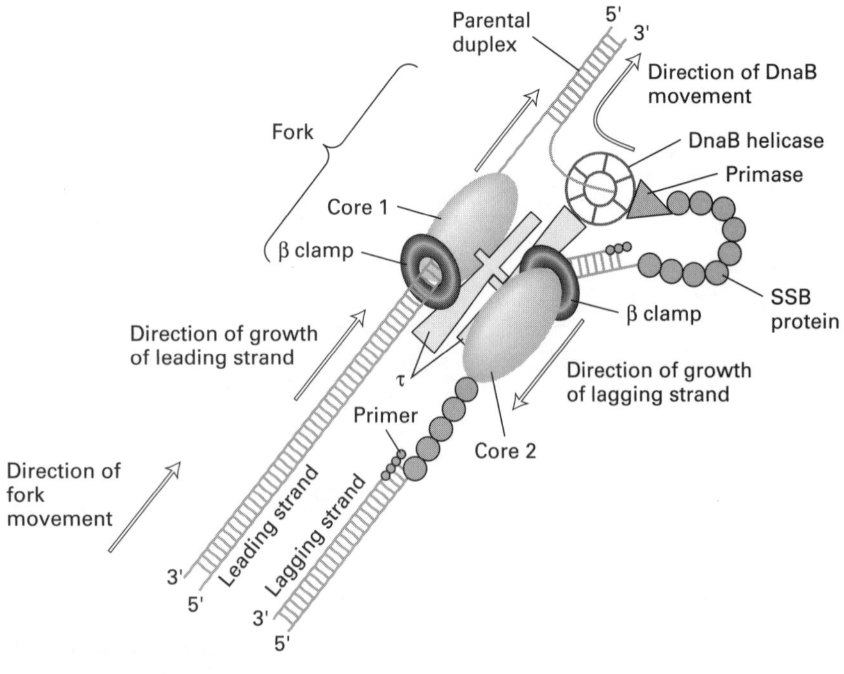

FIGURE 12-11, page 464
Schematic model of the relationship between *E. coli* replication proteins at a growing fork.

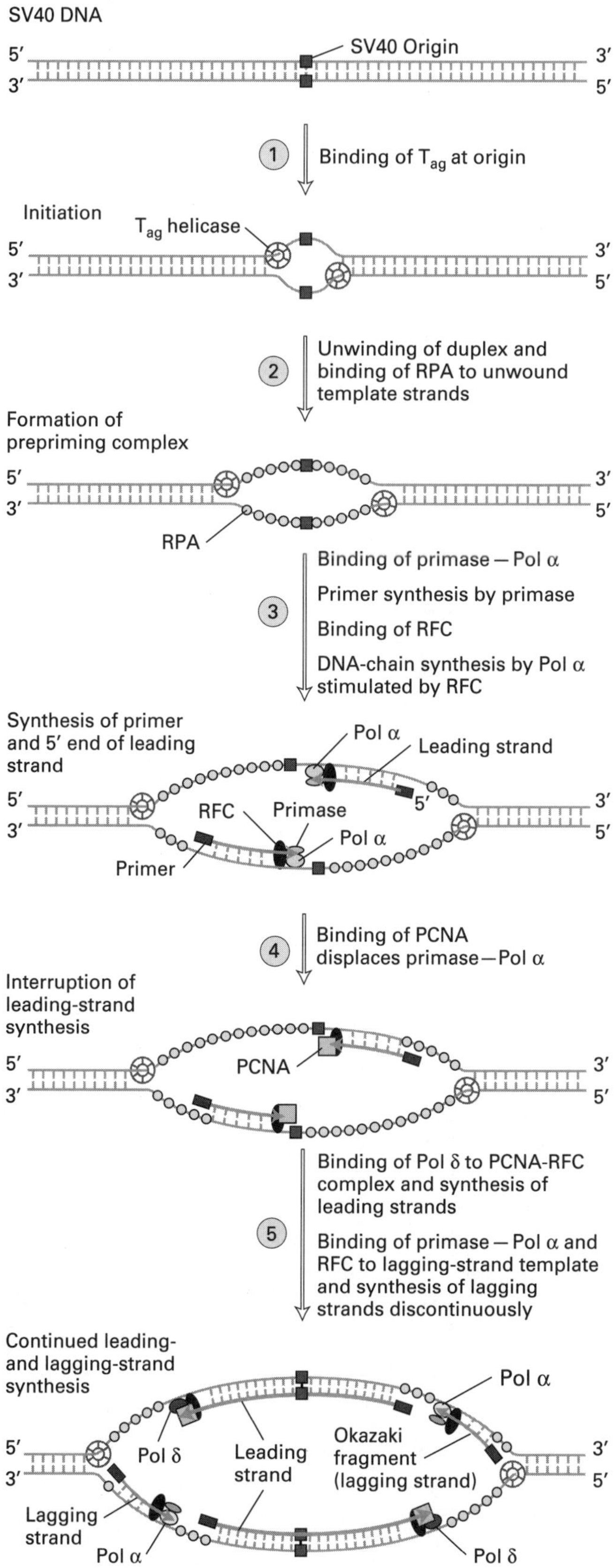

FIGURE 12-12, page 465
Model of in vitro replication of SV40 DNA by eukaryotic enzymes.

MEDIA CONNECTIONS
Telomere Replication

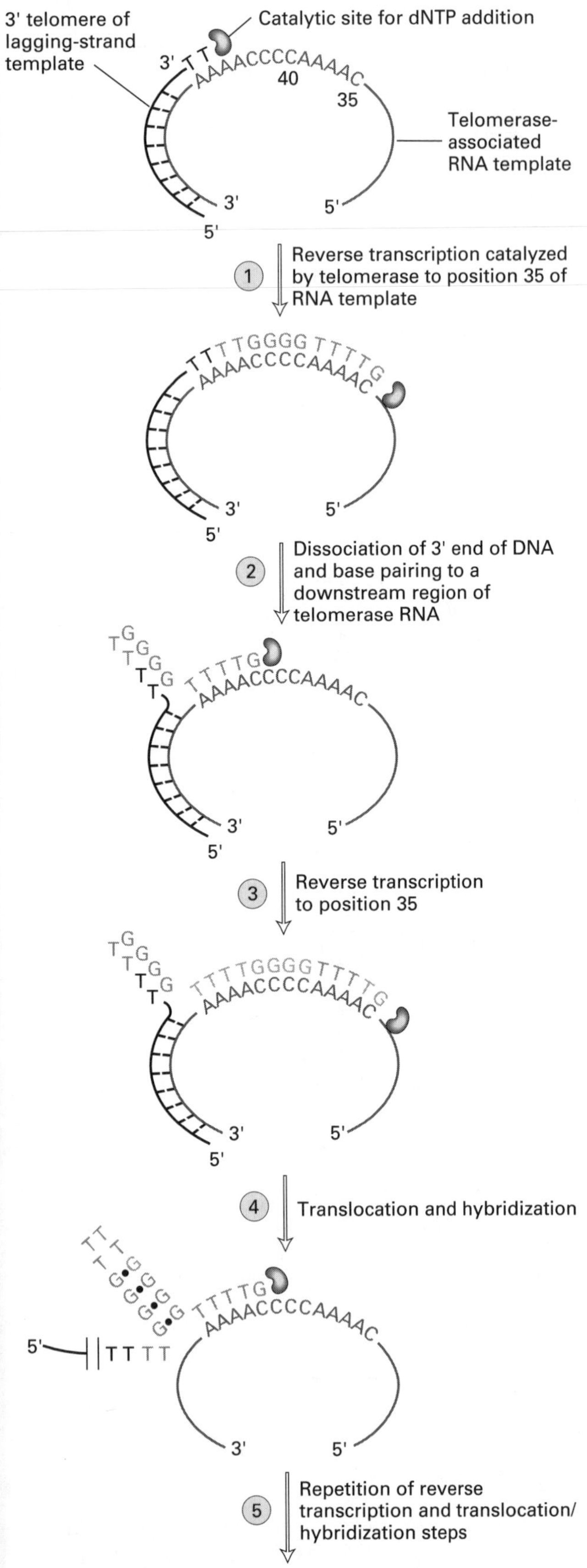

FIGURE 12-13, page 466
Mechanism of action of telomerase.

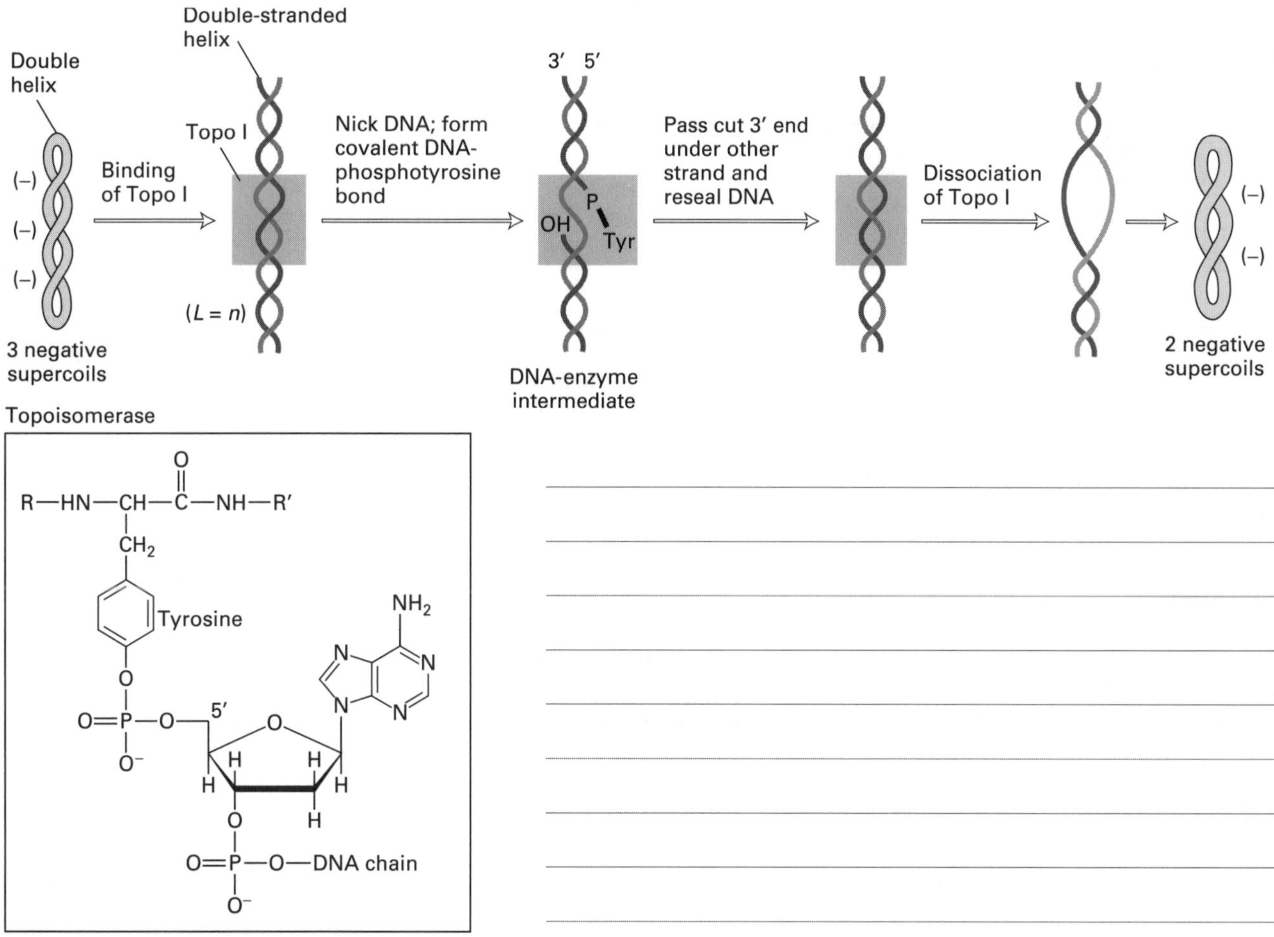

FIGURE 12-14, page 468
Action of *E. coli* type I topoisomerase (Topo I).

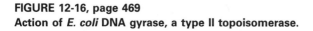

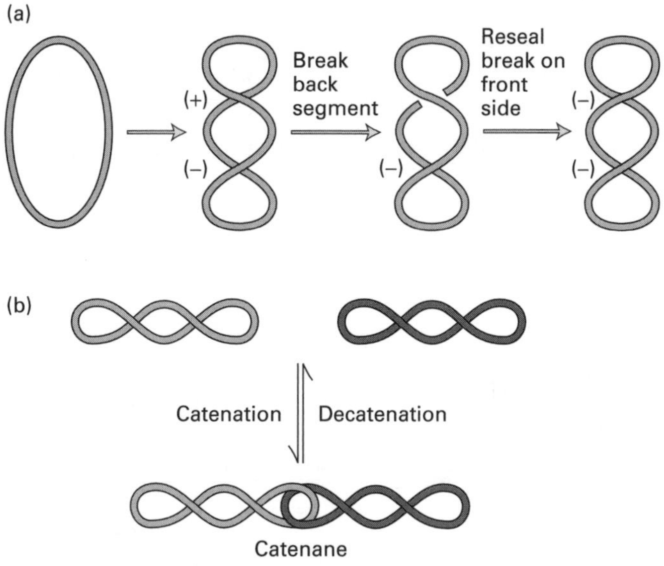

FIGURE 12-16, page 469
Action of *E. coli* DNA gyrase, a type II topoisomerase.

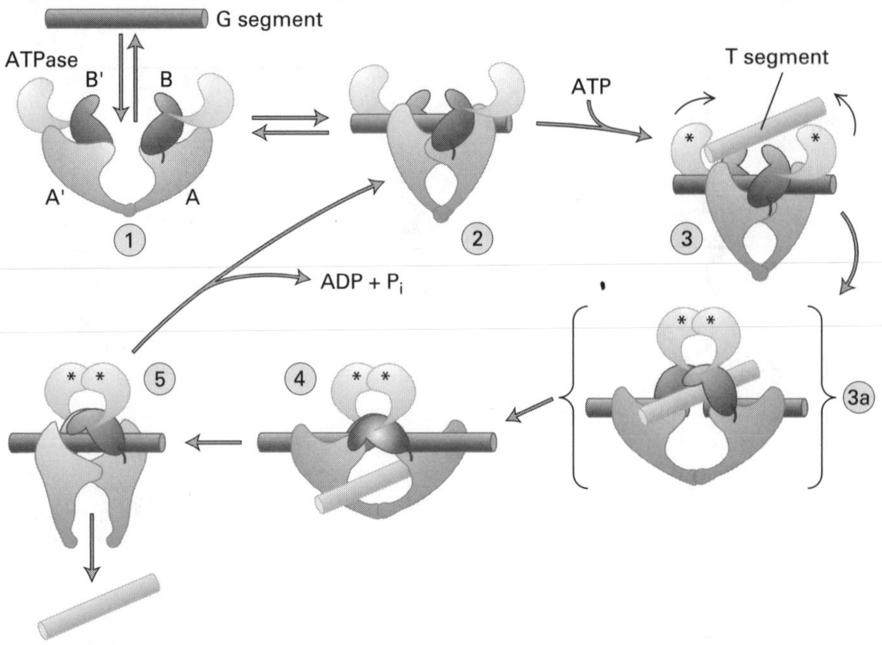

FIGURE 12-17, page 470
Molecular model for the catalytic activity of *E. coli* topoisomerase II (DNA gyrase).

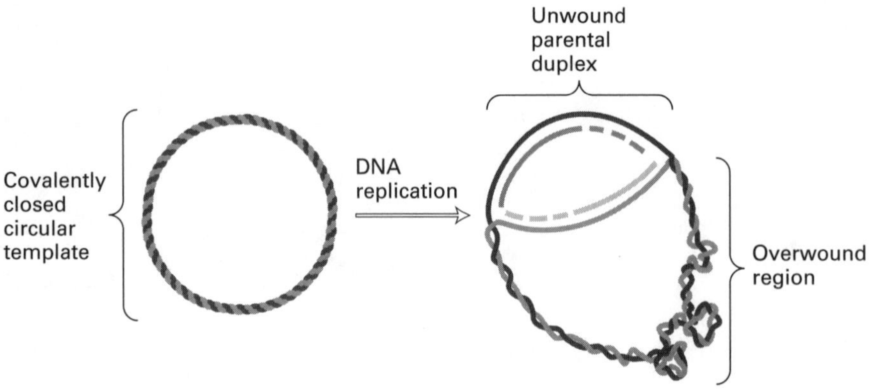

FIGURE 12-18, page 470
Movement of the growing fork during DNA replication induces formation of positive supercoils in the duplex DNA ahead of the fork.

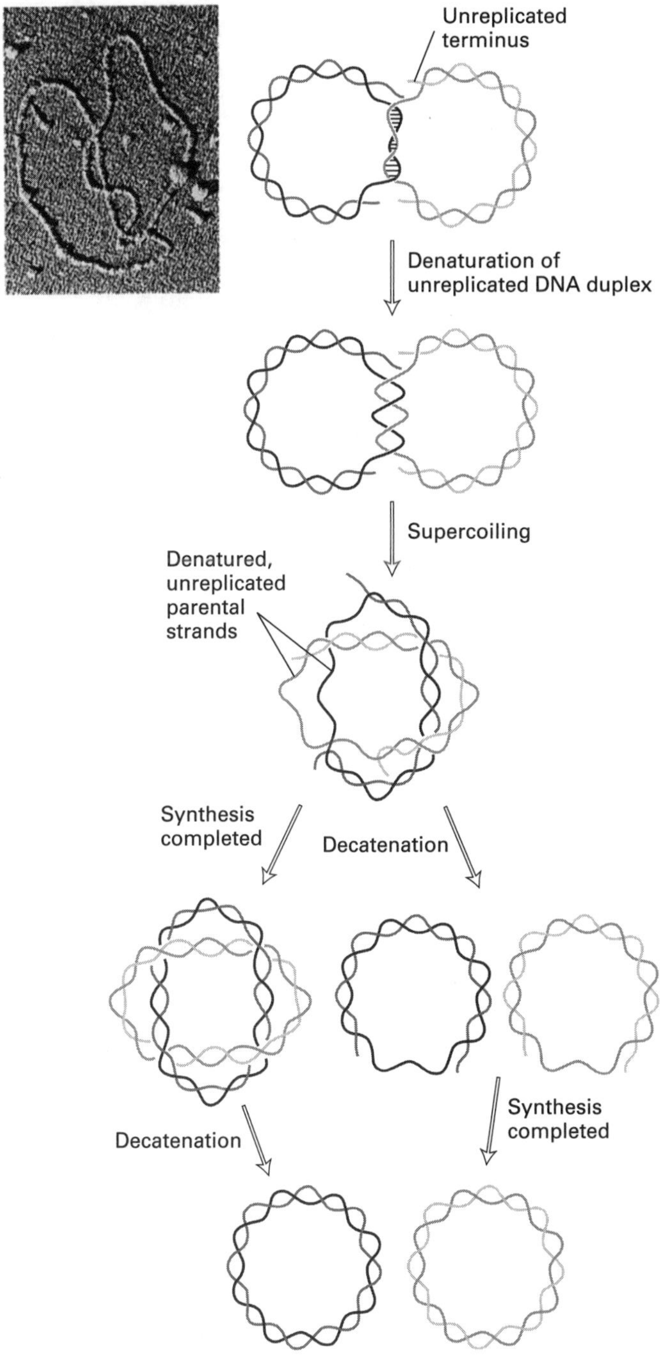

FIGURE 12-19, page 471
Completion of replication of circular DNA molecules.

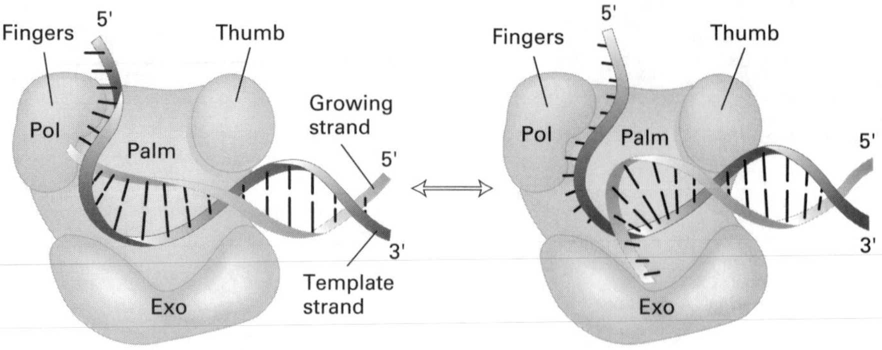

FIGURE 12-21, page 474
Schematic model of the proofreading function of DNA polymerases.

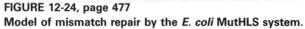

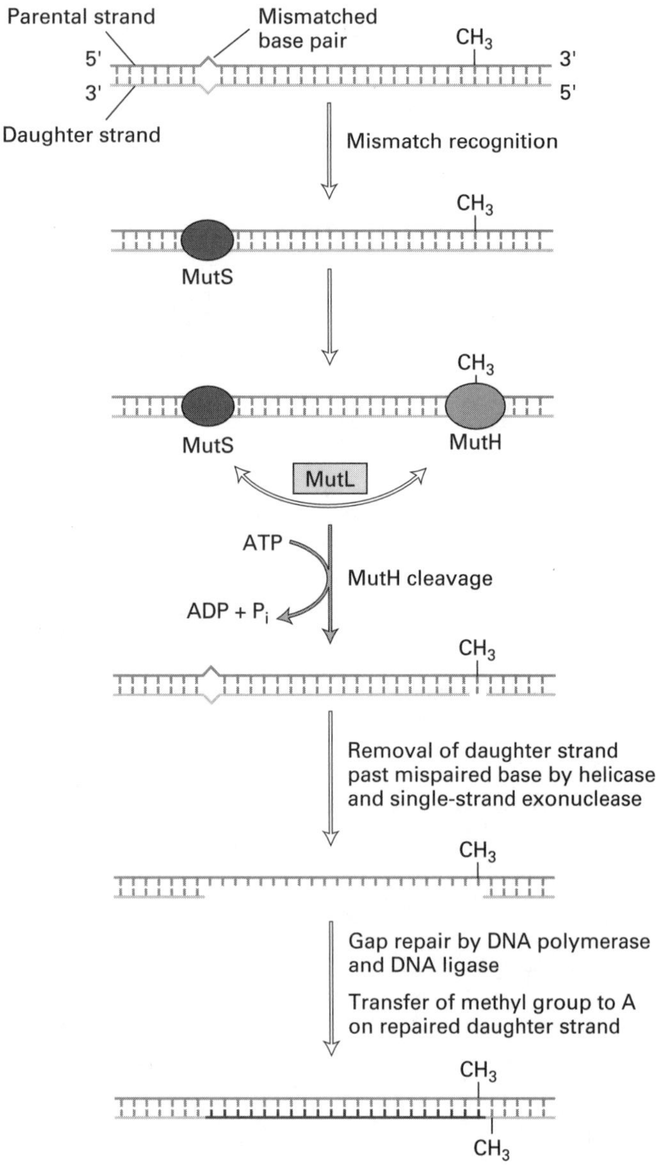

FIGURE 12-24, page 477
Model of mismatch repair by the *E. coli* MutHLS system.

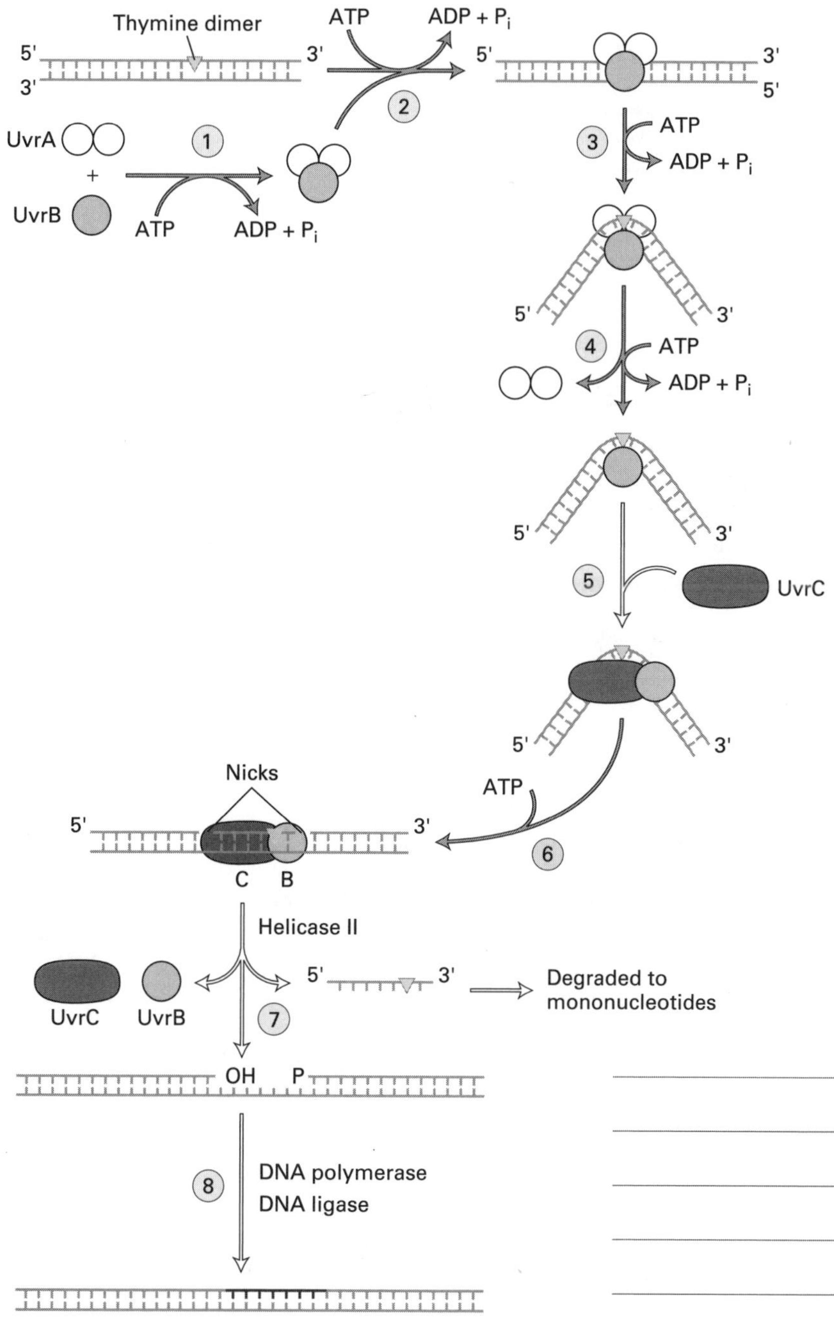

FIGURE 12-26, page 479
Excision repair of DNA by *E. coli* UvrABC mechanism.

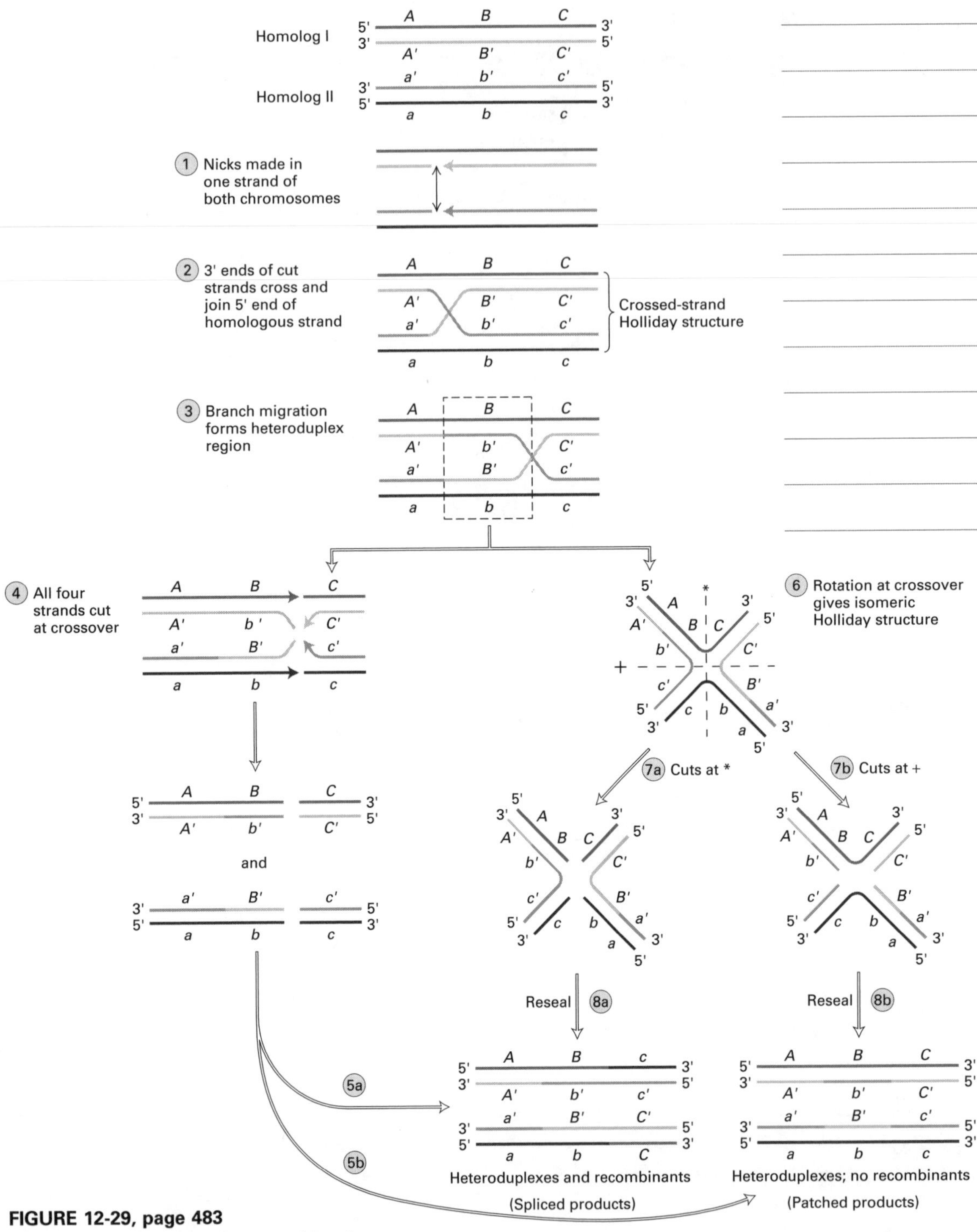

FIGURE 12-29, page 483
Holliday model of genetic recombination.

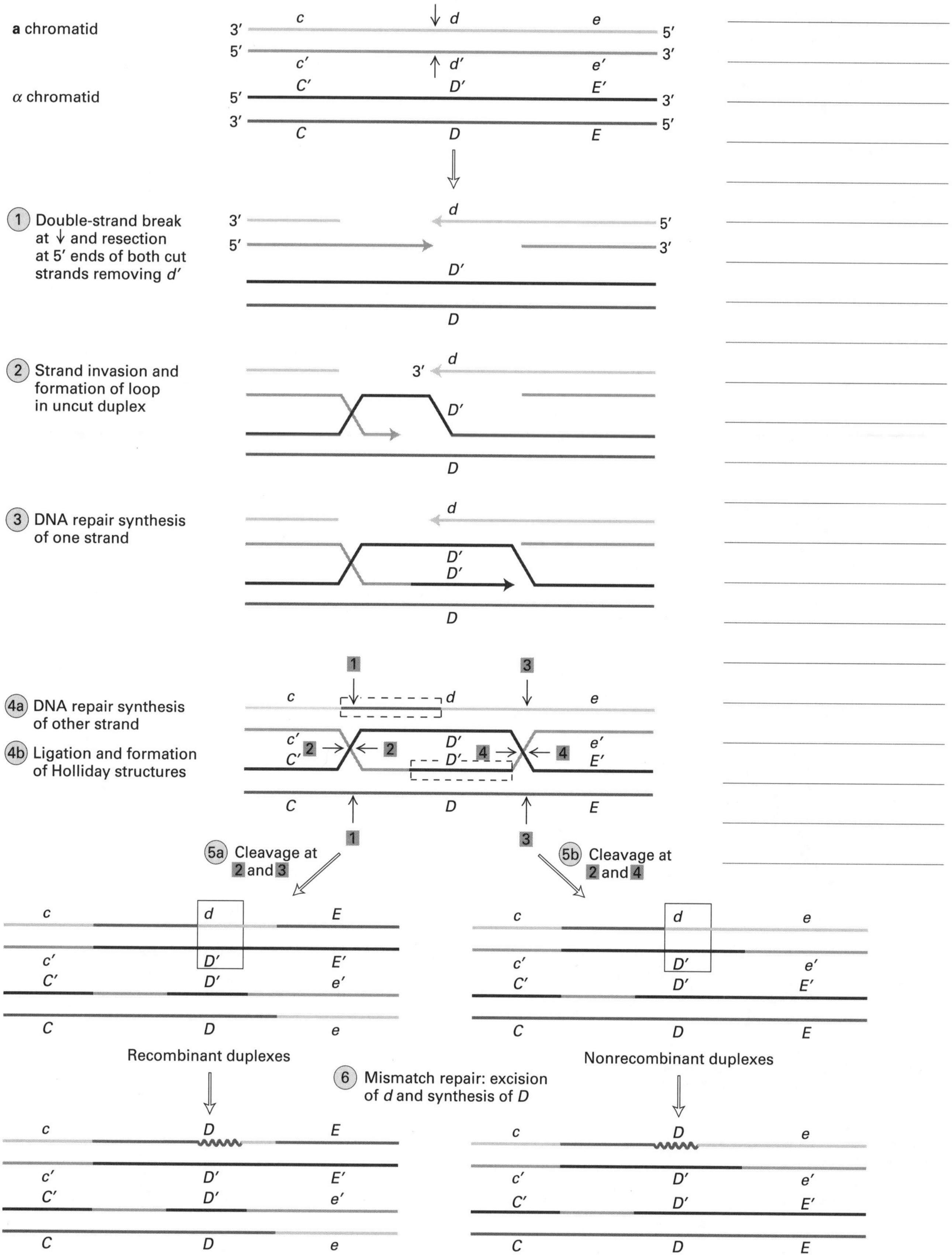

FIGURE 12-31, page 485
Double-strand break model of meiotic recombination developed from studies in the yeast *S. cerevisiae*.

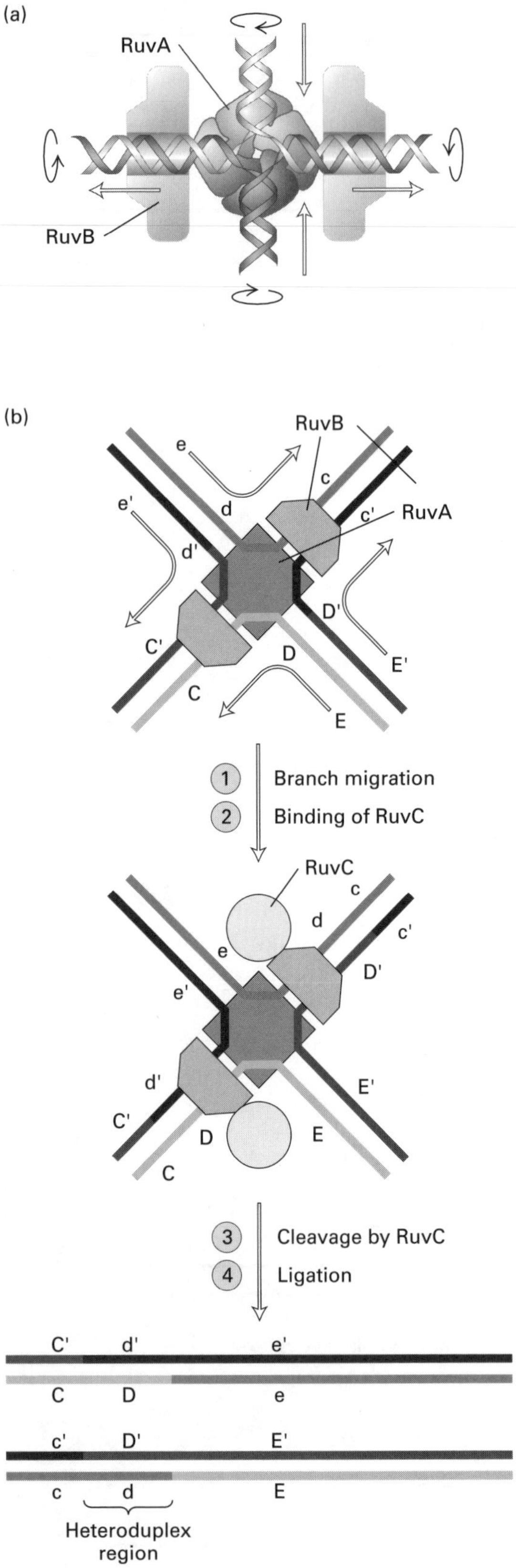

FIGURE 12-35, page 488
Action of *E. coli* proteins in branch migration and resolution of Holliday junctions.

Additional Notes

Regulation of the Eukaryotic Cell Cycle

13

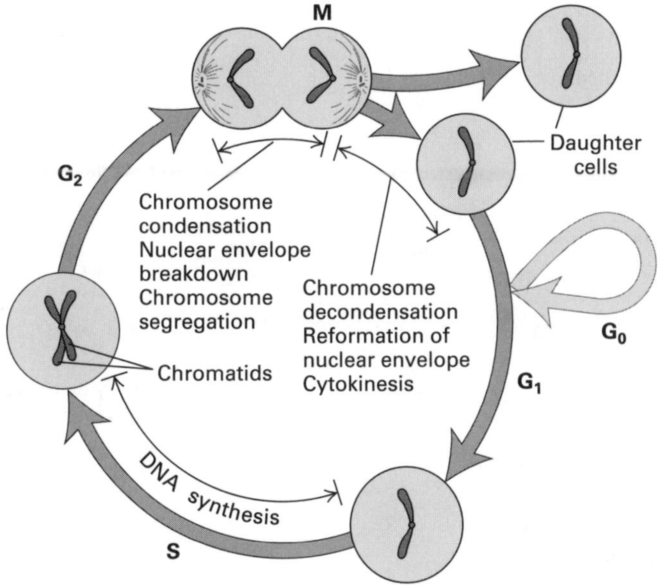

FIGURE 13-1, page 496
The fate of a single parental chromosome throughout the eukaryotic cell cycle.

MEDIA CONNECTIONS
Cell Cycle Control

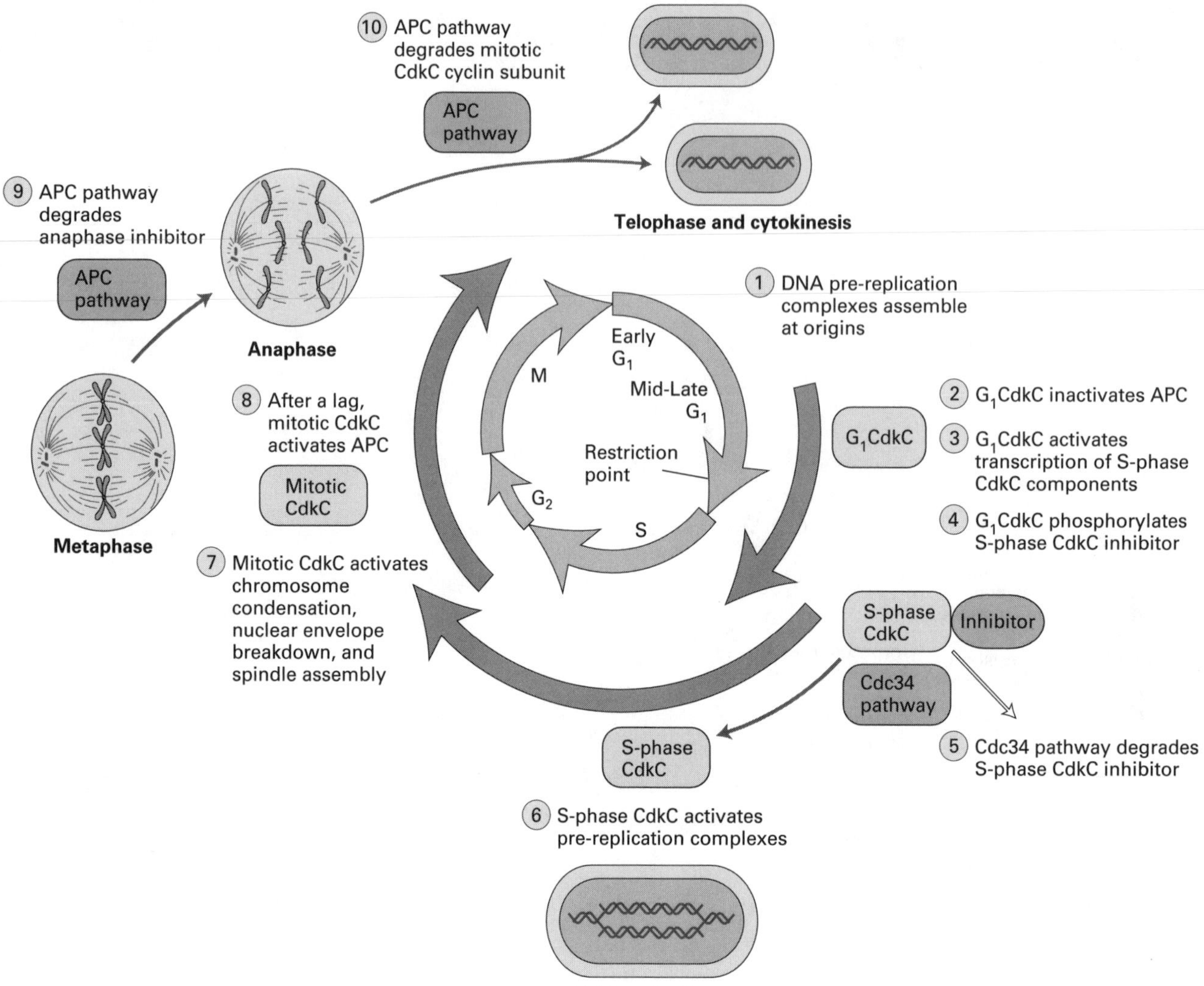

(10) APC pathway degrades mitotic CdkC cyclin subunit
APC pathway
Telophase and cytokinesis
(9) APC pathway degrades anaphase inhibitor
APC pathway
Anaphase
Metaphase
(8) After a lag, mitotic CdkC activates APC
Mitotic CdkC
(7) Mitotic CdkC activates chromosome condensation, nuclear envelope breakdown, and spindle assembly
M
Early G_1
Mid-Late G_1
Restriction point
G_2
S
(1) DNA pre-replication complexes assemble at origins
G_1CdkC
(2) G_1CdkC inactivates APC
(3) G_1CdkC activates transcription of S-phase CdkC components
(4) G_1CdkC phosphorylates S-phase CdkC inhibitor
S-phase CdkC
Inhibitor
Cdc34 pathway
(5) Cdc34 pathway degrades S-phase CdkC inhibitor
S-phase CdkC
(6) S-phase CdkC activates pre-replication complexes
DNA replication

FIGURE 13-2, page 497
Current model for regulation of the eukaryotic cell cycle.

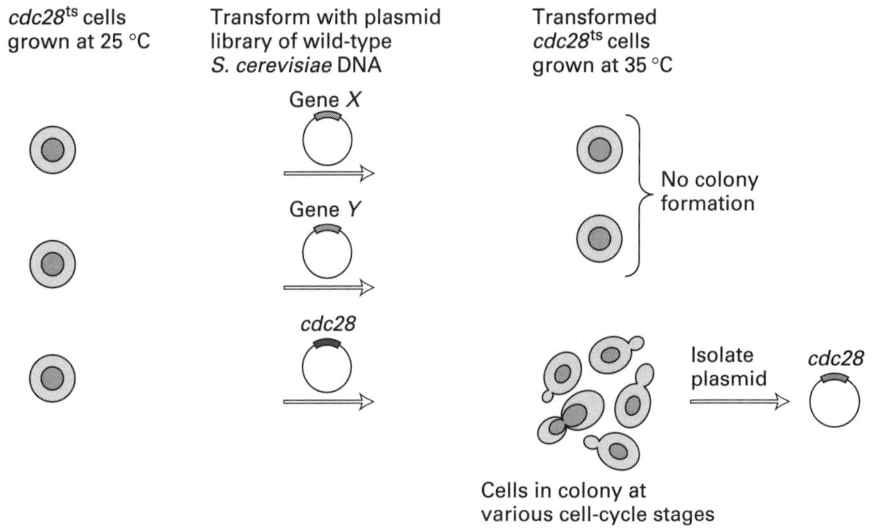

FIGURE 13-4, page 499
Isolation of wild-type cell-division cycle *(CDC)* genes from *S. cerevisiae* cells carrying temperature-sensitive mutations in these genes.

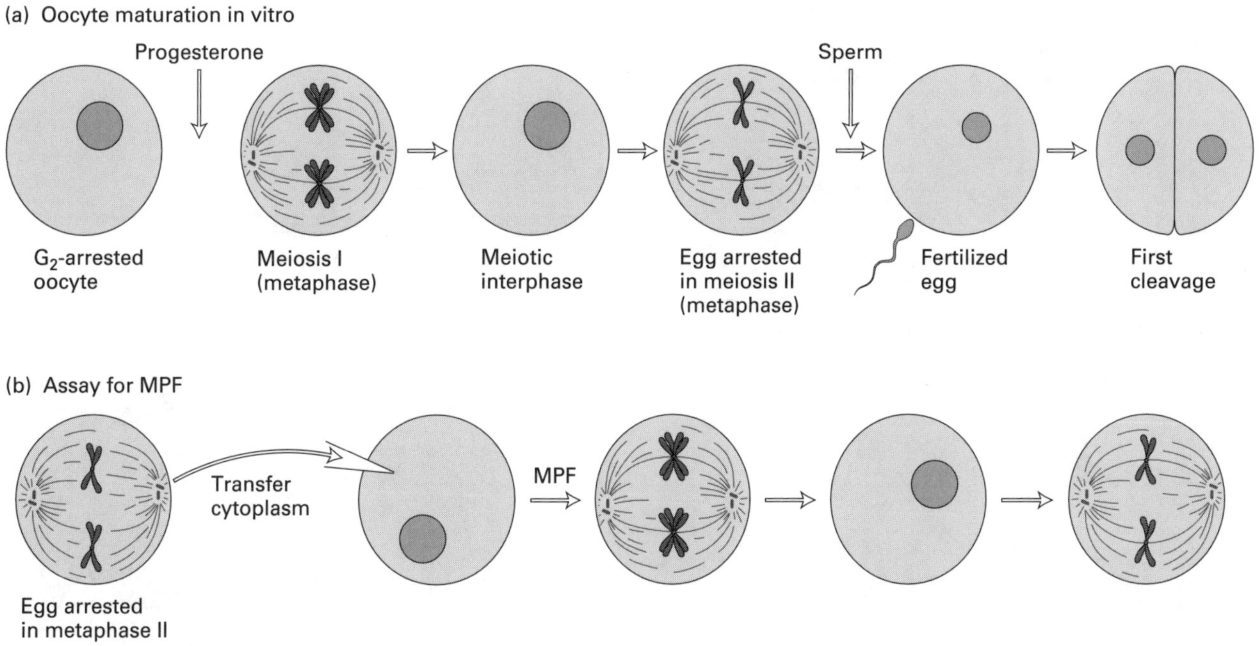

FIGURE 13-5, page 501
In vitro maturation of *Xenopus* oocytes and assay of maturation-promoting factor (MPF).

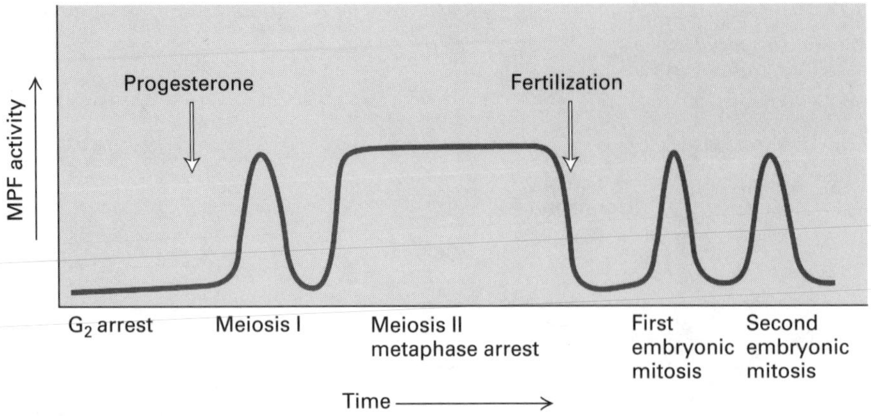

FIGURE 13-6, page 501
Oscillation of MPF activity during meiotic and mitotic cell cycles of *Xenopus* oocytes and early frog embryos.

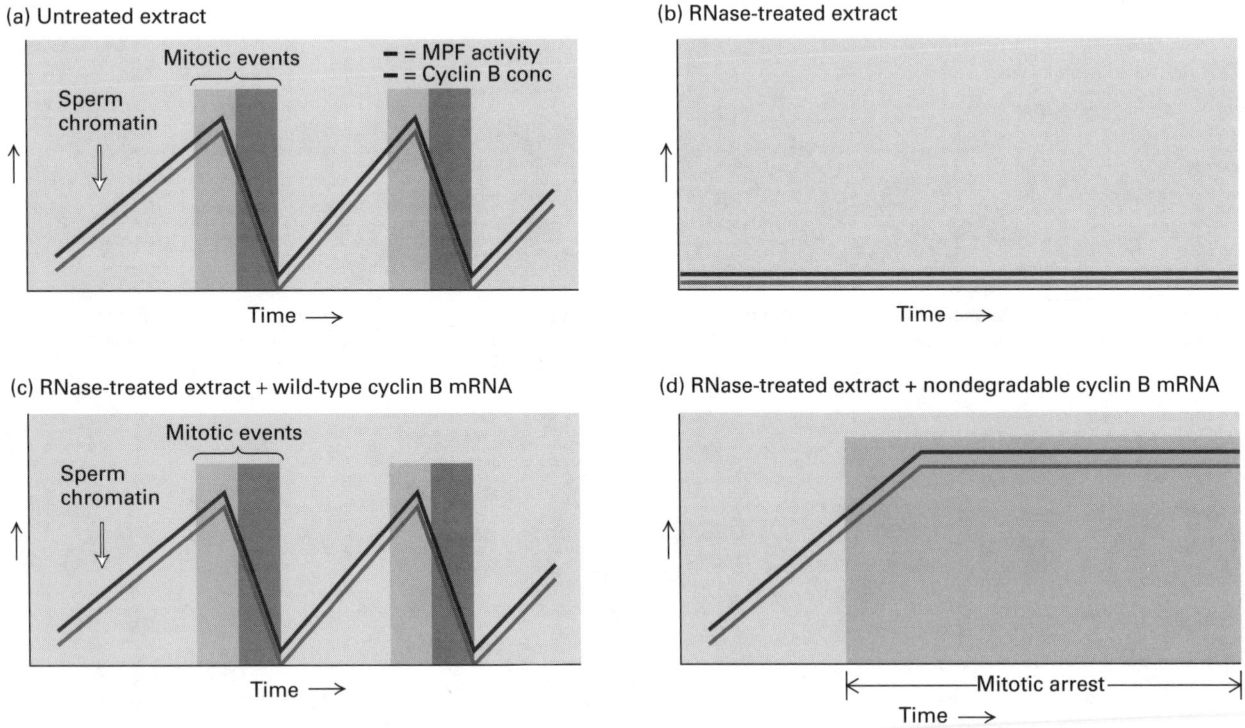

FIGURE 13-7, page 503
Experimental demonstration that the synthesis and degradation of cyclin B are required for the cycling of MPF activity and mitotic events in *Xenopus* egg extracts.

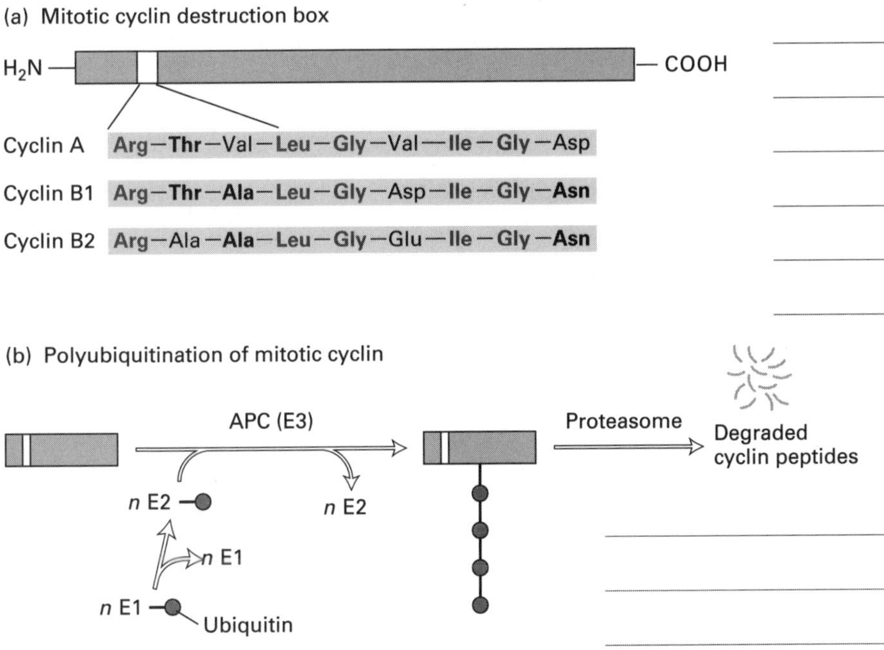

FIGURE 13-8, page 504
Polyubiquitination of mitotic cyclins.

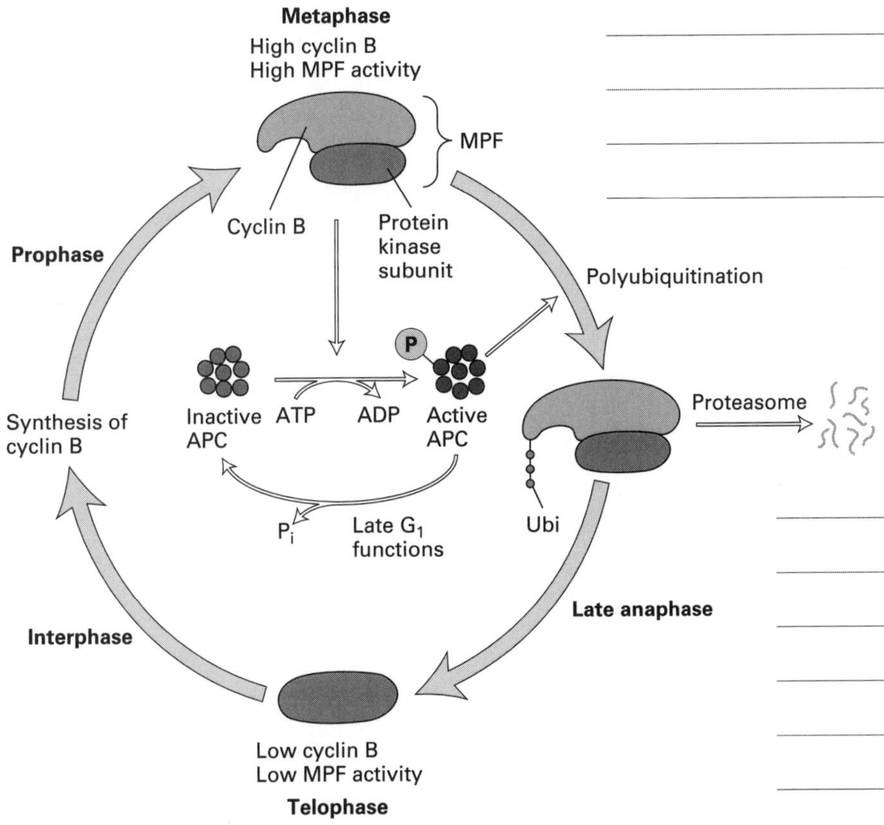

FIGURE 13-9, page 505
Regulation of mitotic cyclin levels in cycling cells.

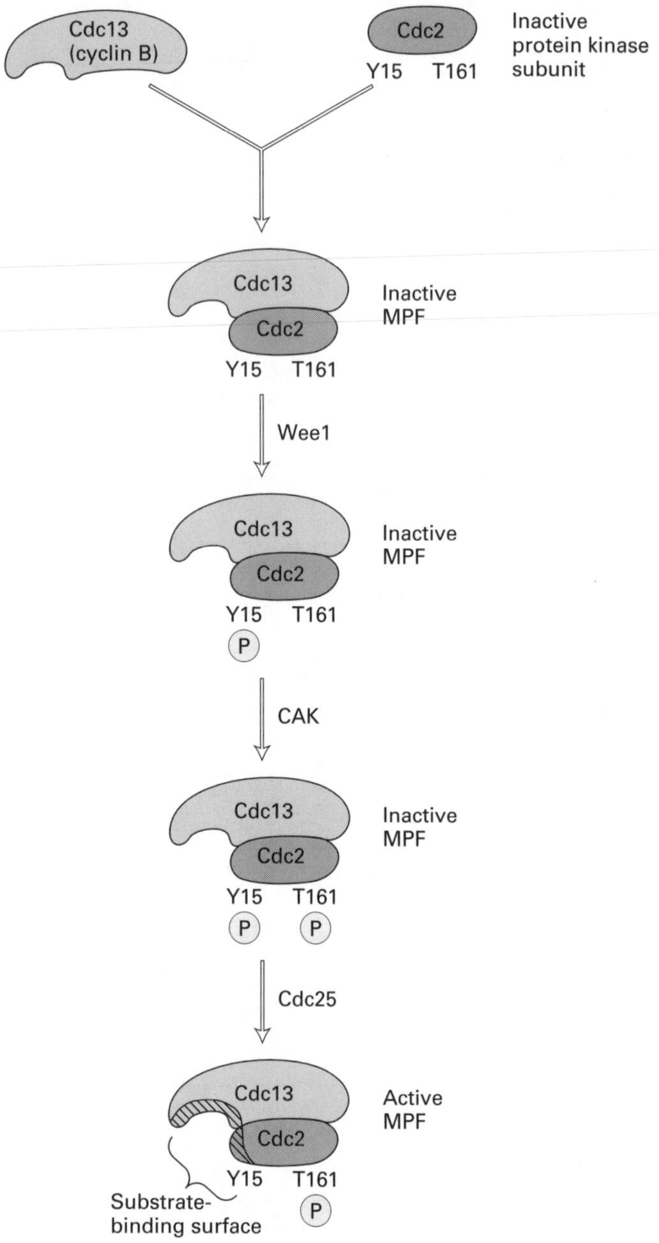

FIGURE 13-13, page 508
Regulation of MPF protein kinase activity in *S. pombe* by Cdc13 (cyclin B), Wee1, CAK (Cdc2-activating kinase), and Cdc25.

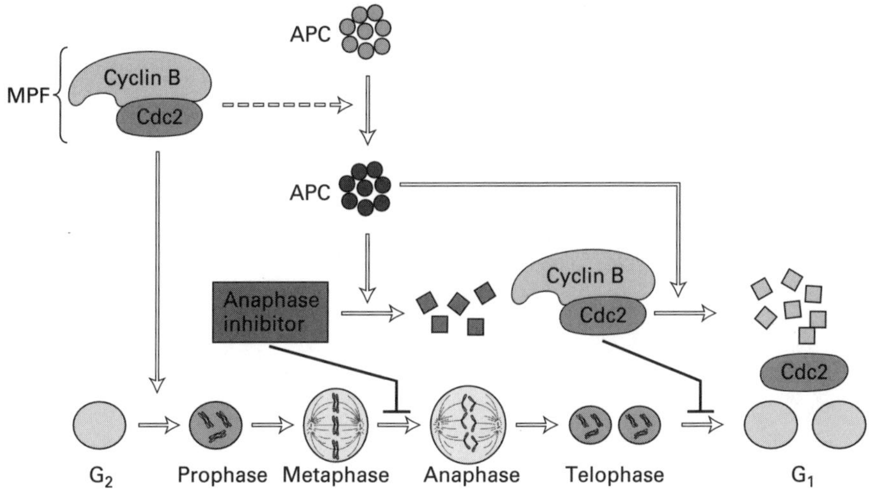

FIGURE 13-19, page 516
Control of entry into anaphase and exit from mitosis by the anaphase-promoting complex (APC), which directs the degradation of at least two classes of proteins.

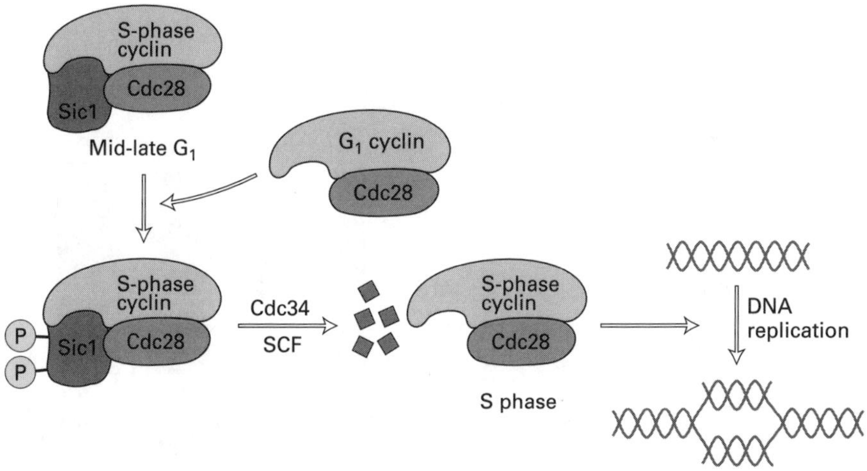

FIGURE 13-25, page 522
Control of the $G_1 \rightarrow S$ phase transition in *S. cerevisiae* by regulated proteolysis of Sic1.

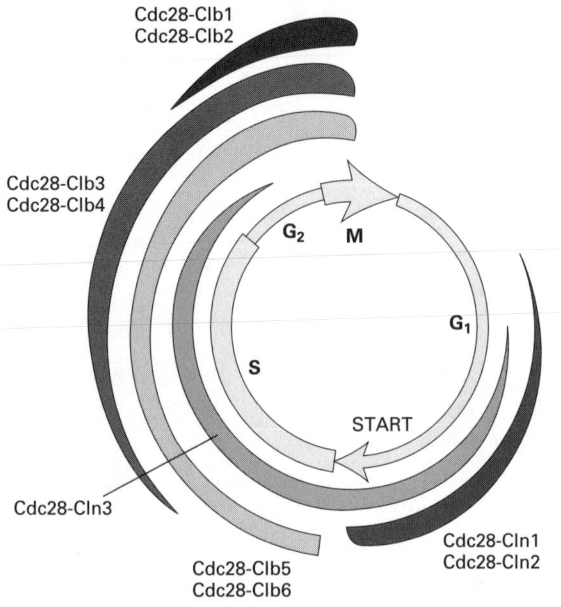

FIGURE 13-26, page 523
Activity of *S. cerevisiae* Cdc28-cyclin complexes through the course of the cell cycle.

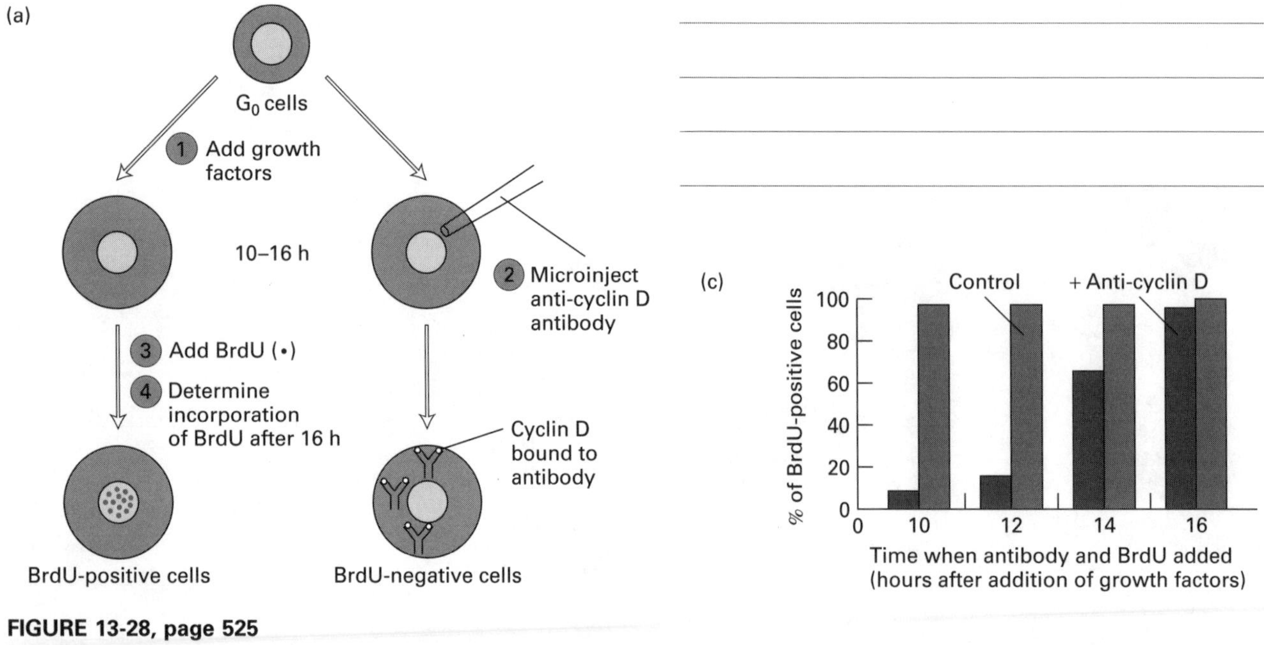

FIGURE 13-28, page 525
Experimental demonstration that cyclin D is required for passage through the restriction point in the mammalian cell cycle.

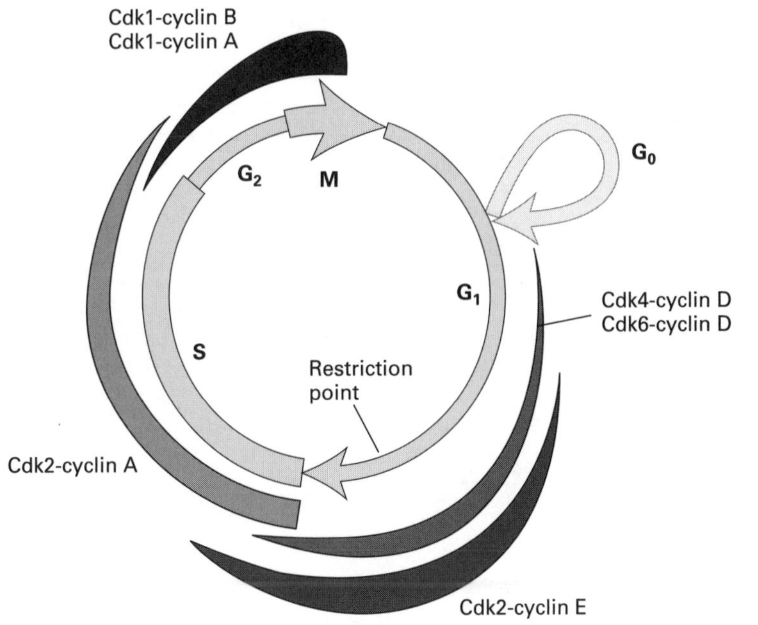

FIGURE 13-29, page 526
Activity of mammalian Cdk-cyclin complexes through the course of the cell cycle in G_0 cells induced to divide by treatment with growth factors.

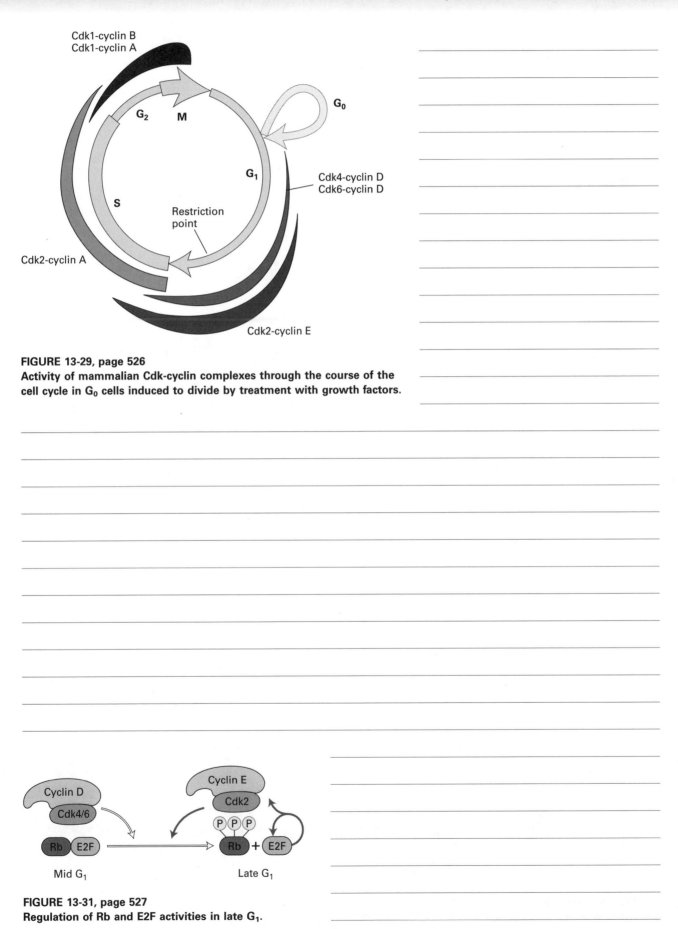

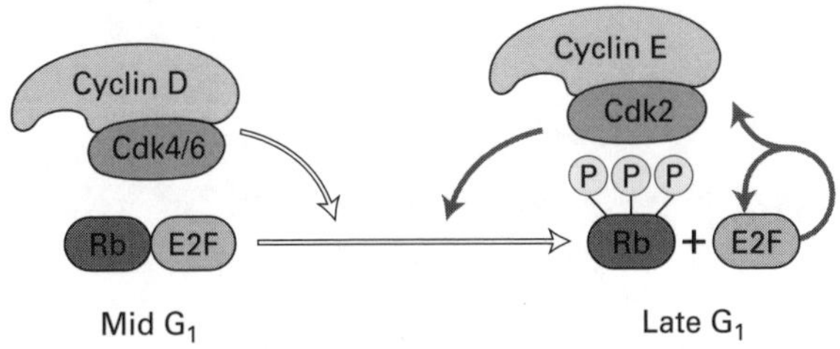

FIGURE 13-31, page 527
Regulation of Rb and E2F activities in late G_1.

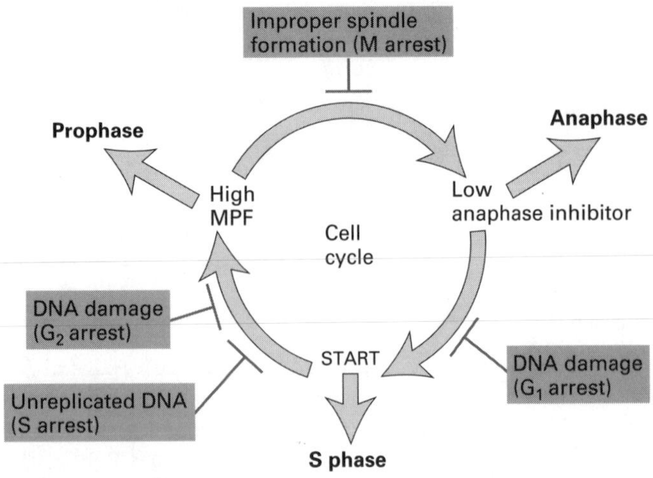

FIGURE 13-34, page 530
Stages at which checkpoint controls can arrest passage through the cell cycle.

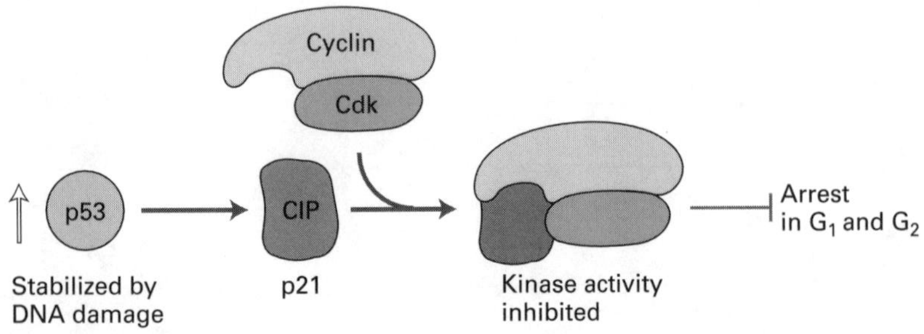

FIGURE 13-36, page 532
p53-induced cell-cycle arrest in response to DNA damage.

Additional Notes

Gene Control in Development

14

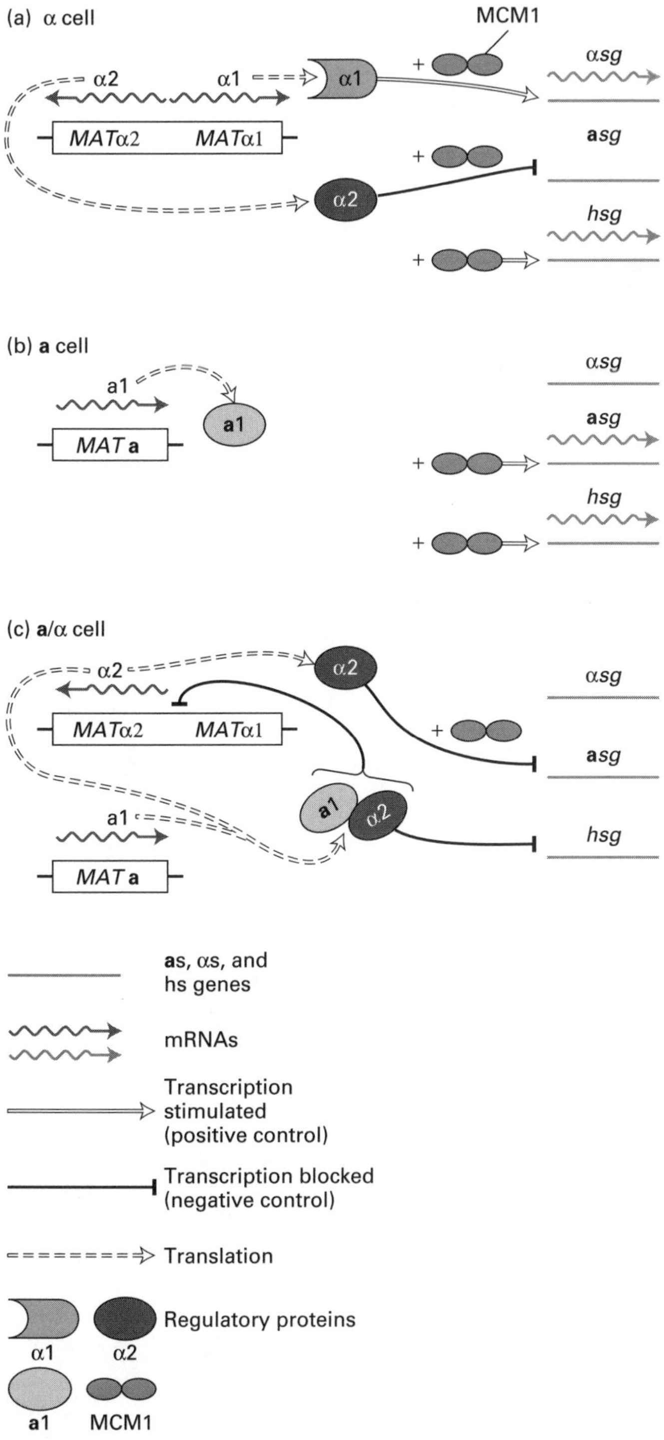

FIGURE 14-1, page 538
Regulation of cell type–specific genes in *S. cerevisiae* by regulatory proteins encoded at the *MAT* locus together with MCM1, a constitutive transcription factor produced by all three cell types.

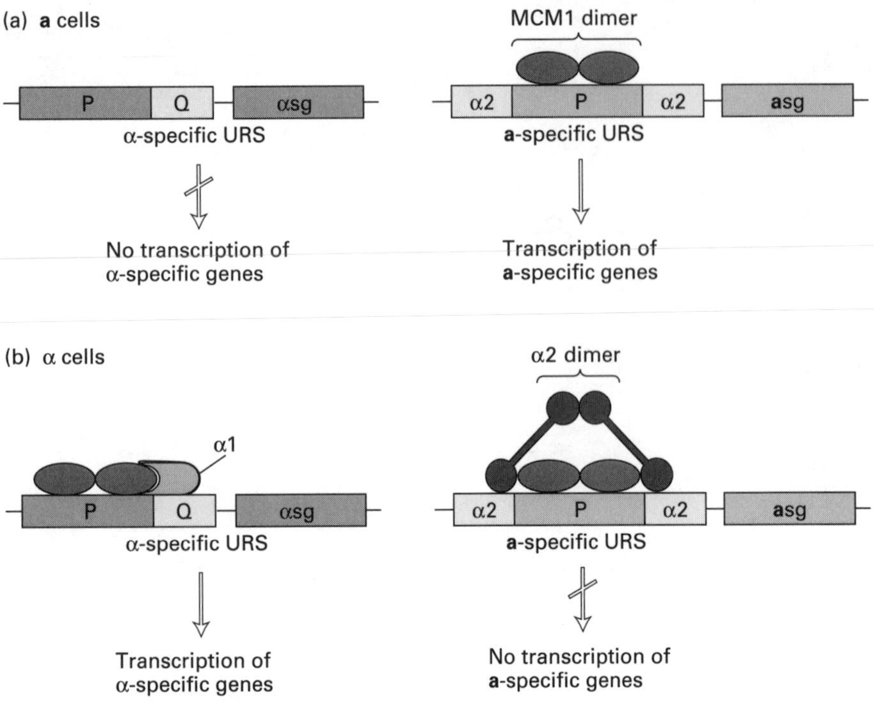

FIGURE 14-2, page 539
Activity of MCM1 in a **and** ***α*** **yeast cells.**

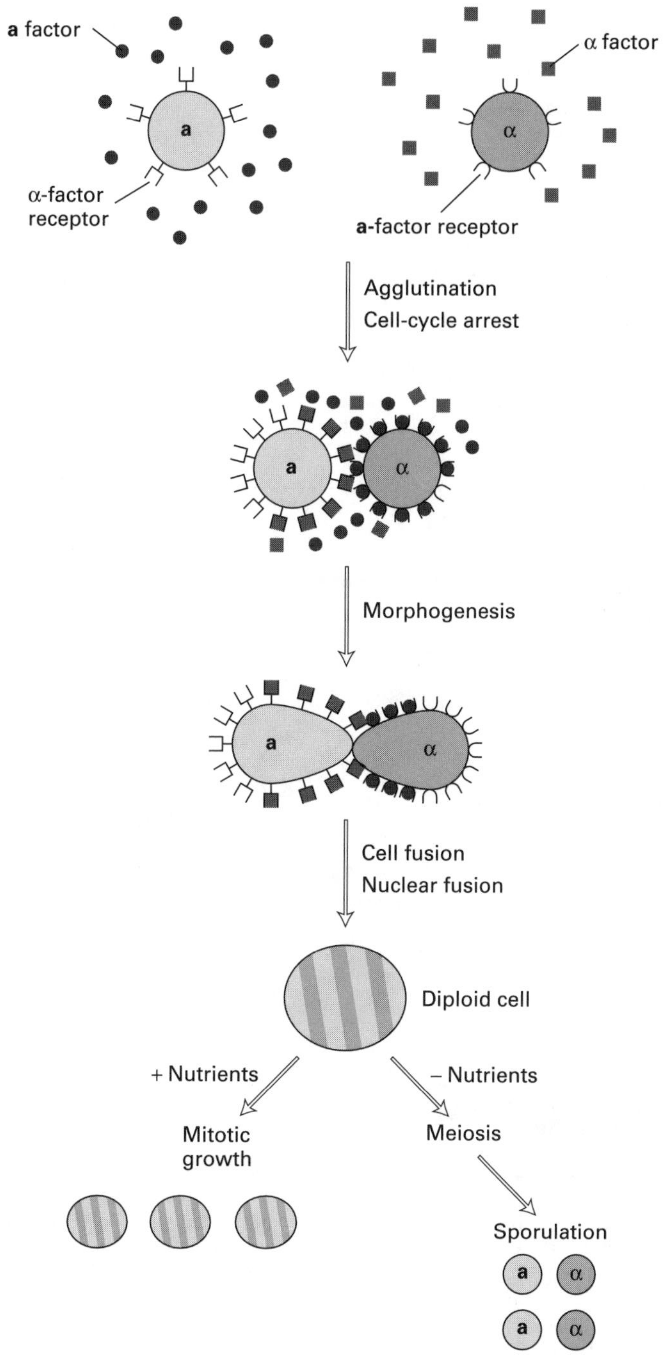

FIGURE 14-4, page 540
Haploid yeast cells produce pheromones, or mating factors, and pheromone receptors.

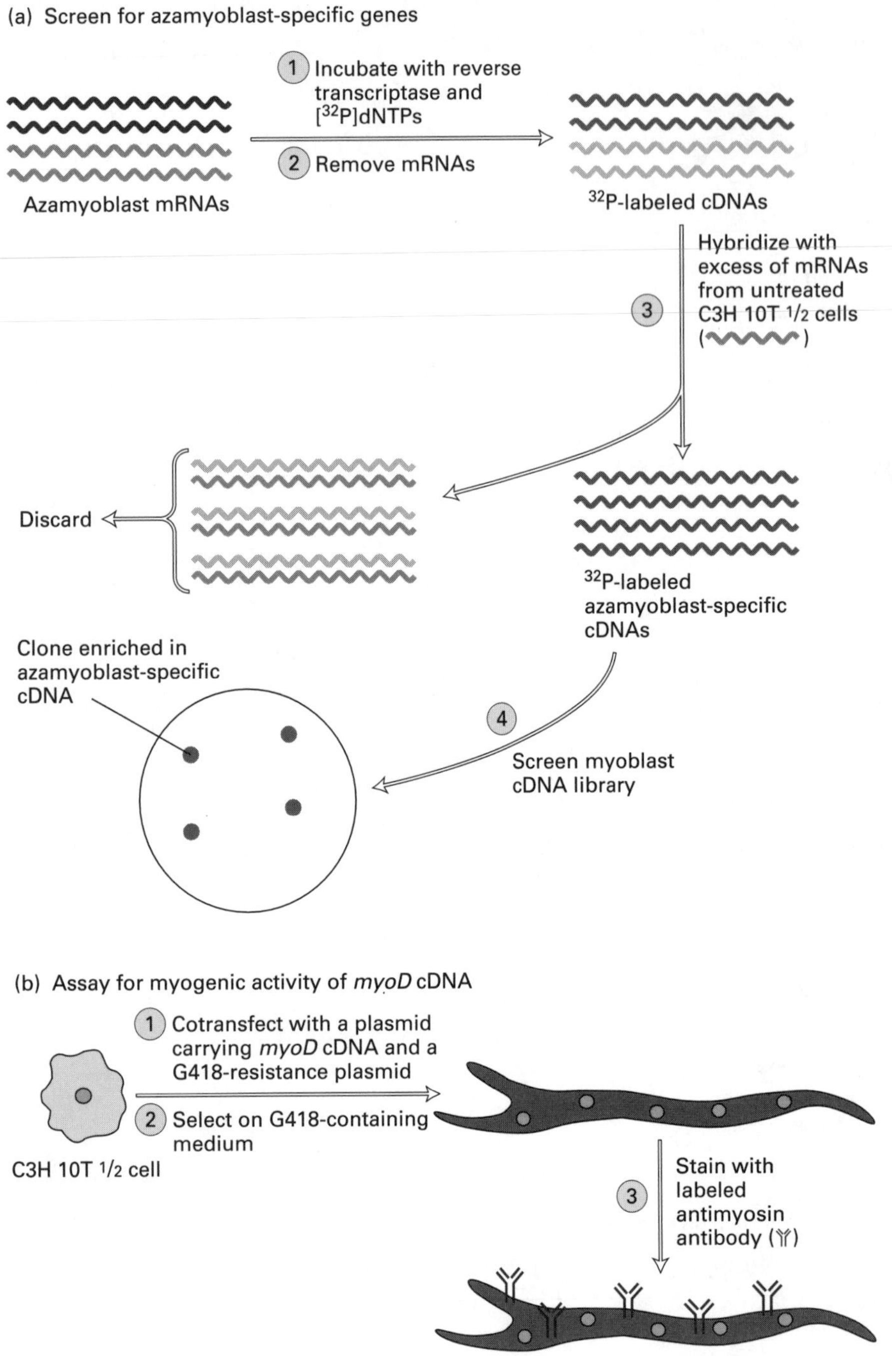
(a) Screen for azamyoblast-specific genes
1 Incubate with reverse transcriptase and [32P]dNTPs
2 Remove mRNAs
Azamyoblast mRNAs
32P-labeled cDNAs
3 Hybridize with excess of mRNAs from untreated C3H 10T 1/2 cells
Discard
32P-labeled azamyoblast-specific cDNAs
Clone enriched in azamyoblast-specific cDNA
4 Screen myoblast cDNA library
(b) Assay for myogenic activity of myoD cDNA
1 Cotransfect with a plasmid carrying myoD cDNA and a G418-resistance plasmid
2 Select on G418-containing medium
C3H 10T 1/2 cell
3 Stain with labeled antimyosin antibody

FIGURE 14-11, page 545
Identification and assay of genes that drive myogenesis.

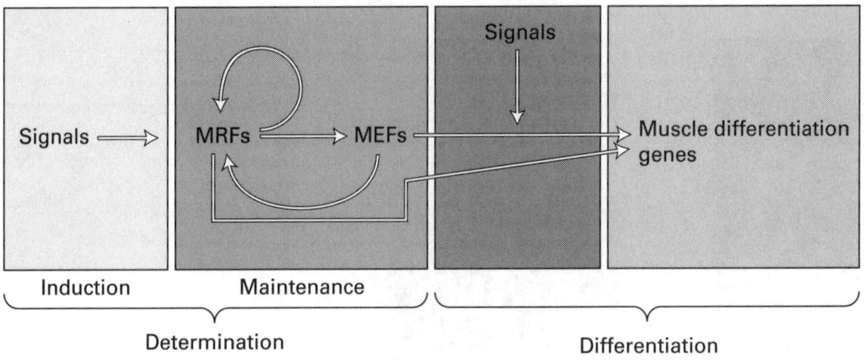

FIGURE 14-16, page 550
Maintenance of the myogenic program.

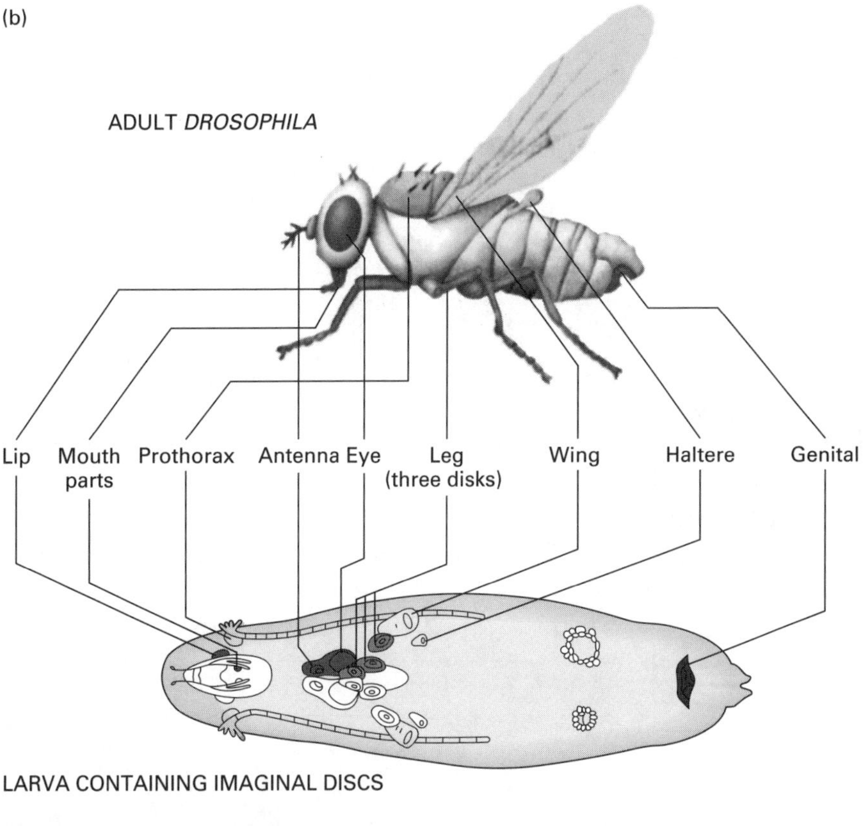

FIGURE 14-21 (b), page 554
The development of *D. melanogaster*.

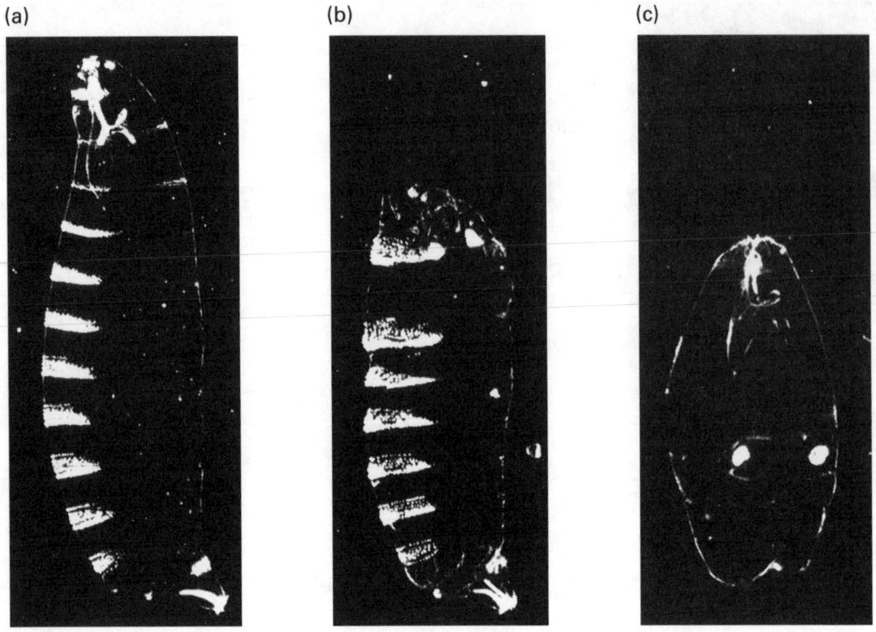

FIGURE 14-24, page 557
Abnormal patterns in the outer (cuticular) structures of the *Drosophila* embryo resulting from mutations in two of the four maternal gene systems that regulate axis determination during early embryogenesis.

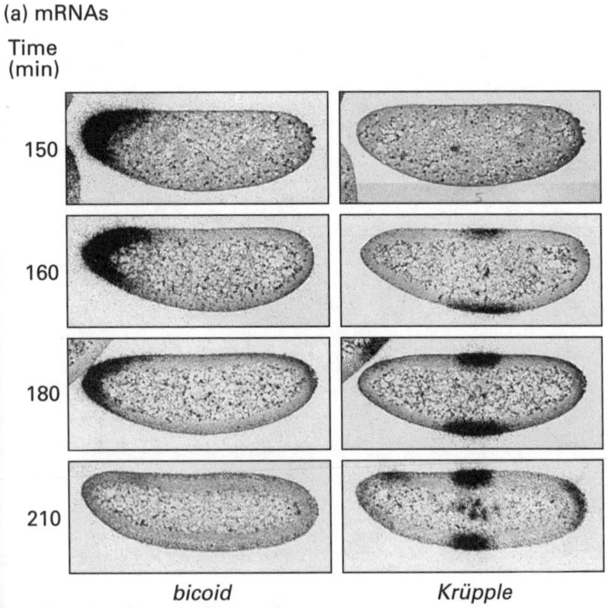

MEDIA CONNECTIONS
Gene Control in Embryonic Development

FIGURE 14-25 (a), page 558
Localization of developmentally important gene products in early *Drosophila* embryos.

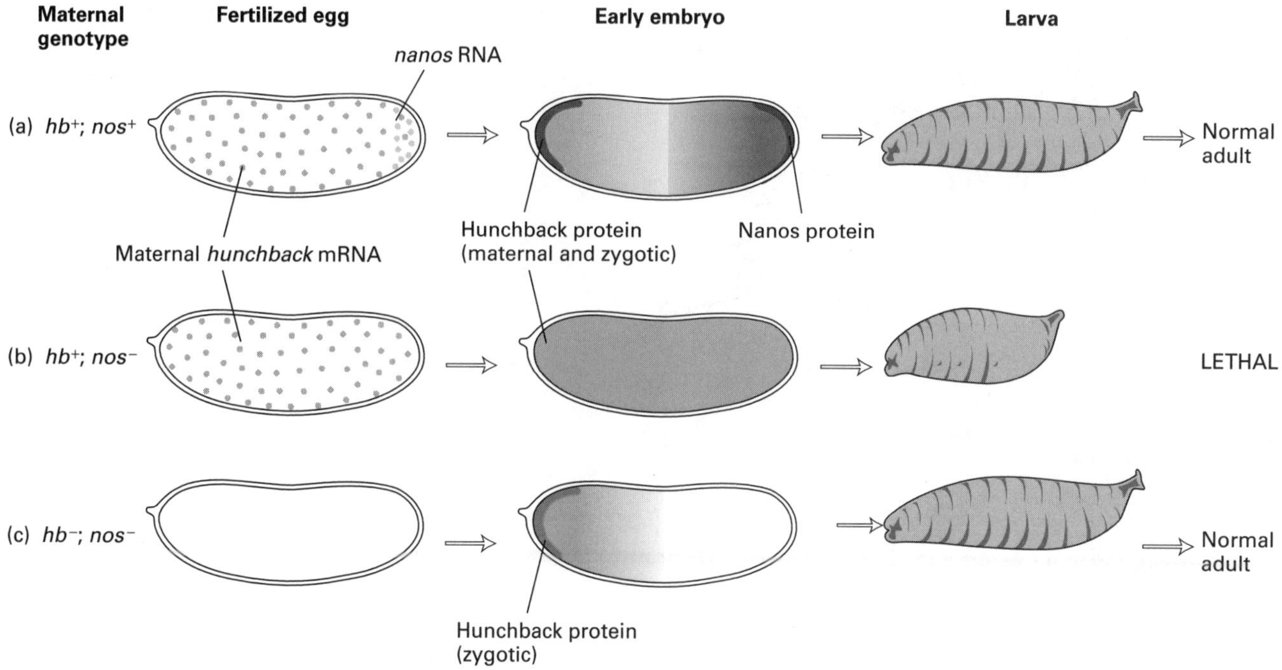

FIGURE 14-27, page 560
Role of maternally derived Nanos protein in excluding Hunchback protein from the posterior region of *Drosophila* embryos.

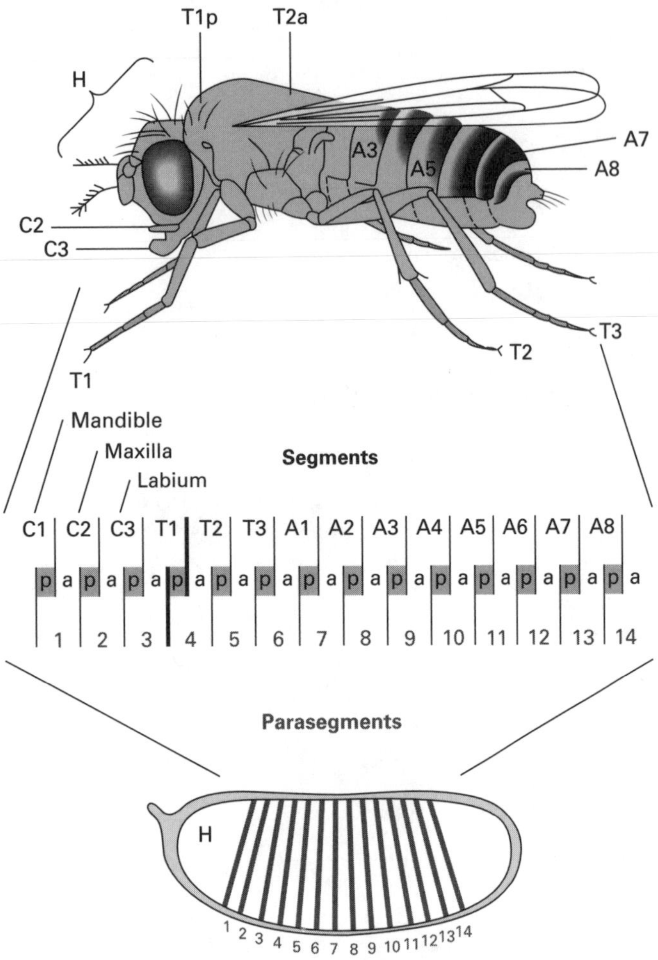

FIGURE 14-30, page 562
The relationship between segments in the adult fly and parasegments, which are developmental units corresponding to the domains of activity of selector genes.

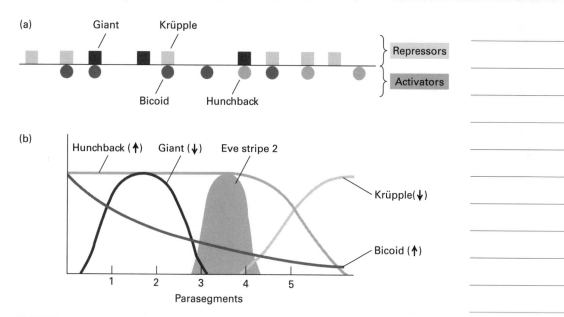

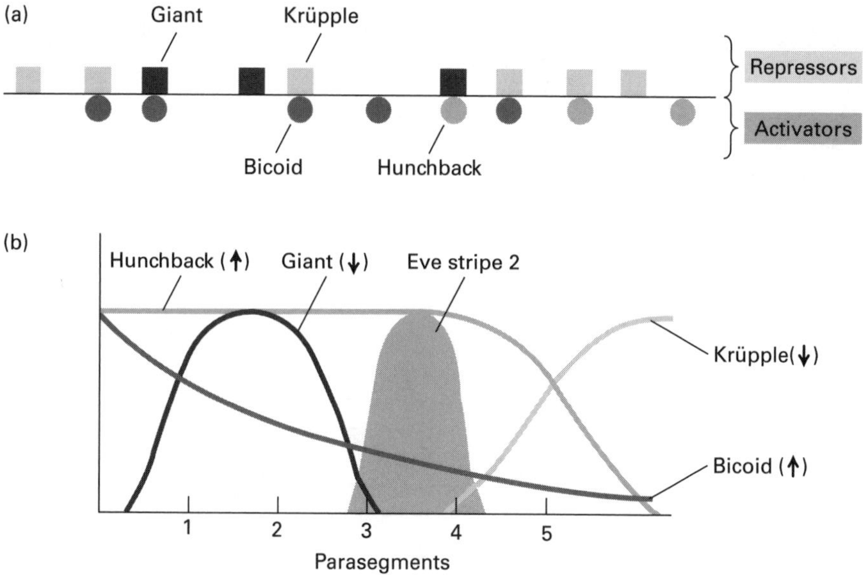

FIGURE 14-31, page 563
Expression of the Even-skipped (Eve) stripe 2 in the *Drosophila* embryo.

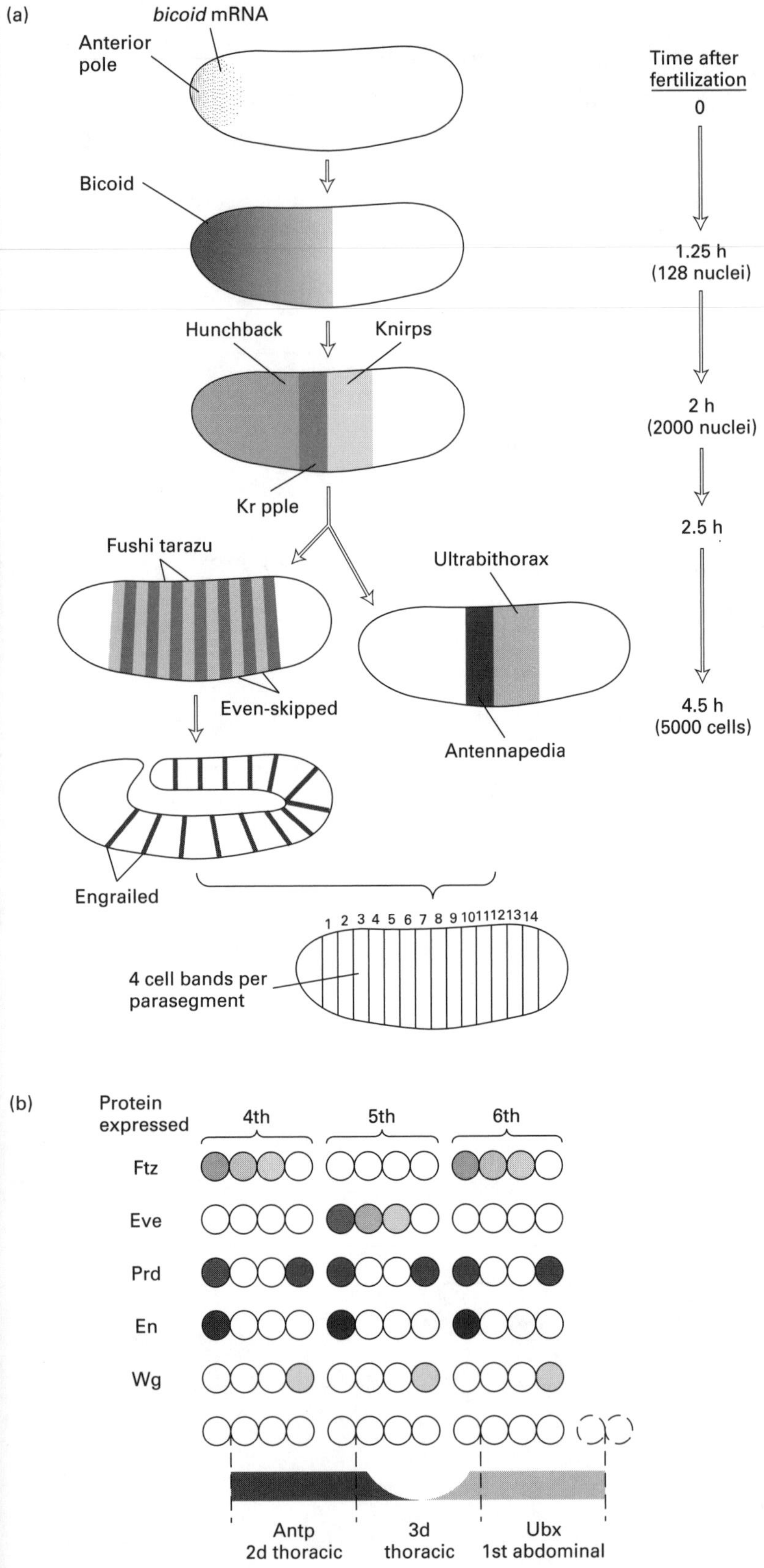

FIGURE 14-32, page 564
Summary of sequential expression of various genes during early development of the *Drosophila* embryo and localization of their gene products within the embryo.

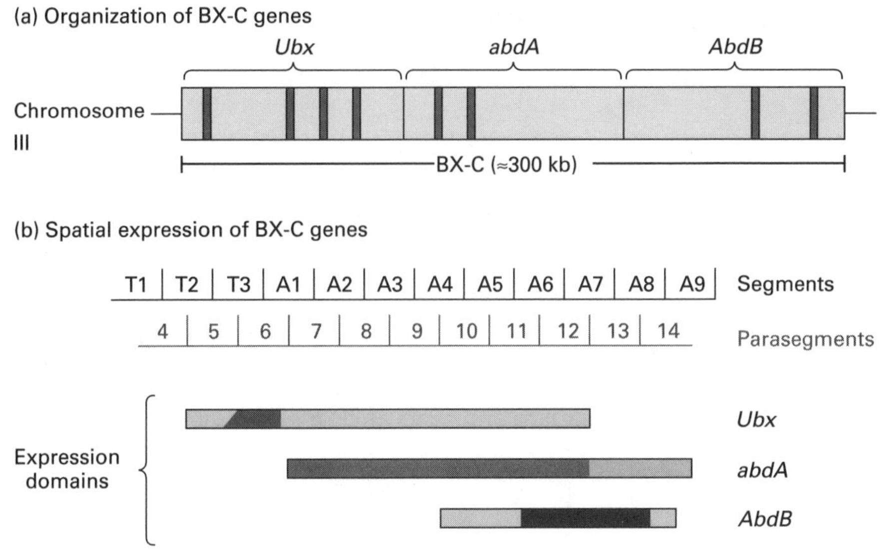

FIGURE 14-34, page 565
(a) Organization of genes within the bithorax complex (BX-C) on *Drosophila* chromosome III. (b) Expression patterns of the products (mRNAs and/or proteins) of the *Ultrabithorax (Ubx), abdominal A (abdA),* and *Abdominal B (AbdB)* genes in parasegments 4–14.

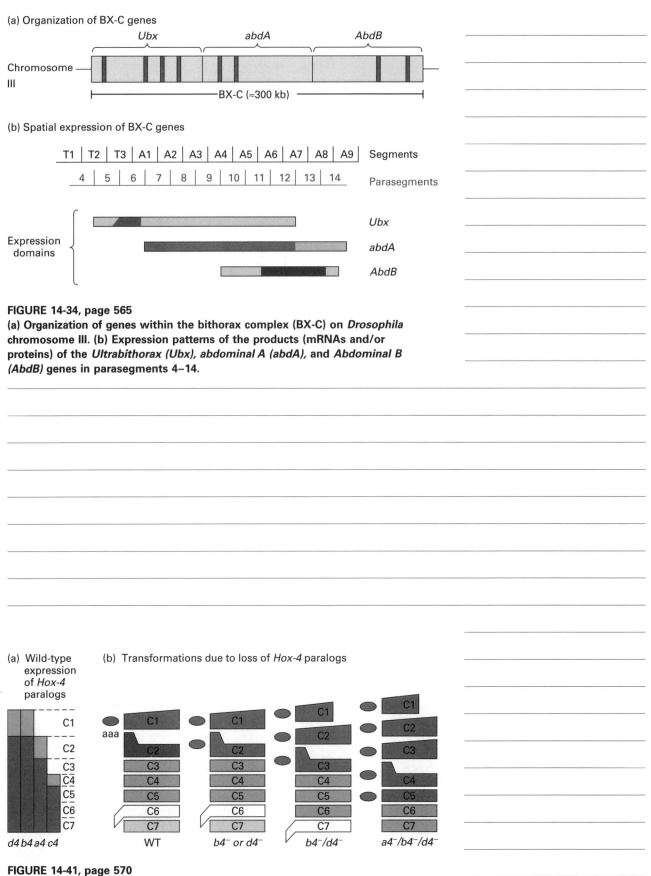

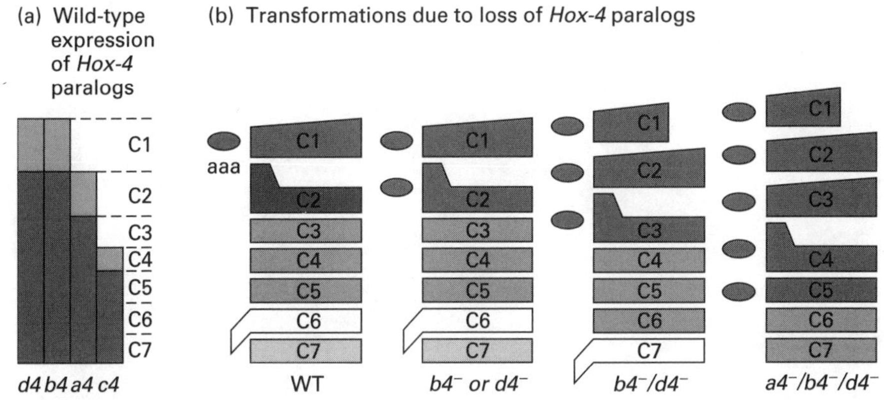

FIGURE 14-41, page 570
Effect of loss-of-function mutations in *Hox-4* paralogs on development of cervical vertebrae in mice.

Additional Notes

Transport across Cell Membranes

15

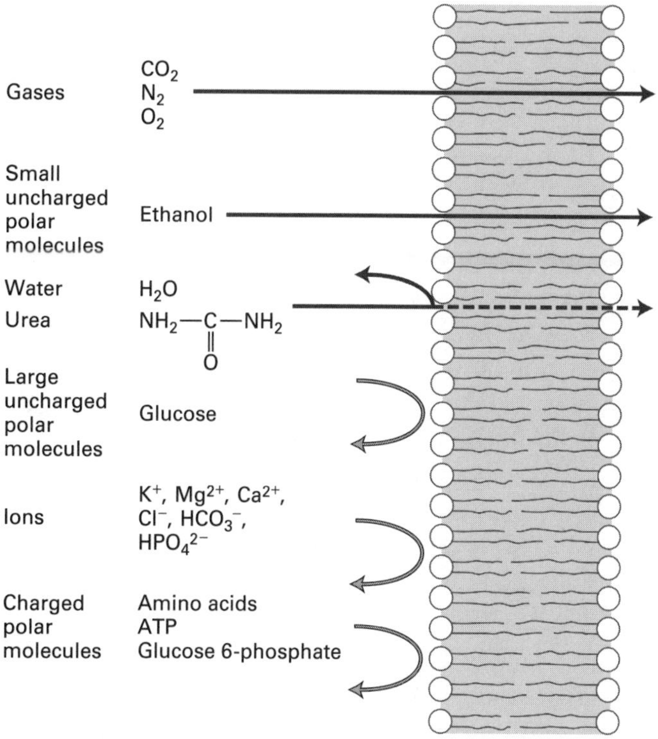

FIGURE 15-1, page 579
A pure artificial phospholipid bilayer is permeable to small hydrophobic molecules and small uncharged polar molecules.

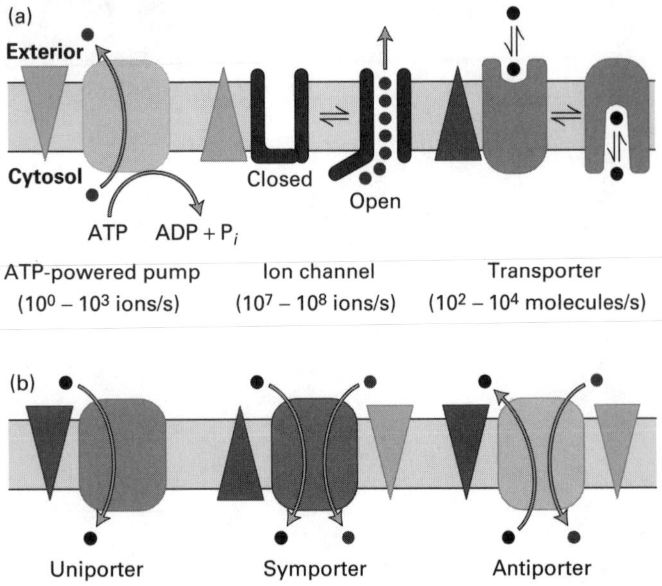

FIGURE 15-3, page 581
Schematic diagrams illustrating action of membrane transport proteins.

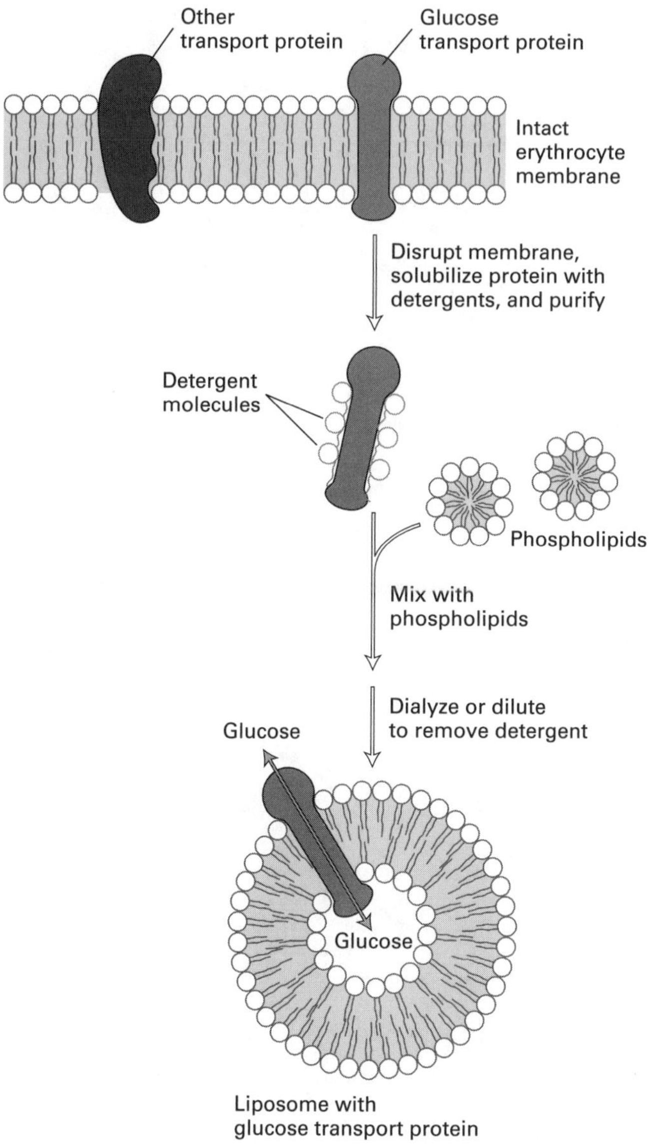

FIGURE 15-4, page 582
Liposomes containing a single type of transport protein can be used to investigate properties of the transport process.

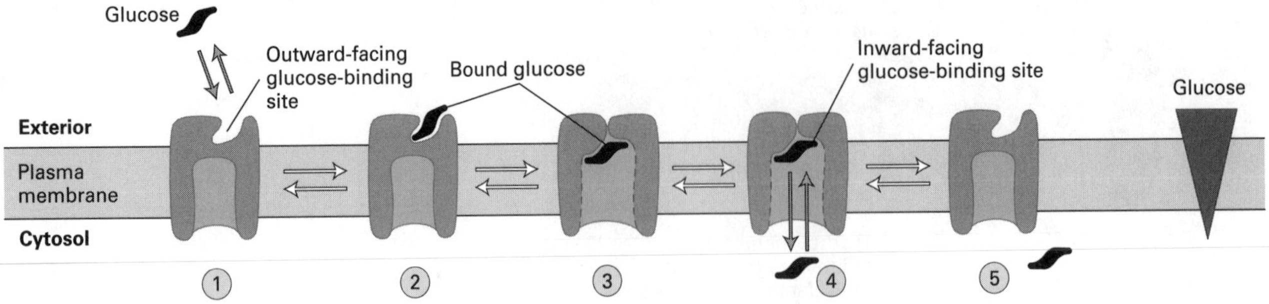

FIGURE 15-7, page 584
Model of the mechanism of uniport transport by GLUT1, which is believed to shuttle between two conformational states.

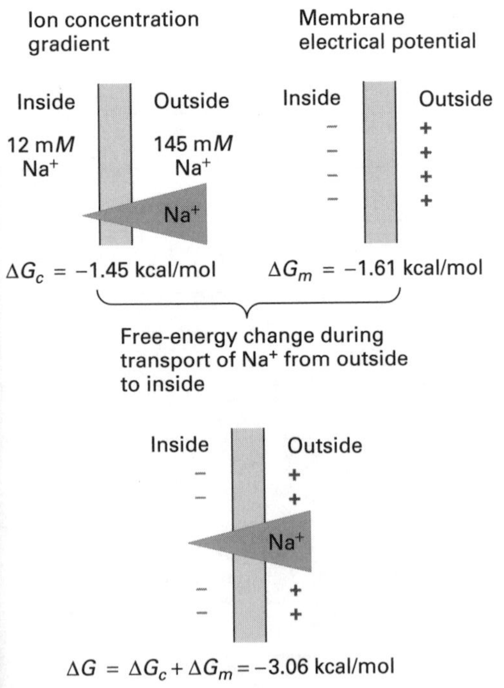

FIGURE 15-9, page 588
Transmembrane forces acting on Na^+ ions.

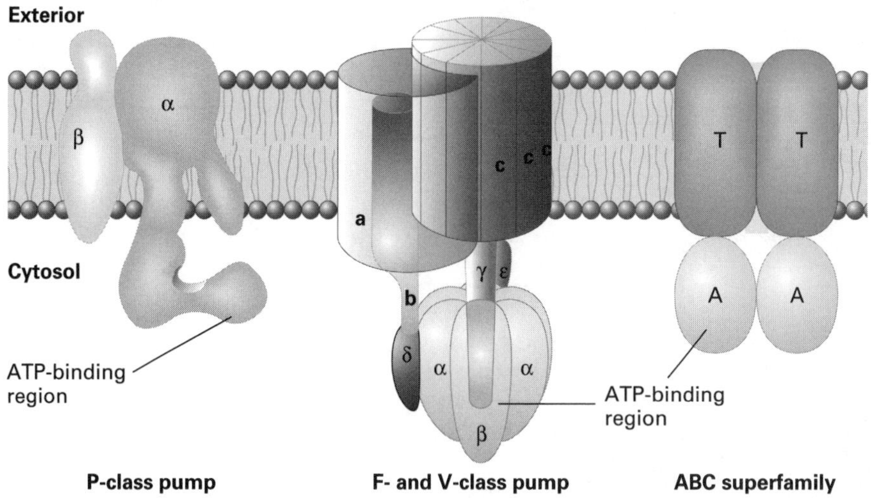

FIGURE 15-10, page 589
The four classes of ATP-powered transport proteins.

MEDIA CONNECTIONS
Biological Energy Interconversions

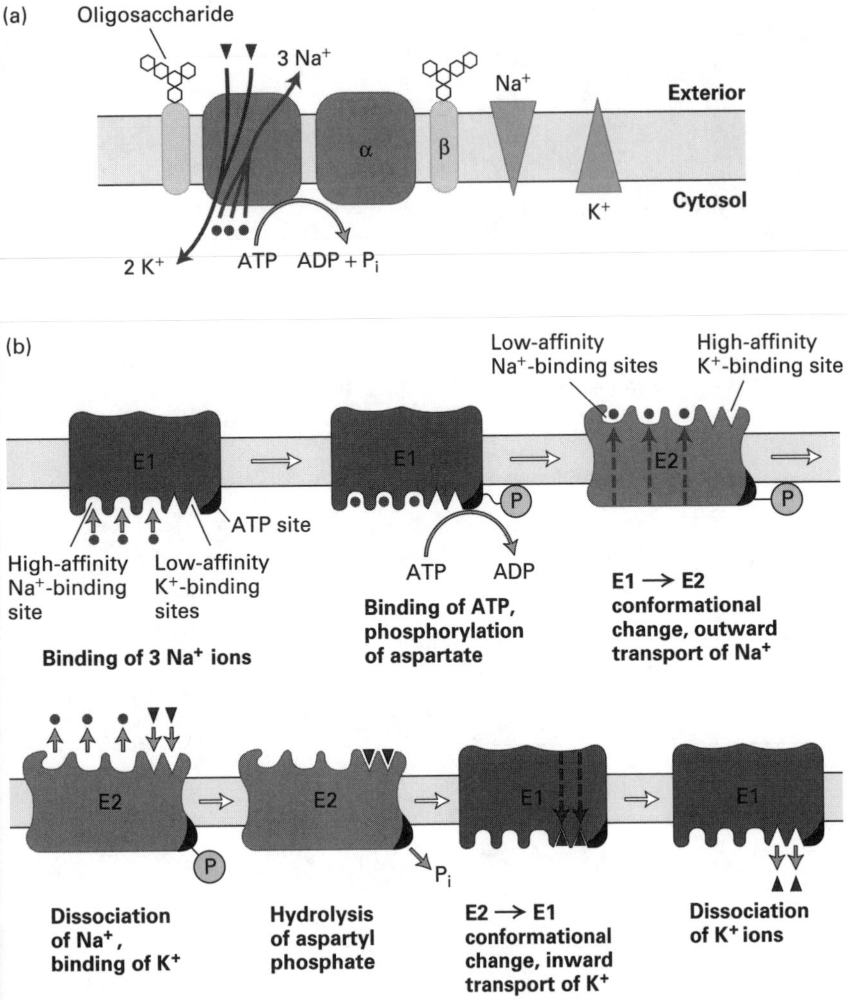

FIGURE 15-13, page 593
Models for the structure and function of the Na^+/K^+ ATPase in the plasma membrane.

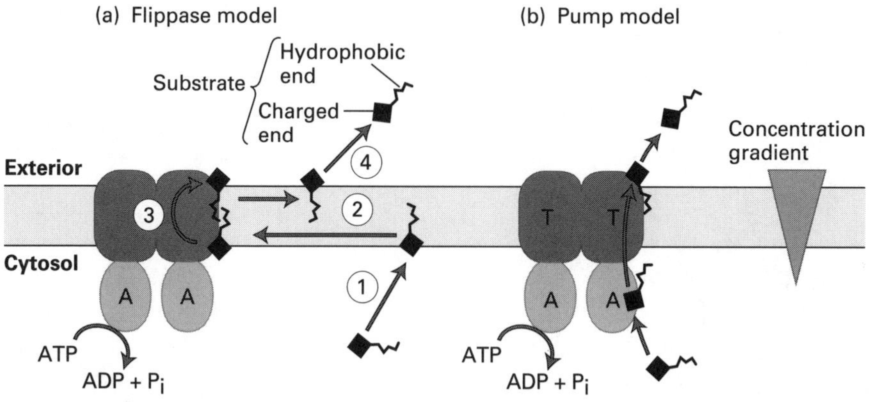

FIGURE 15-17, page 596
Possible mechanisms of action of the MDR1 protein.

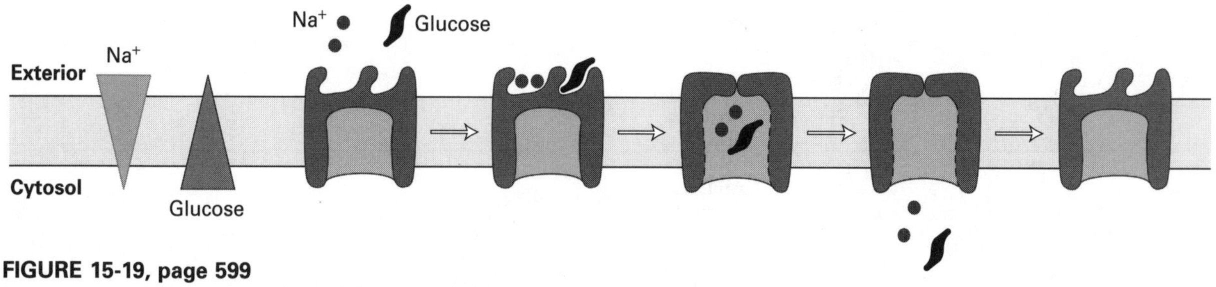

FIGURE 15-19, page 599
Proposed model for operation of the two-Na$^+$/one-glucose symporter.

MEDIA CONNECTIONS
Biological Energy Interconversions

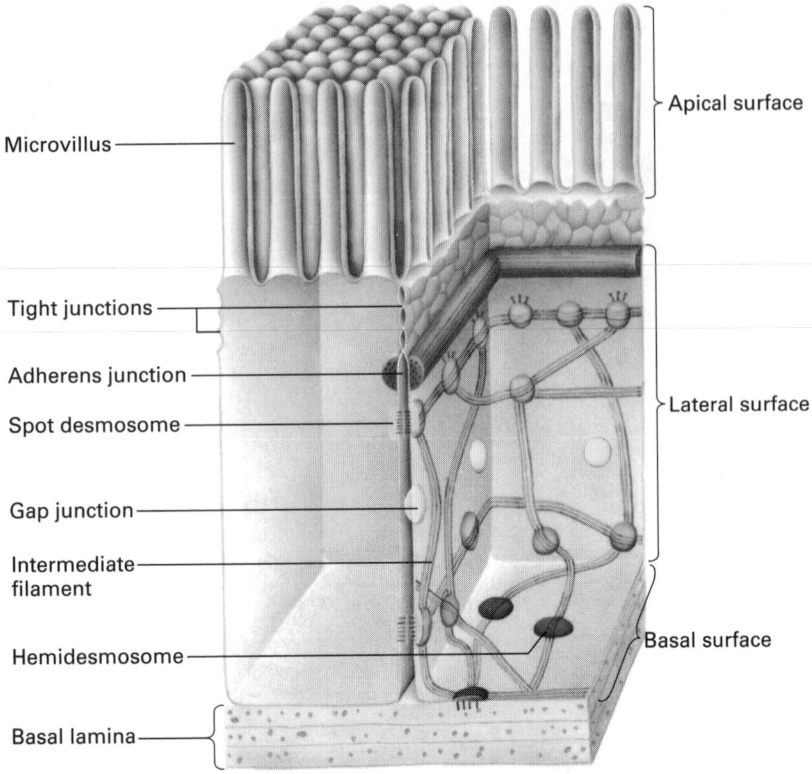

FIGURE 15-23, page 603
Schematic diagram of epithelial cells lining the small intestine and the principal types of cell junctions that connect them.

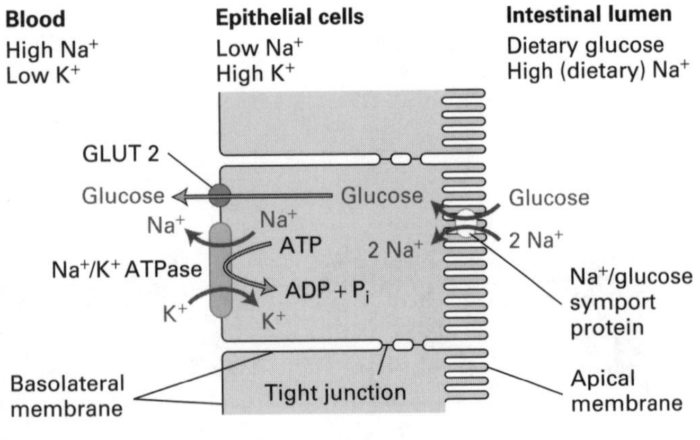

FIGURE 15-25, page 604
Transport of glucose from the intestinal lumen into the blood.

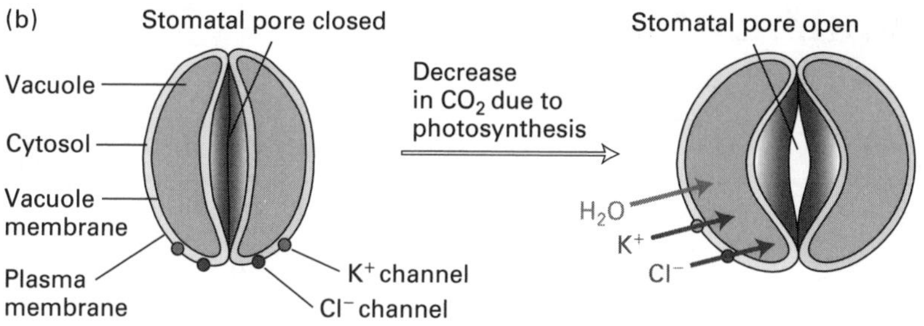

FIGURE 15-34 (b), page 611
The opening and closing of stomata.

Additional Notes

Cellular Energetics: Glycolysis, Aerobic Oxidation, and Photosynthesis

16

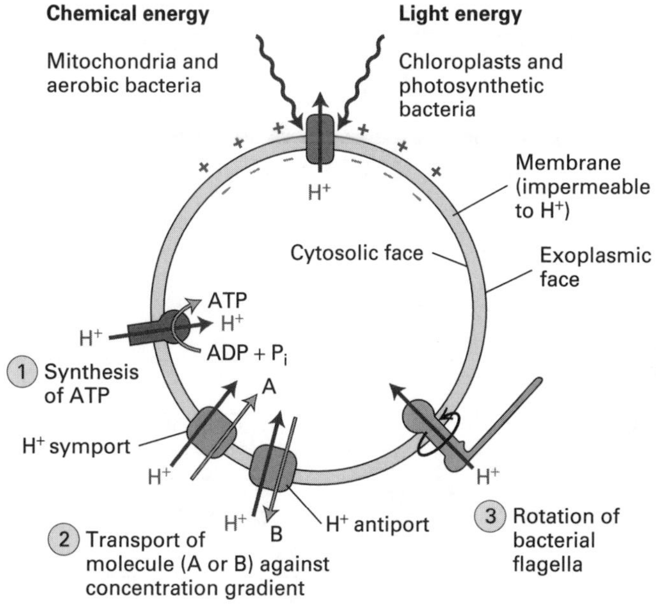

FIGURE 16-1, page 617
Chemiosmotic coupling.

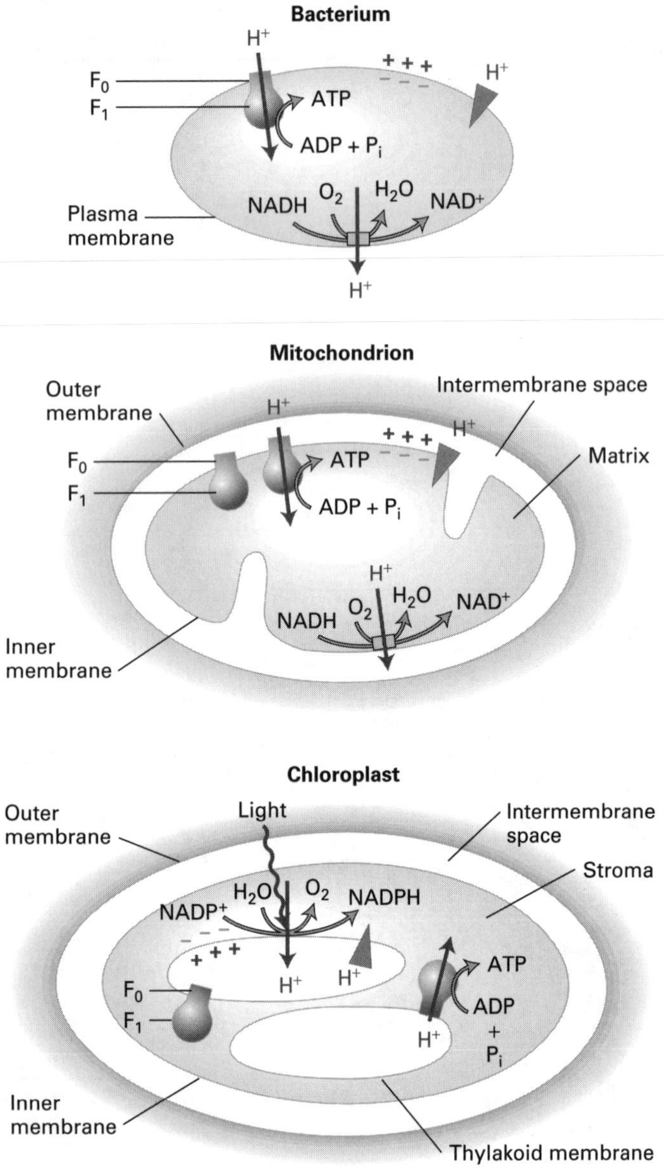

FIGURE 16-2, page 618
Membrane orientation and the direction of proton movement in bacteria, mitochondria, and chloroplasts.

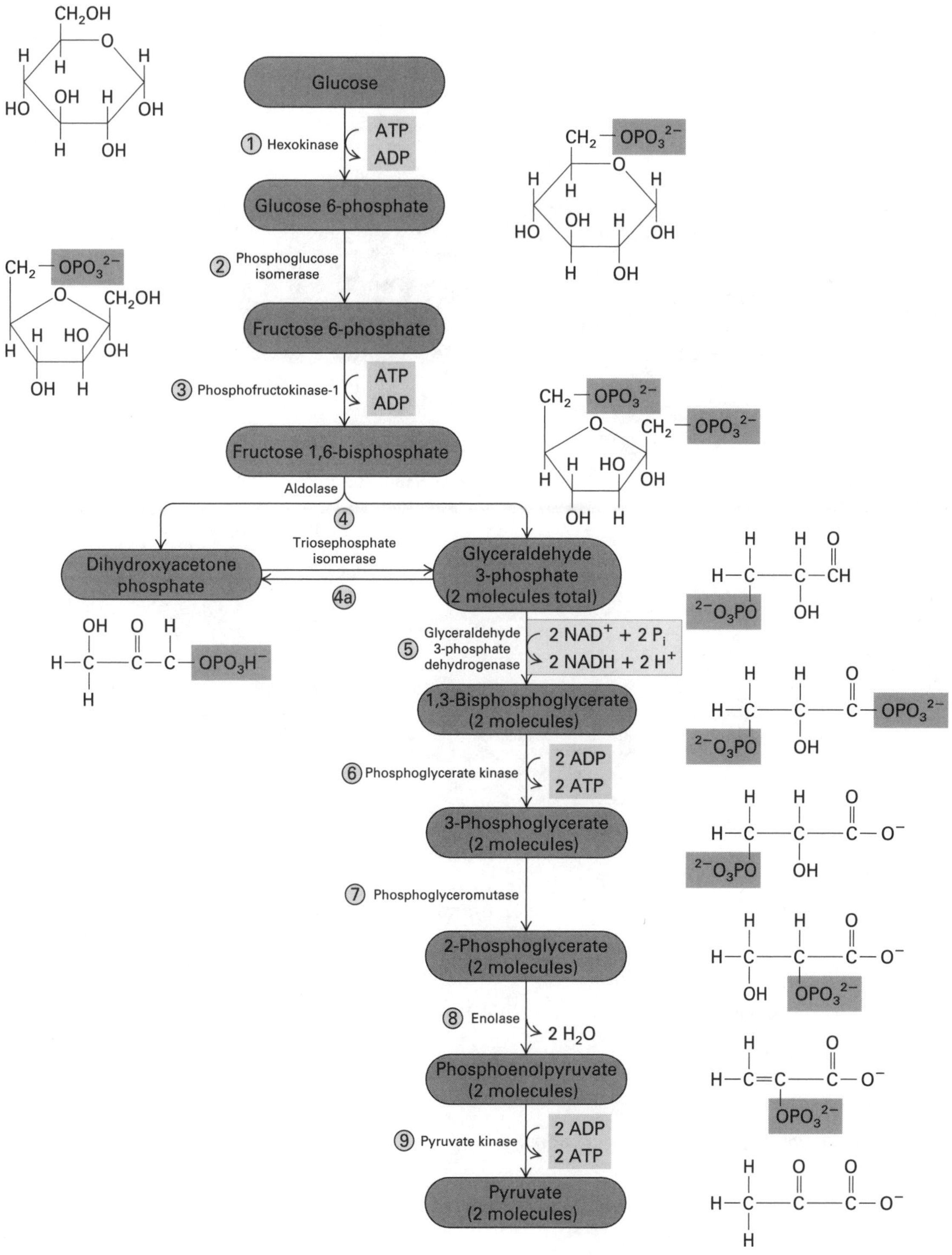

FIGURE 16-3, page 620
The glycolytic pathway by which glucose is degraded to pyruvic acid.

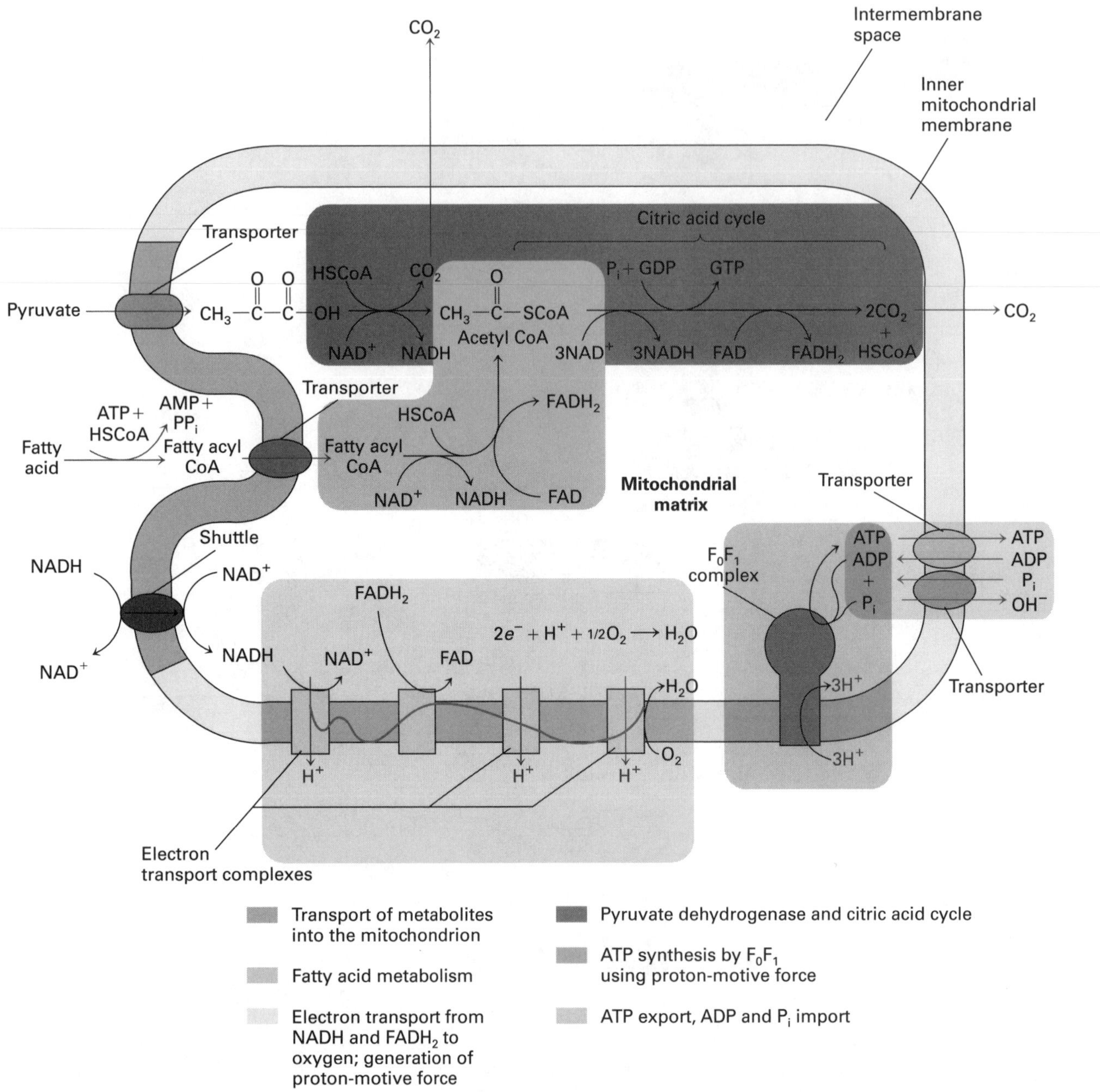

FIGURE 16-9, page 624
Summary of the aerobic oxidation of pyruvate in mitochondria.

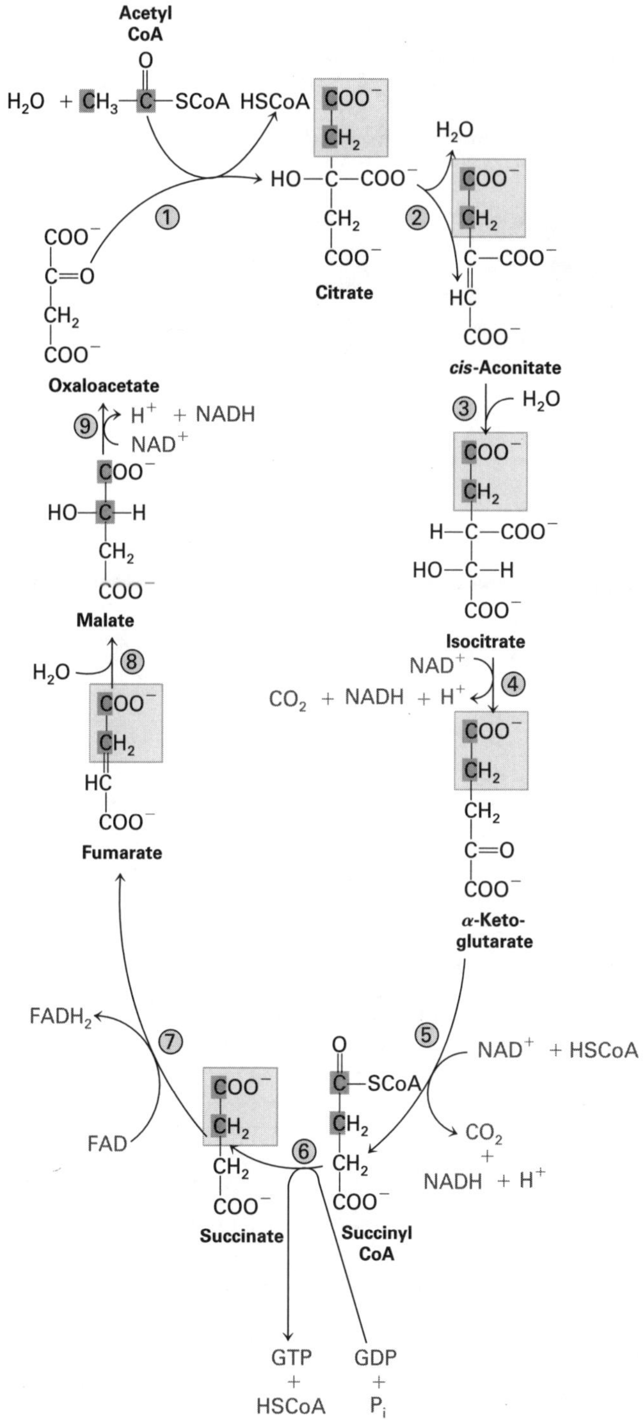

FIGURE 16-12, page 626
The citric acid cycle, in which acetyl groups transferred from acetyl CoA are oxidized to CO_2.

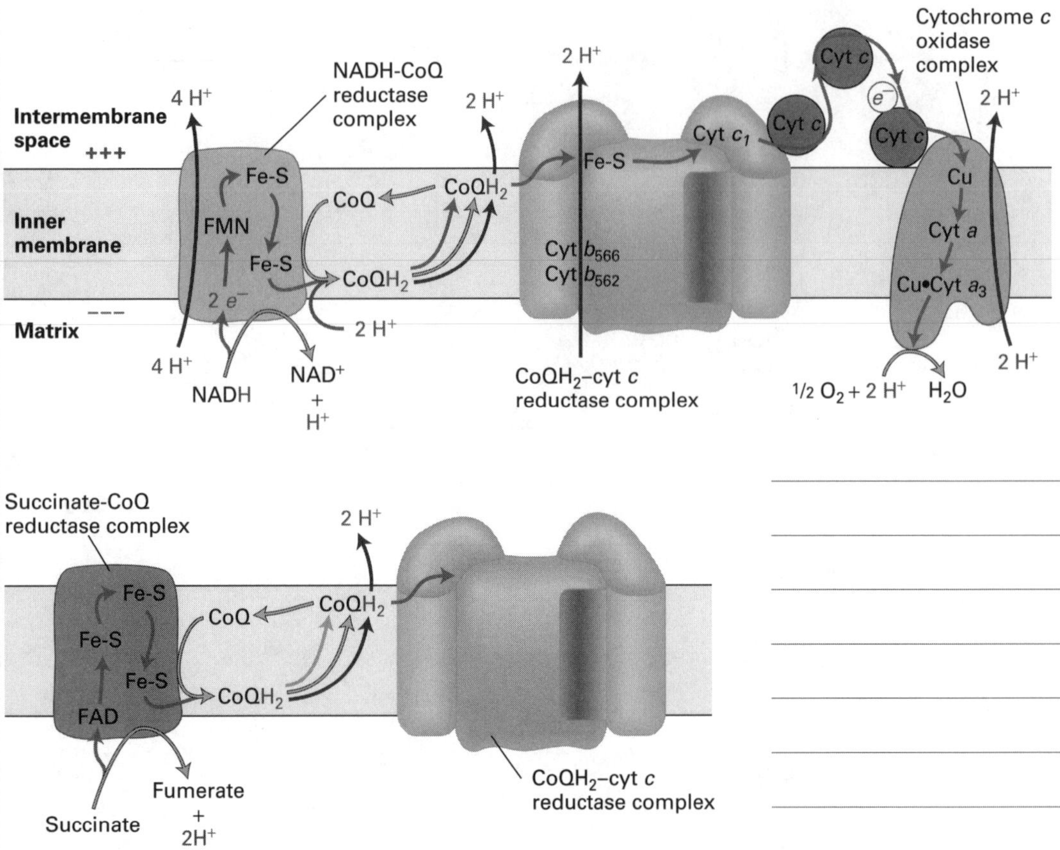

FIGURE 16-19, page 636
The pathway of electron transport (light gray arrows) and proton transport (dark gray arrows) in the inner mitochondrial membrane.

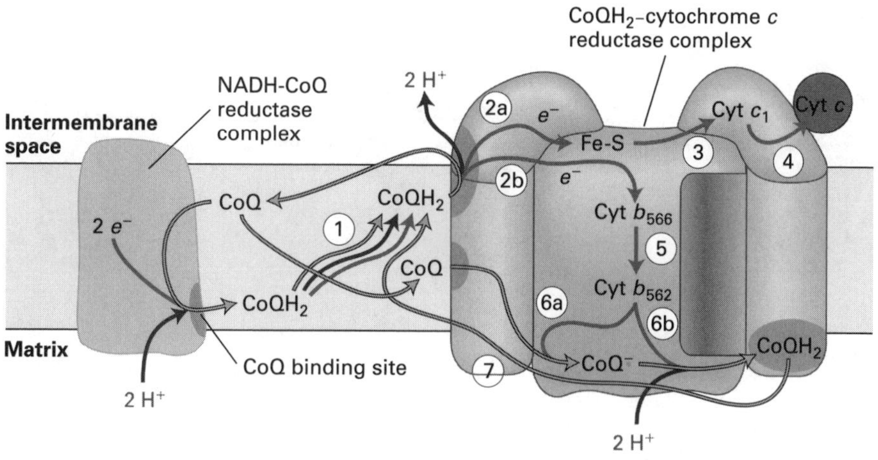

FIGURE 16-25, page 641
The proton-motive Q cycle.

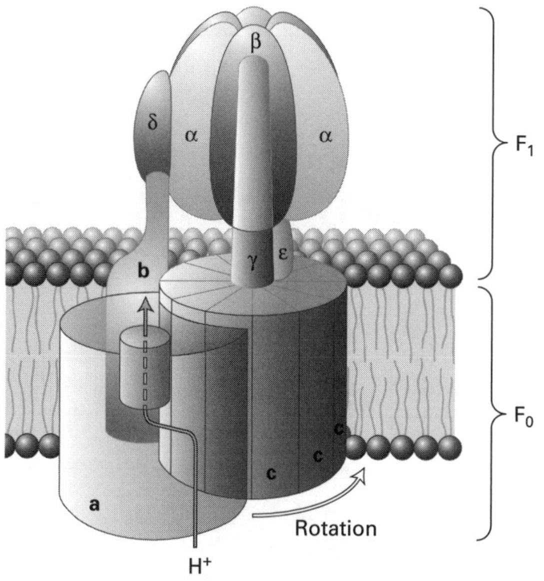

FIGURE 16-28, page 644
Model of the structure of ATP synthase (the F_0F_1 ATPase complex) in the bacterial plasma membrane.

MEDIA CONNECTIONS
ATP Synthesis

MEDIA CONNECTIONS
ATP Synthesis

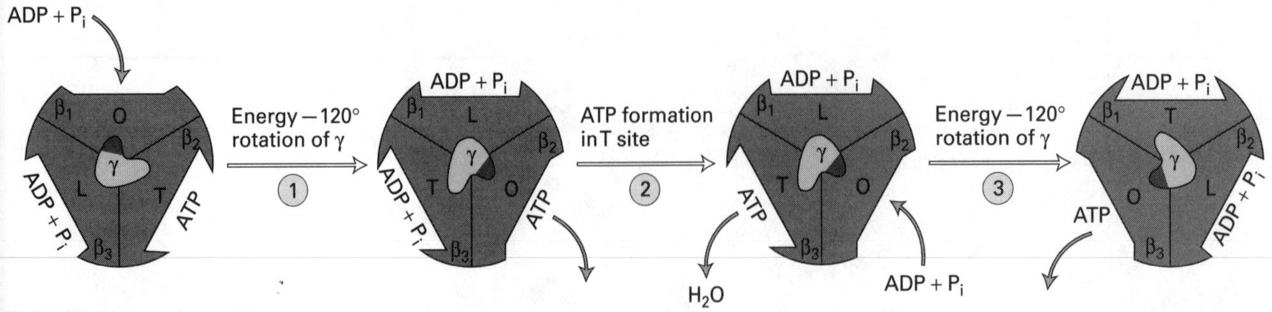

FIGURE 16-30, page 645
The binding-change mechanism of ATP synthesis from ADP and P_i by the F_0F_1 complex.

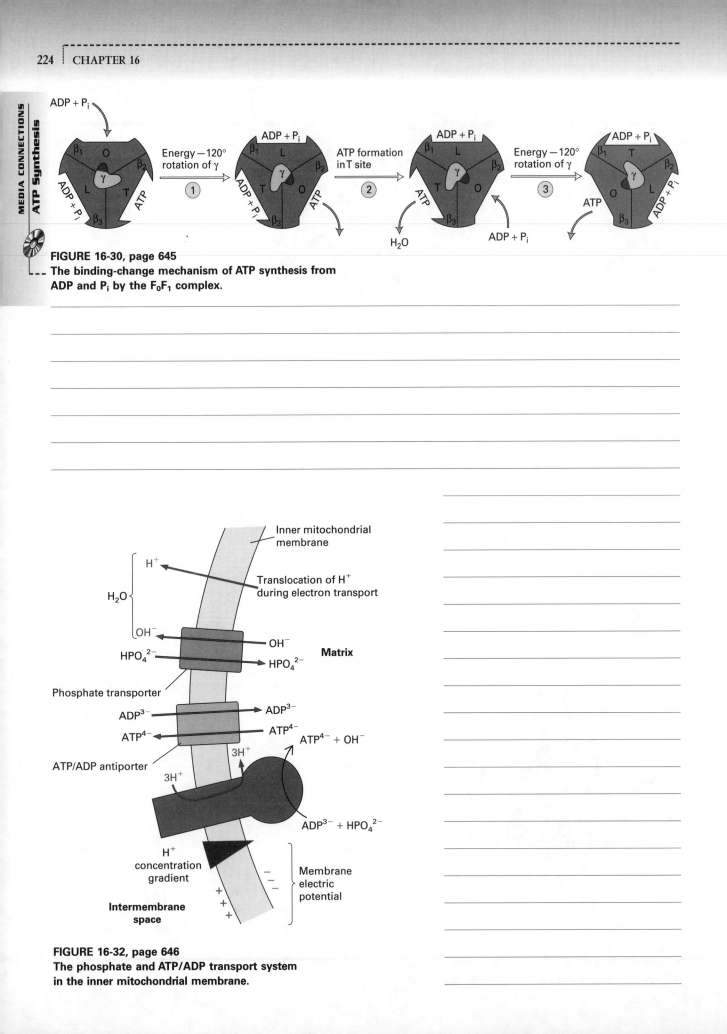

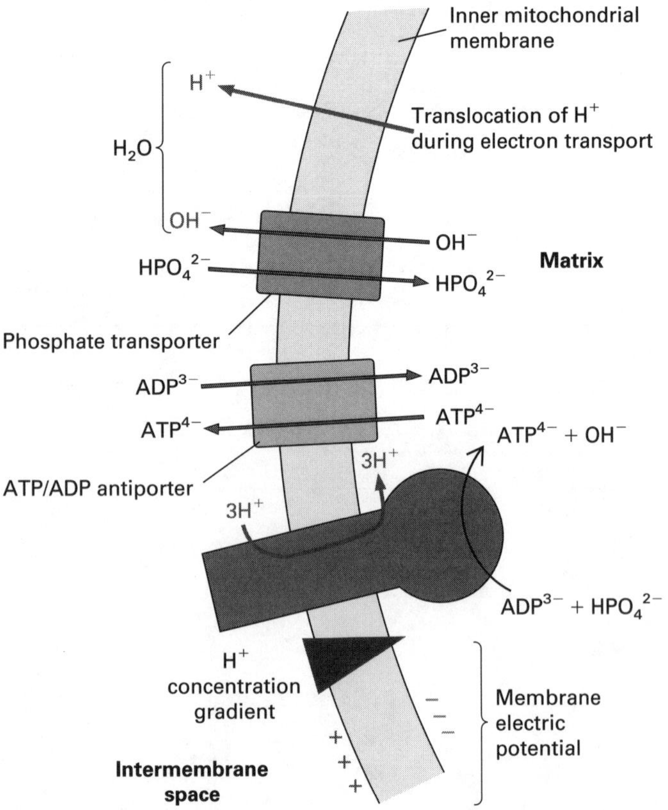

FIGURE 16-32, page 646
The phosphate and ATP/ADP transport system in the inner mitochondrial membrane.

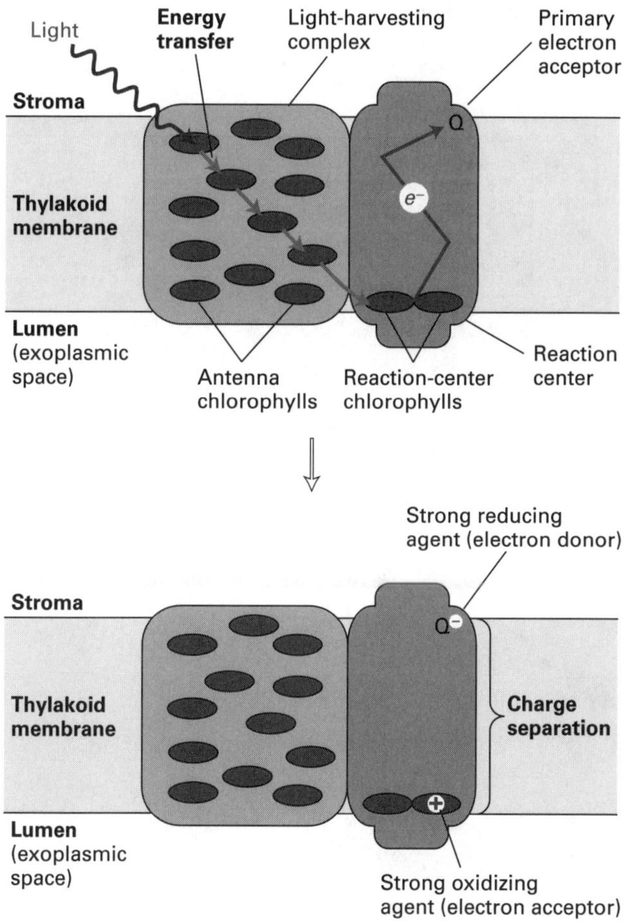

FIGURE 16-38, page 653
The primary event in photosynthesis.

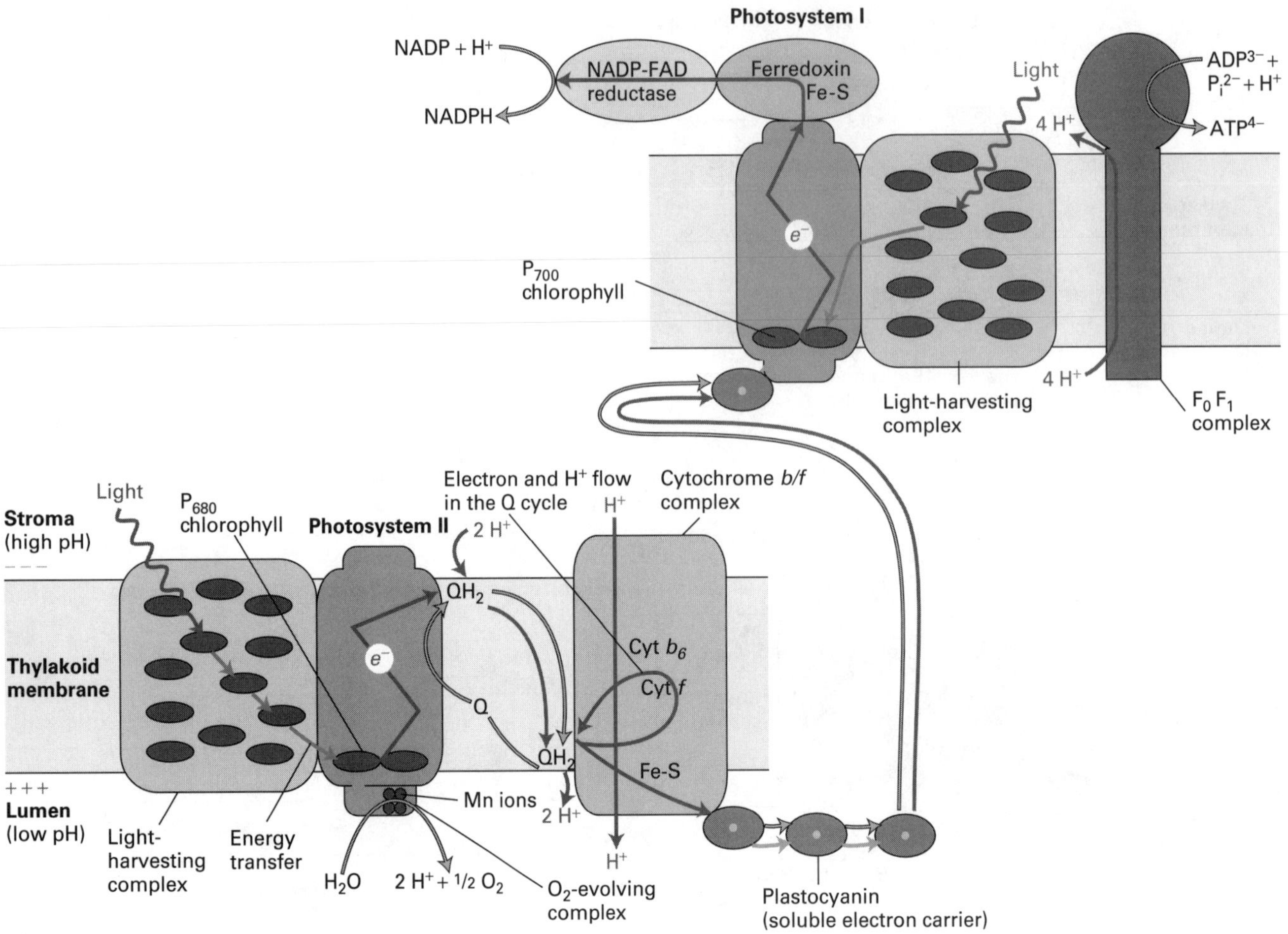

FIGURE 16-42, page 658
Summary of photosynthesis in plants, which utilize two photosystems, PSI and PSII, during linear electron flow.

Additional Notes

Protein Sorting: Organelle Biogenesis and Protein Secretion

17

MEDIA CONNECTIONS
Protein Sorting

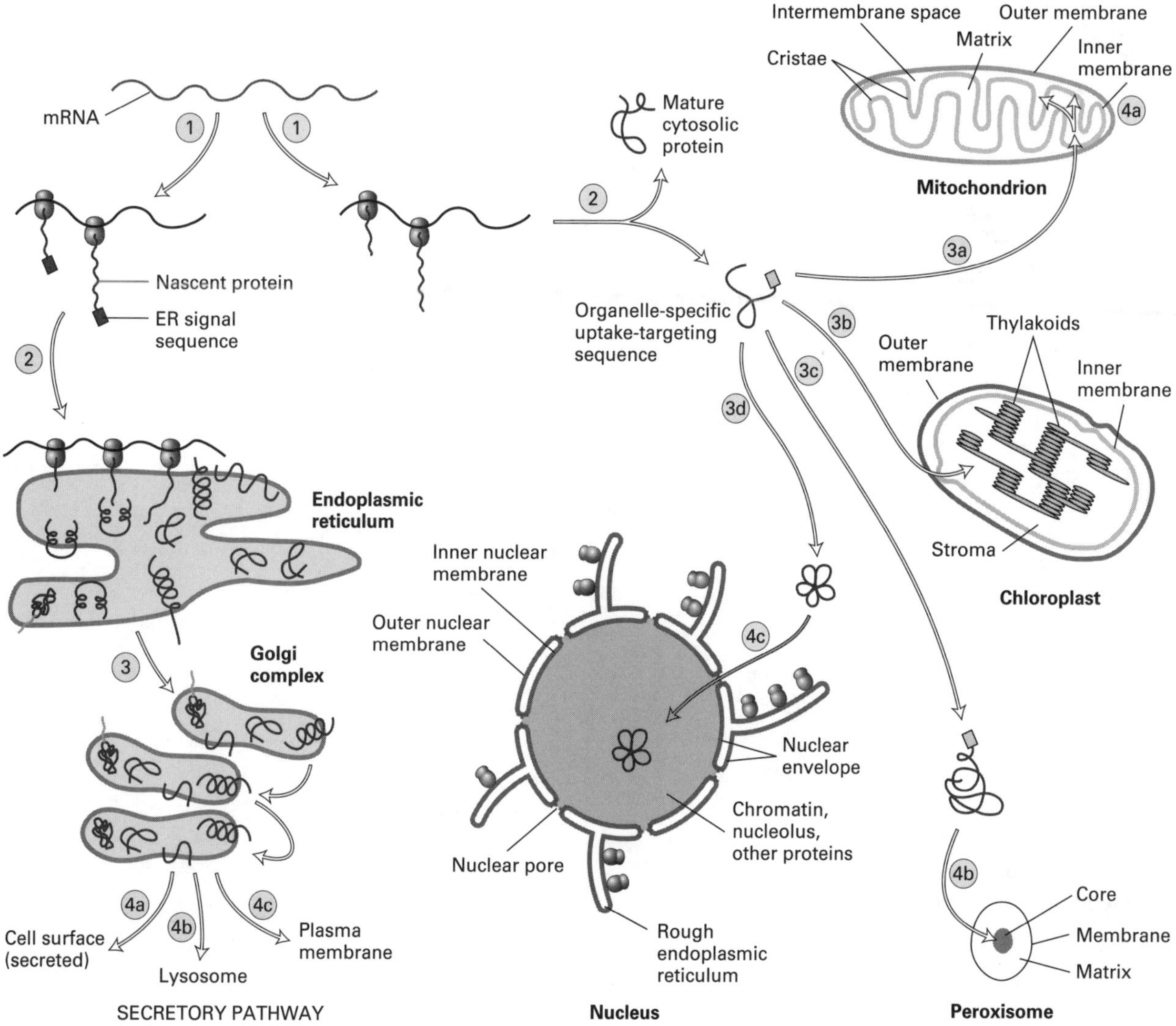

FIGURE 17-1, page 676
Overview of sorting of nuclear-encoded proteins in eukaryotic cells.

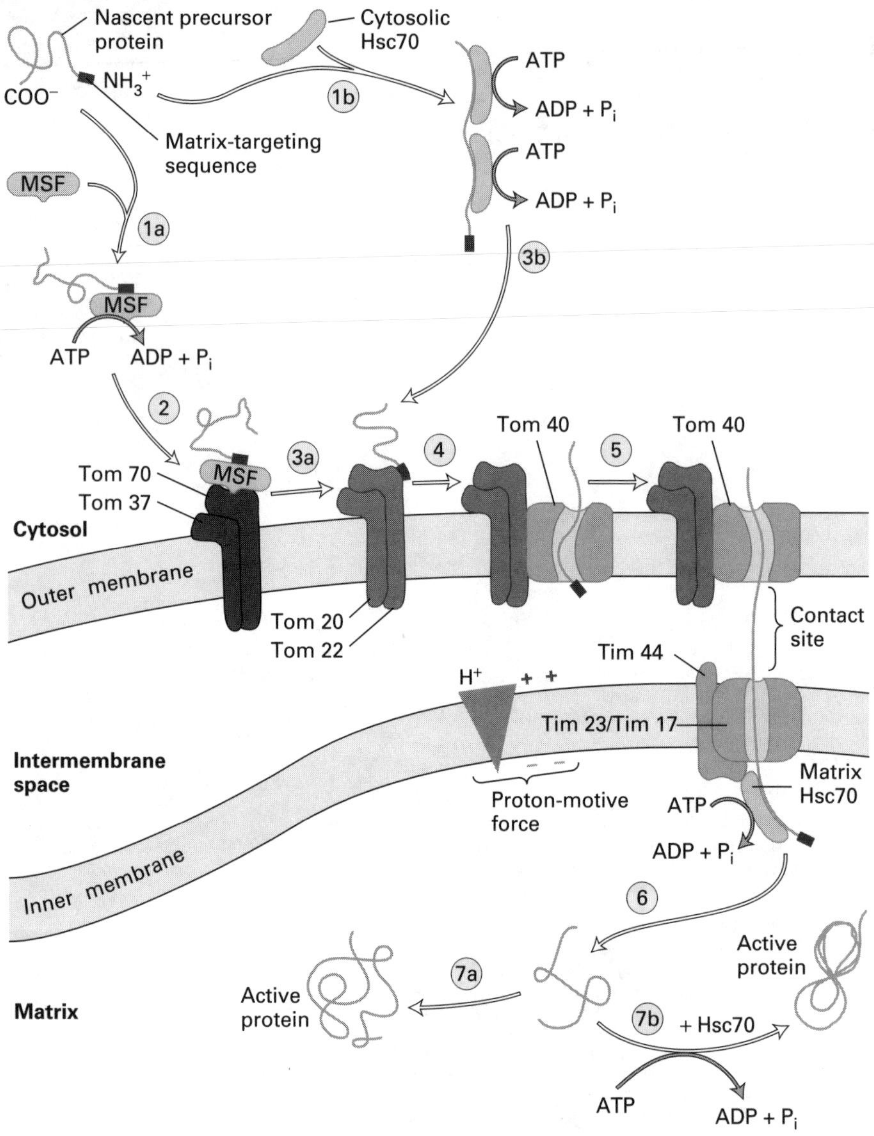

FIGURE 17-4, page 681
Protein import into the mitochondrial matrix.

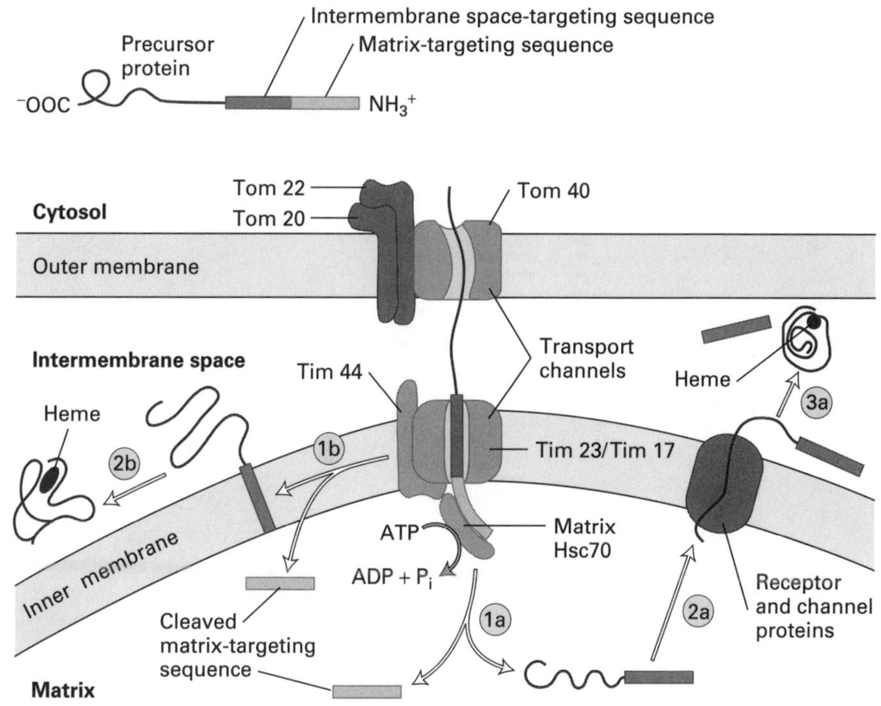

FIGURE 17-6, page 684
Two pathways by which different proteins are transported from the cytosol to the mitochondrial intermembrane space.

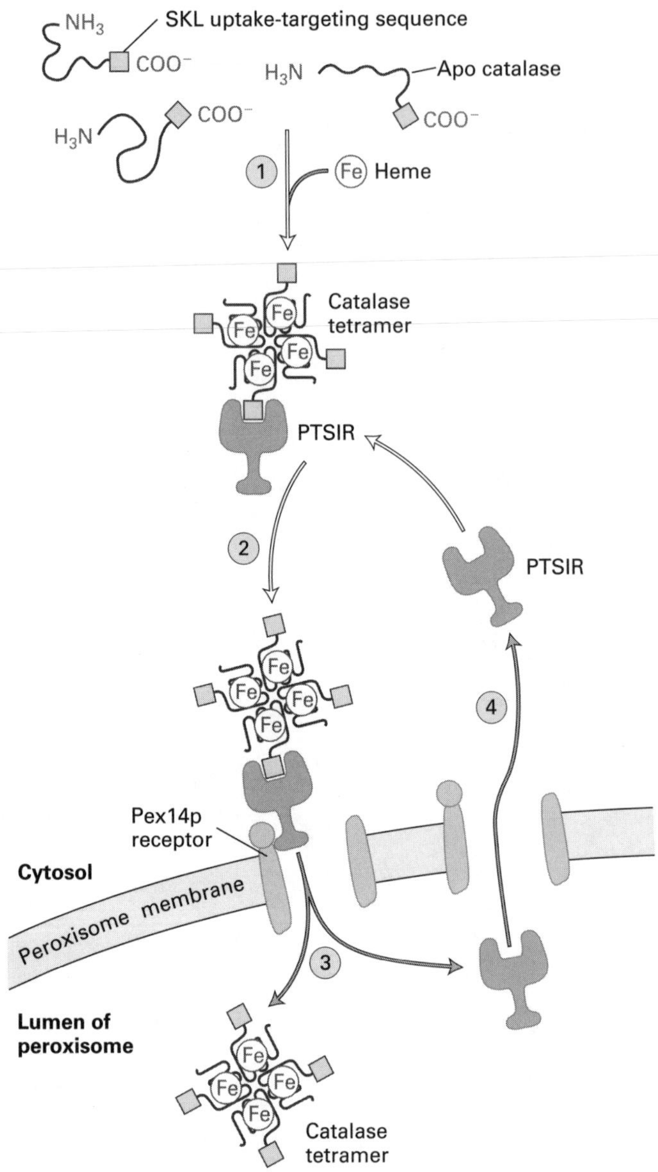

FIGURE 17-10, page 690
Synthesis of catalase and its incorporation into peroxisomes.

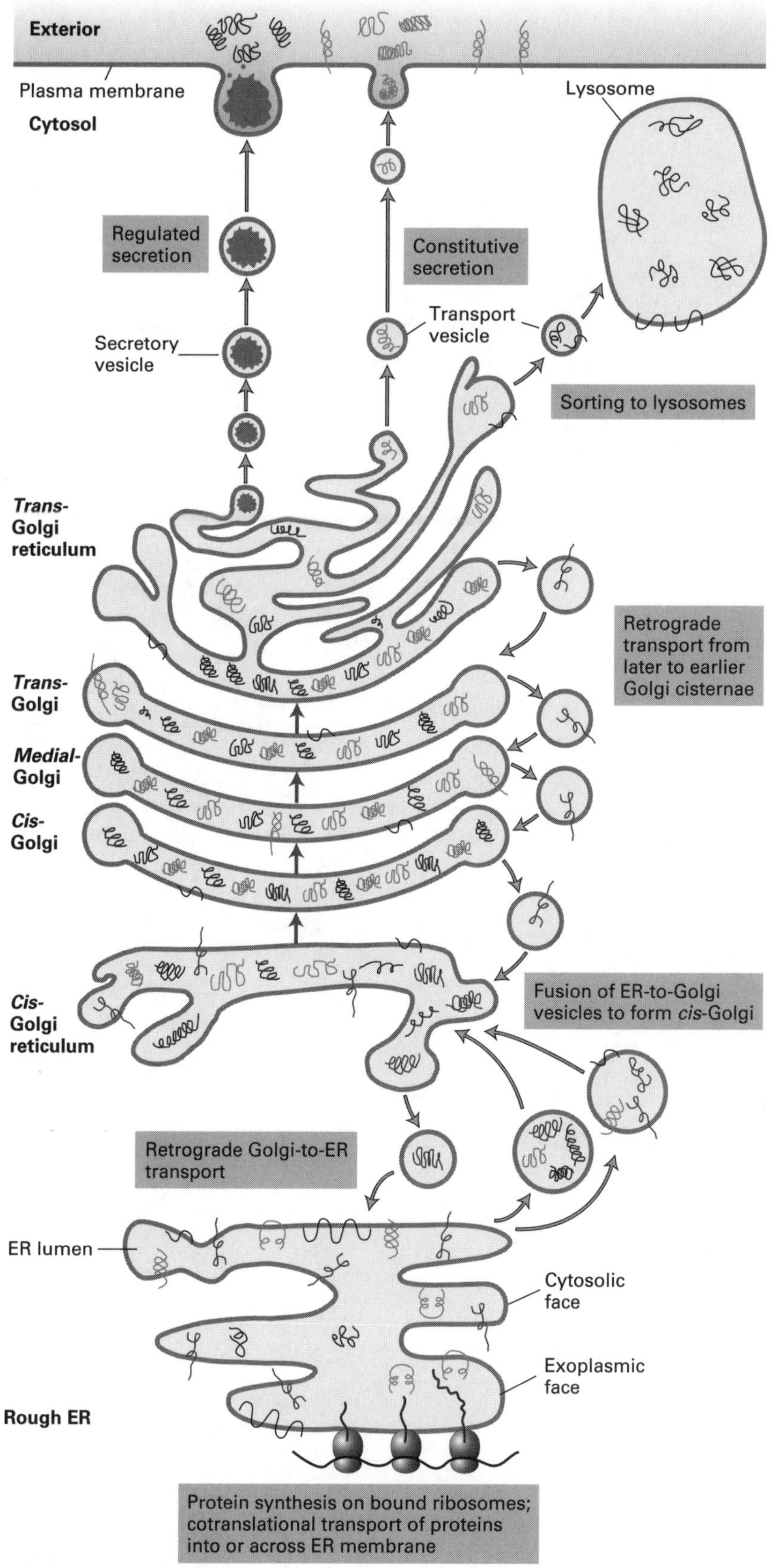

FIGURE 17-13, page 693
The secretory pathway of protein synthesis and sorting.

MEDIA CONNECTIONS
Protein Secretion

MEDIA CONNECTIONS
Synthesis of Secreted and Membrane-Bound Proteins

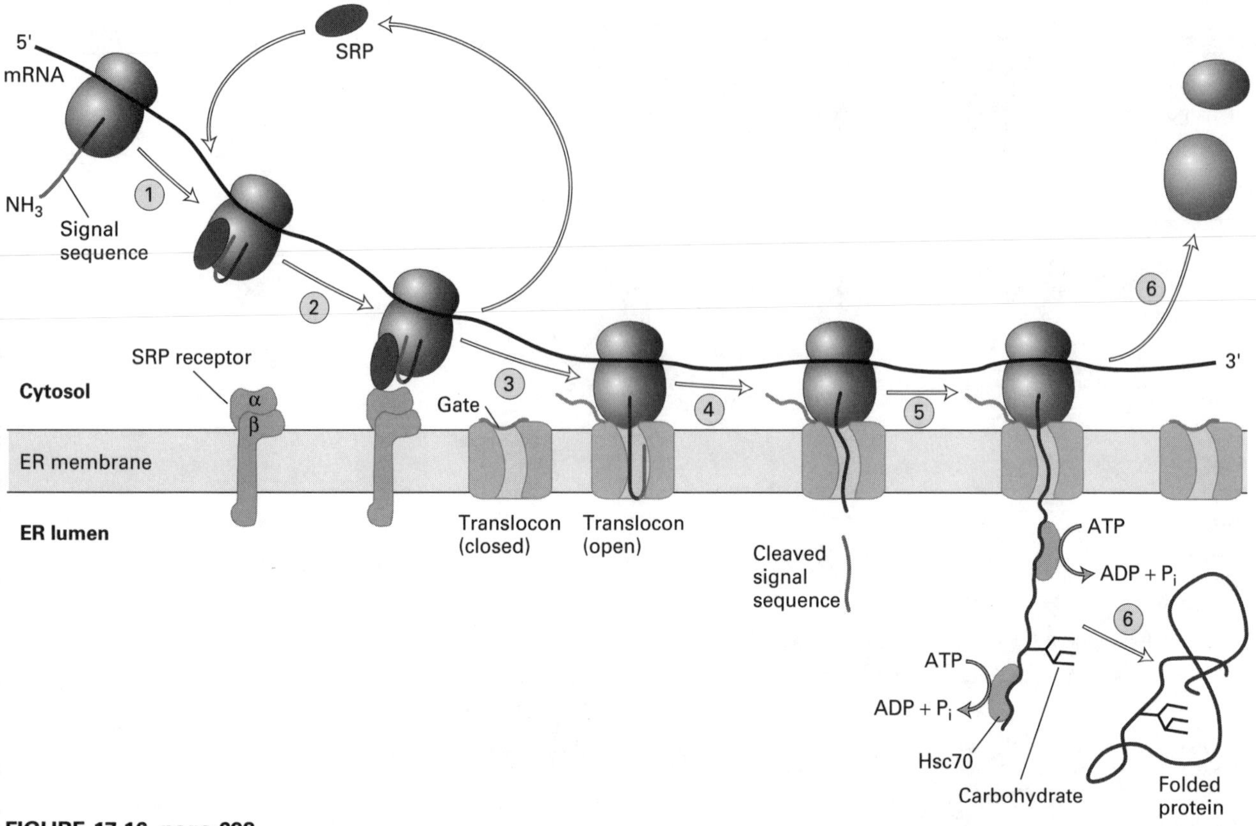

FIGURE 17-16, page 698
Synthesis of secretory proteins on the rough ER.

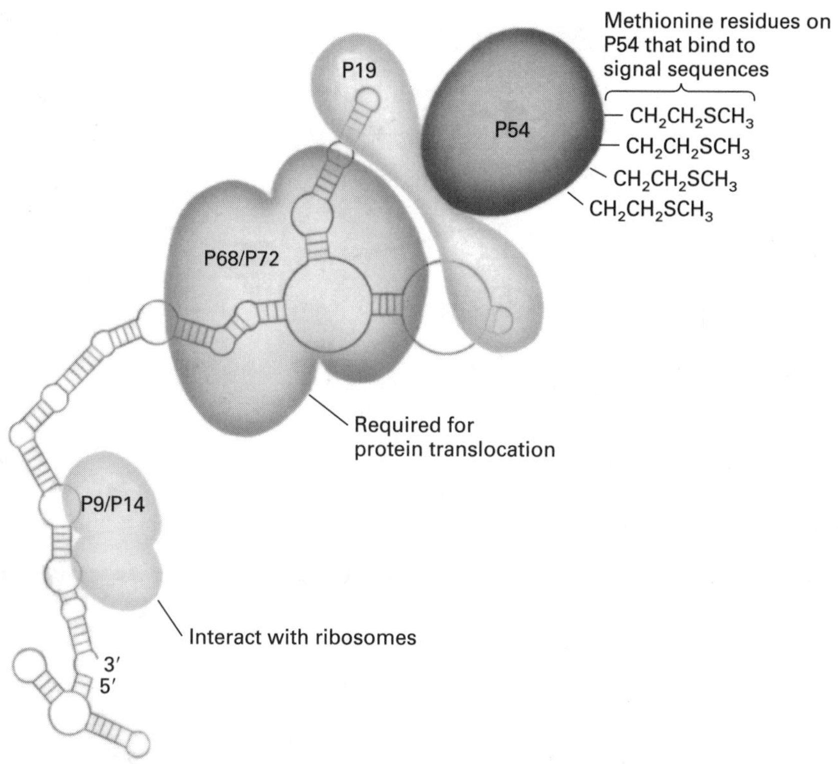

FIGURE 17-17, page 699
Structure of the signal-recognition particle (SRP).

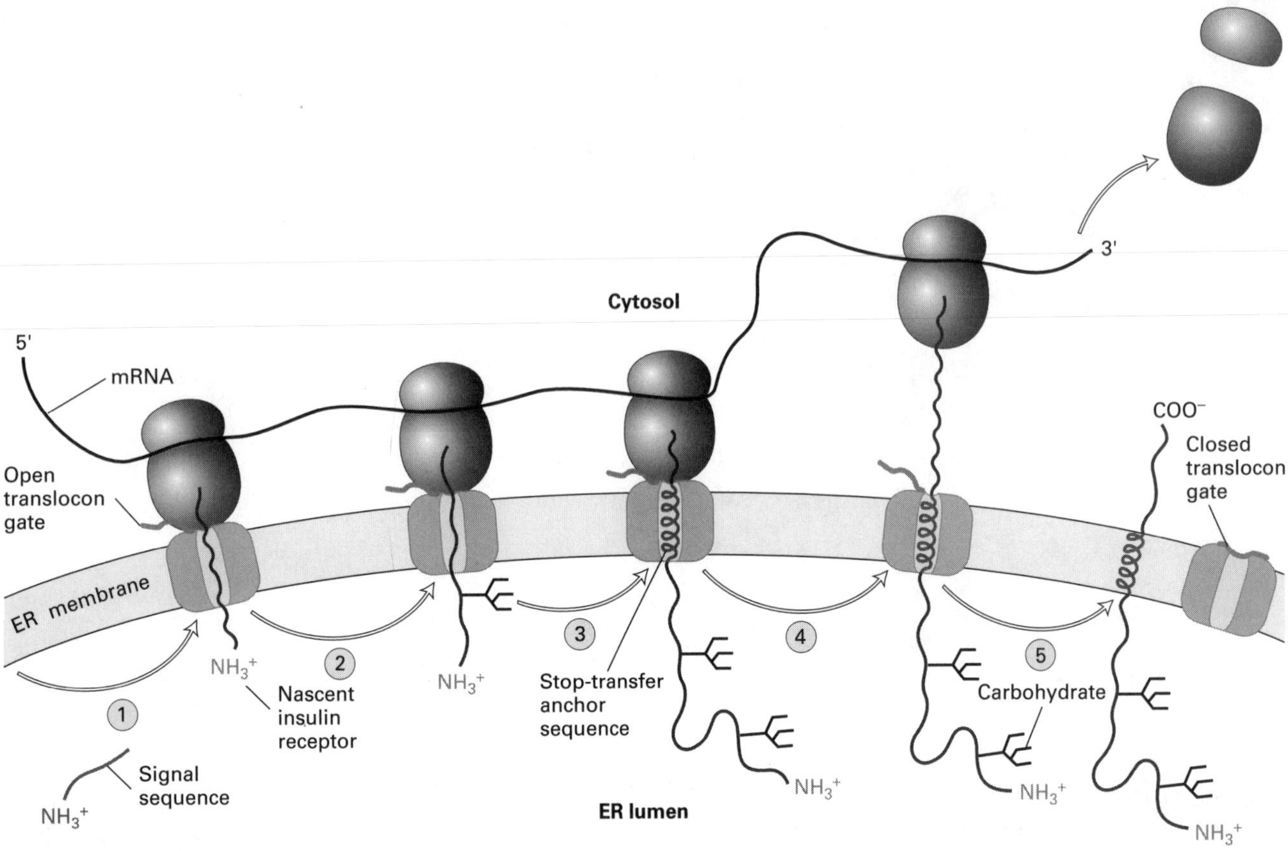

FIGURE 17-22, page 703
Synthesis and insertion into the ER membrane of the insulin receptor and similar proteins that employ a cleaved ER signal sequence and an internal stop-transfer membrane-anchor sequence.

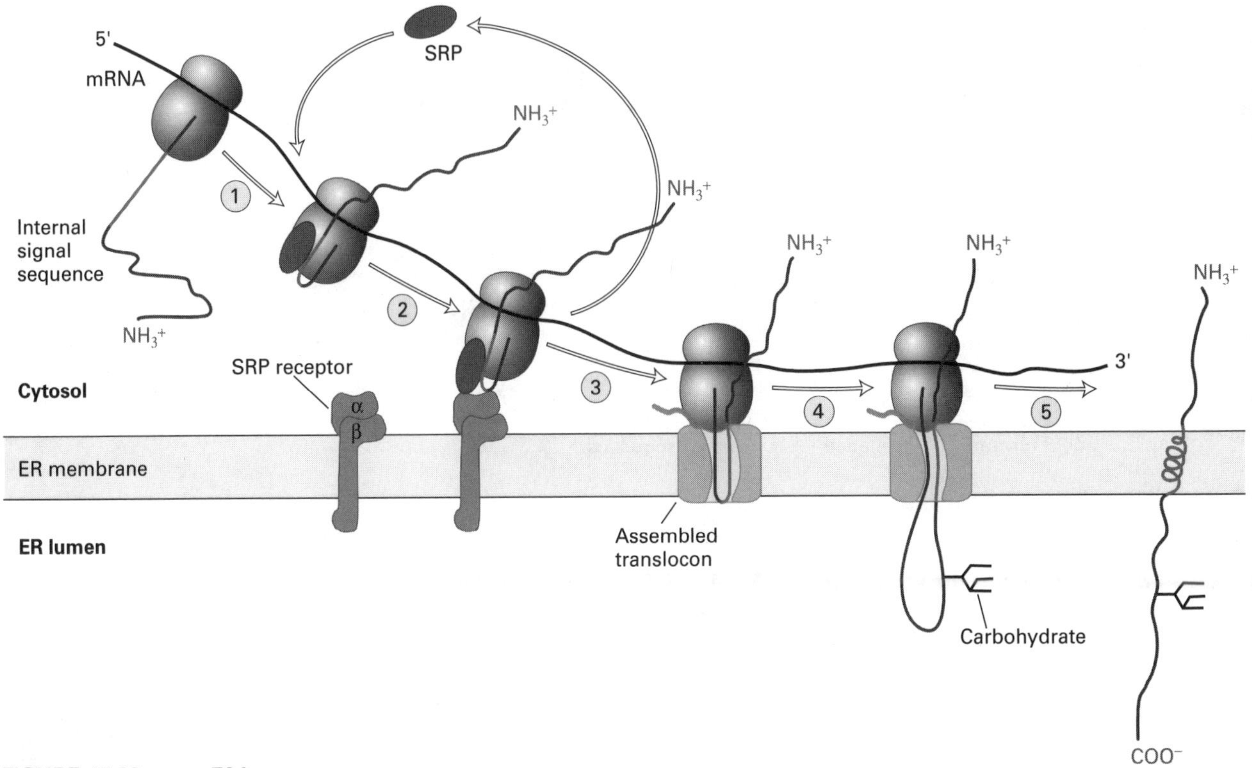

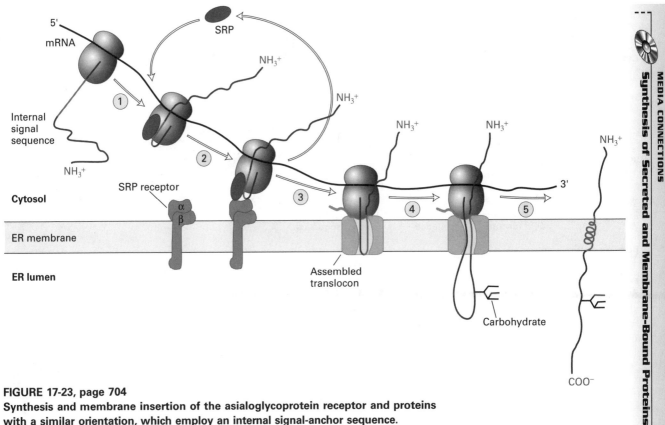

FIGURE 17-23, page 704
Synthesis and membrane insertion of the asialoglycoprotein receptor and proteins with a similar orientation, which employ an internal signal-anchor sequence.

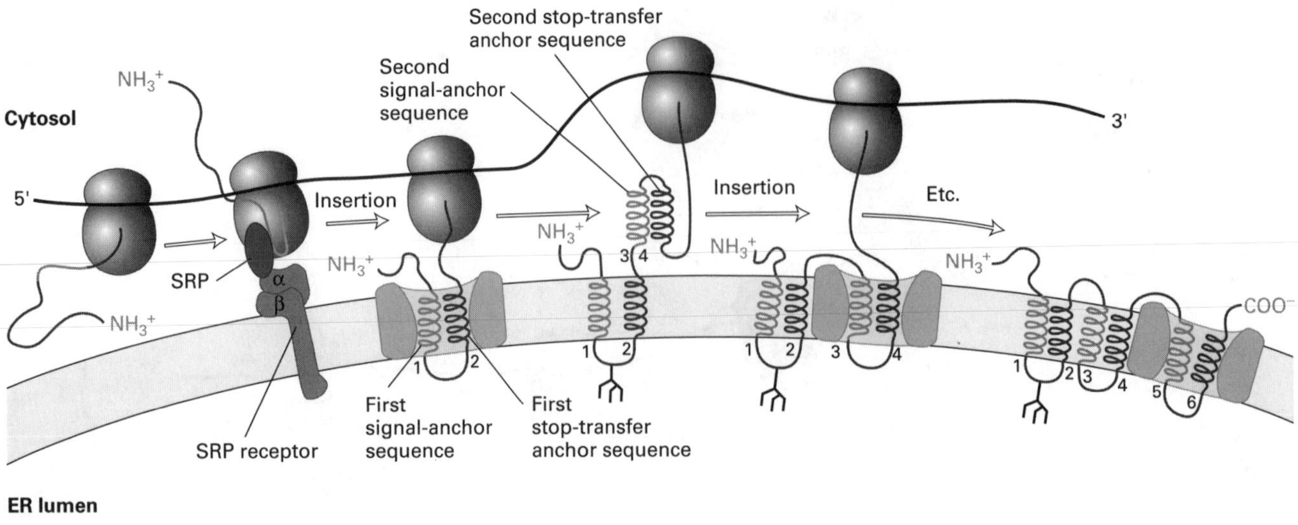

FIGURE 17-24, page 706
Synthesis and insertion into the ER membrane of the GLUT1 glucose transporter and other proteins with multiple transmembrane α-helical segments.

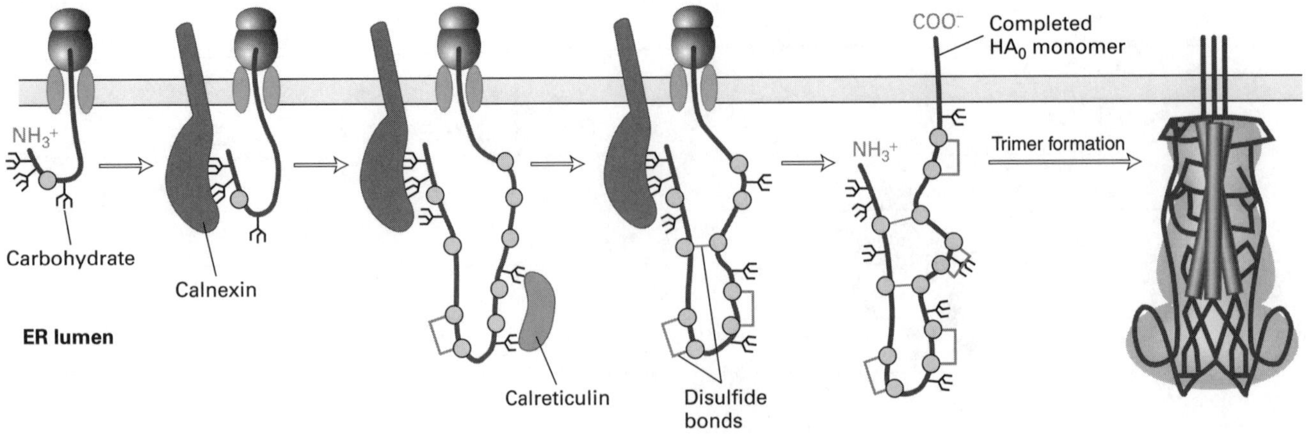

FIGURE 17-27, page 709
Folding of the hemagglutinin (HA) precursor polypeptide HA_0 and formation of an HA_0 trimer within the ER.

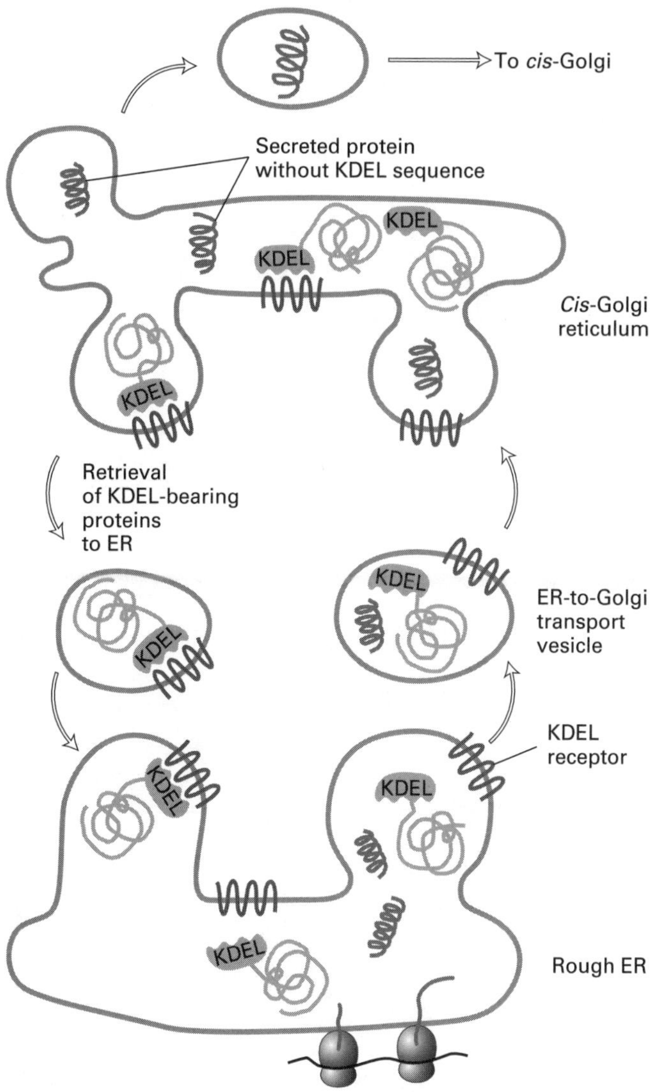

FIGURE 17-29, page 711
Role of the KDEL receptor in the retrieval of ER-resident proteins.

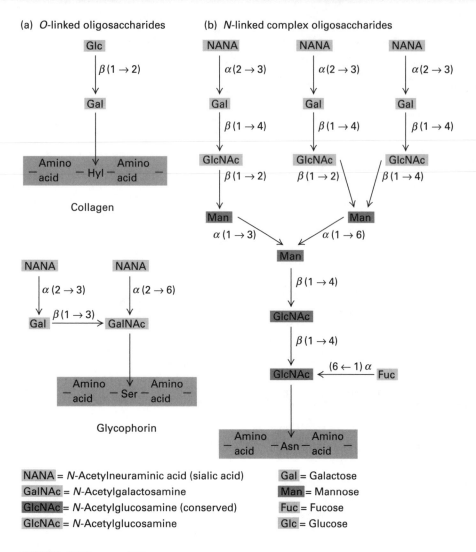

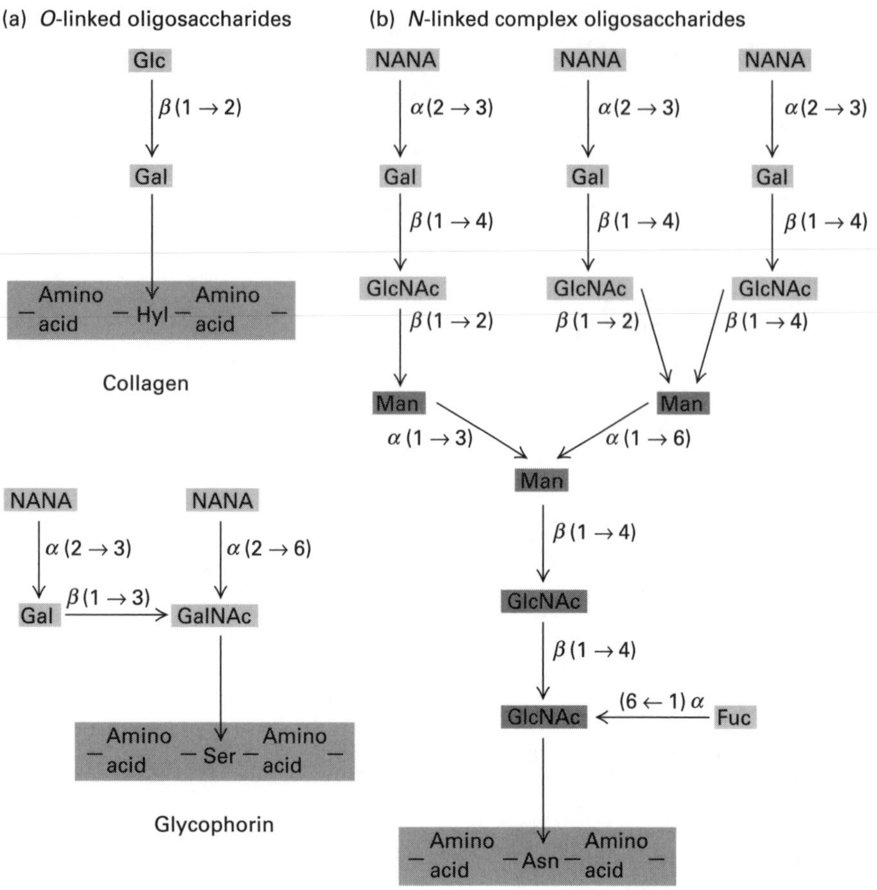

NANA = *N*-Acetylneuraminic acid (sialic acid)
GalNAc = *N*-Acetylgalactosamine
GlcNAc = *N*-Acetylglucosamine (conserved)
GlcNAc = *N*-Acetylglucosamine
Gal = Galactose
Man = Mannose
Fuc = Fucose
Glc = Glucose

FIGURE 17-30, page 713
Structures of typical *O*-linked and *N*-linked oligosaccharides.

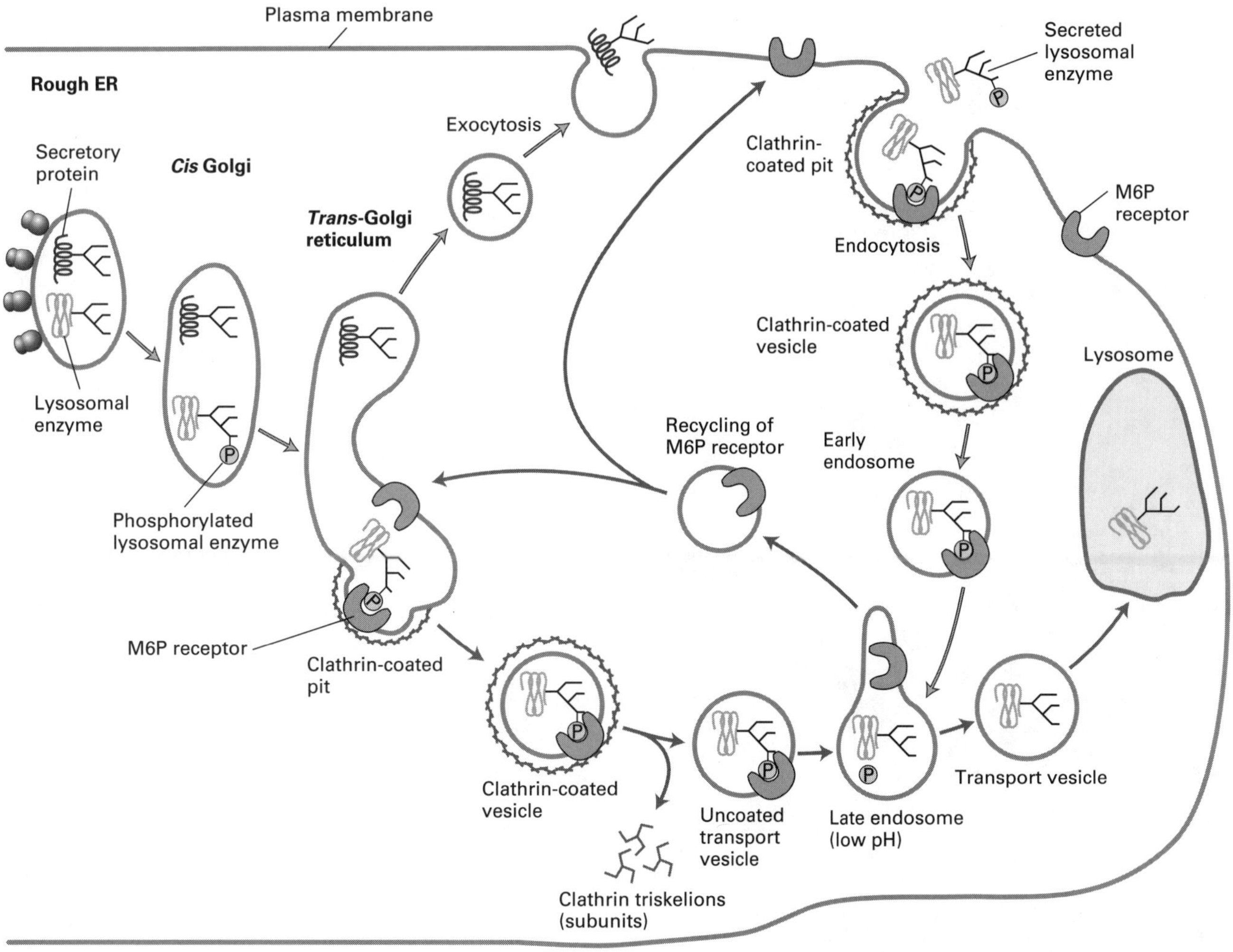

FIGURE 17-40, page 721
The mannose 6-phosphate (M6P) pathway, the major route for targeting lysosomal enzymes to lysosomes.

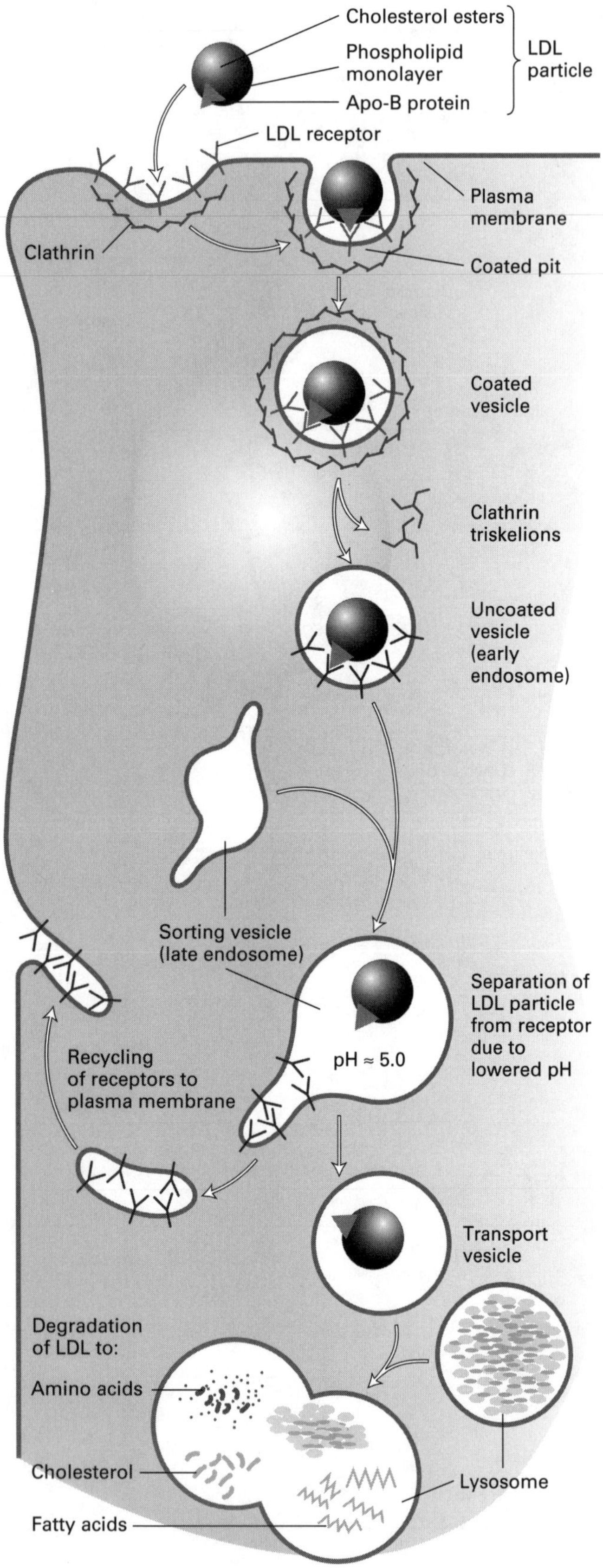

FIGURE 17-46, page 730
Fate of an LDL particle and its receptor after endocytosis.

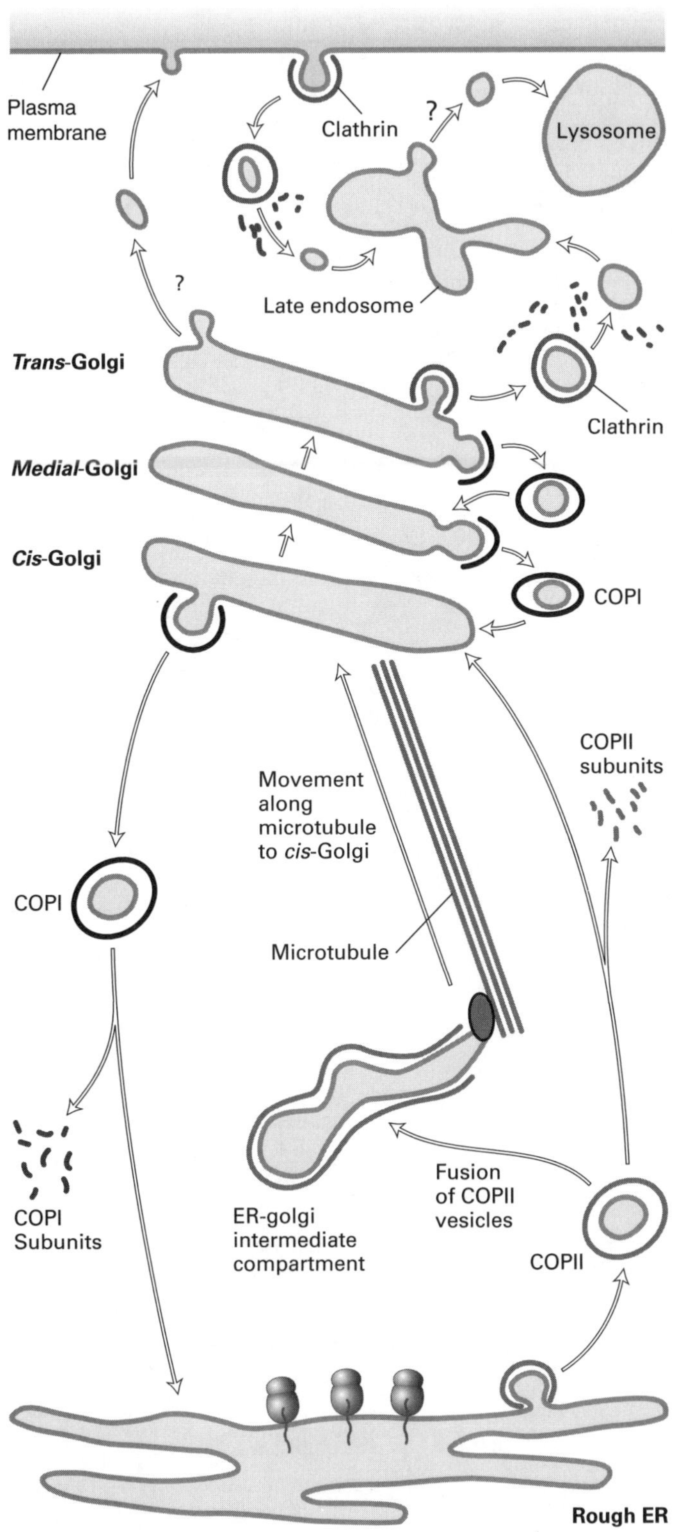

FIGURE 17-50, page 734
Involvement of the three known types of coat proteins —COP I, COP II, and clathrin—in vesicular traffic in the secretory and endocytic pathways.

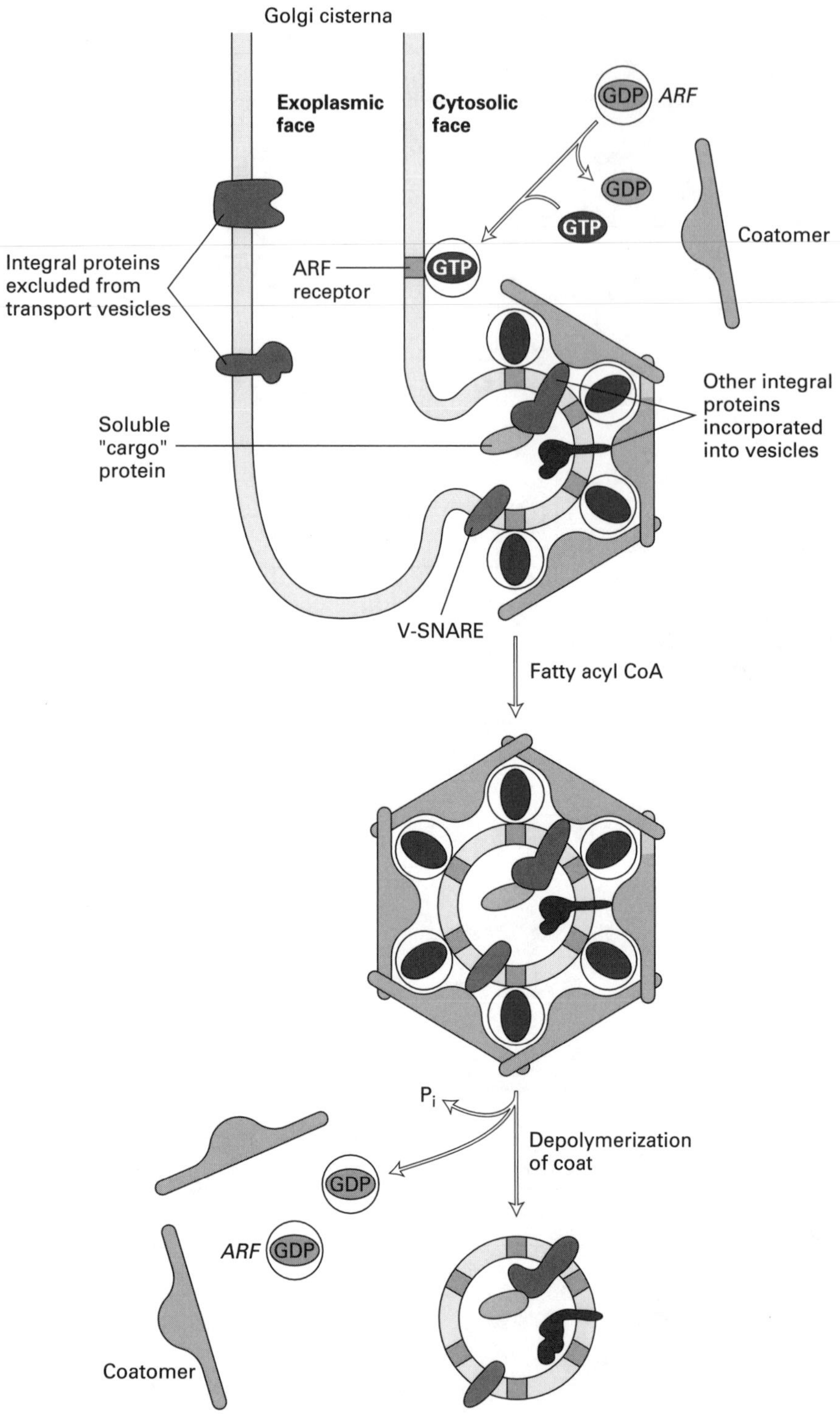

FIGURE 17-58, page 740
Model for formation of COP I–coated vesicles.

Additional Notes

Cell Motility and Shape I: Microfilaments

18

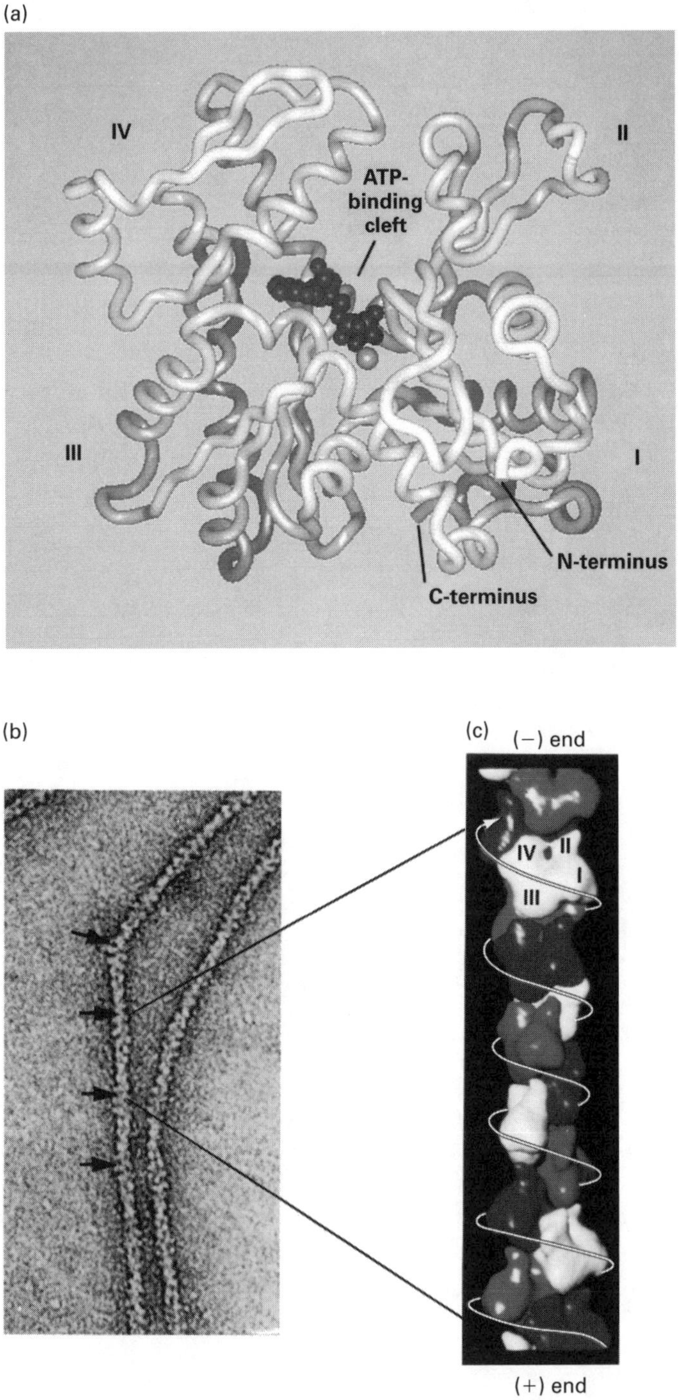

FIGURE 18-2, page 754
Structures of monomeric G-actin and F-actin filament.

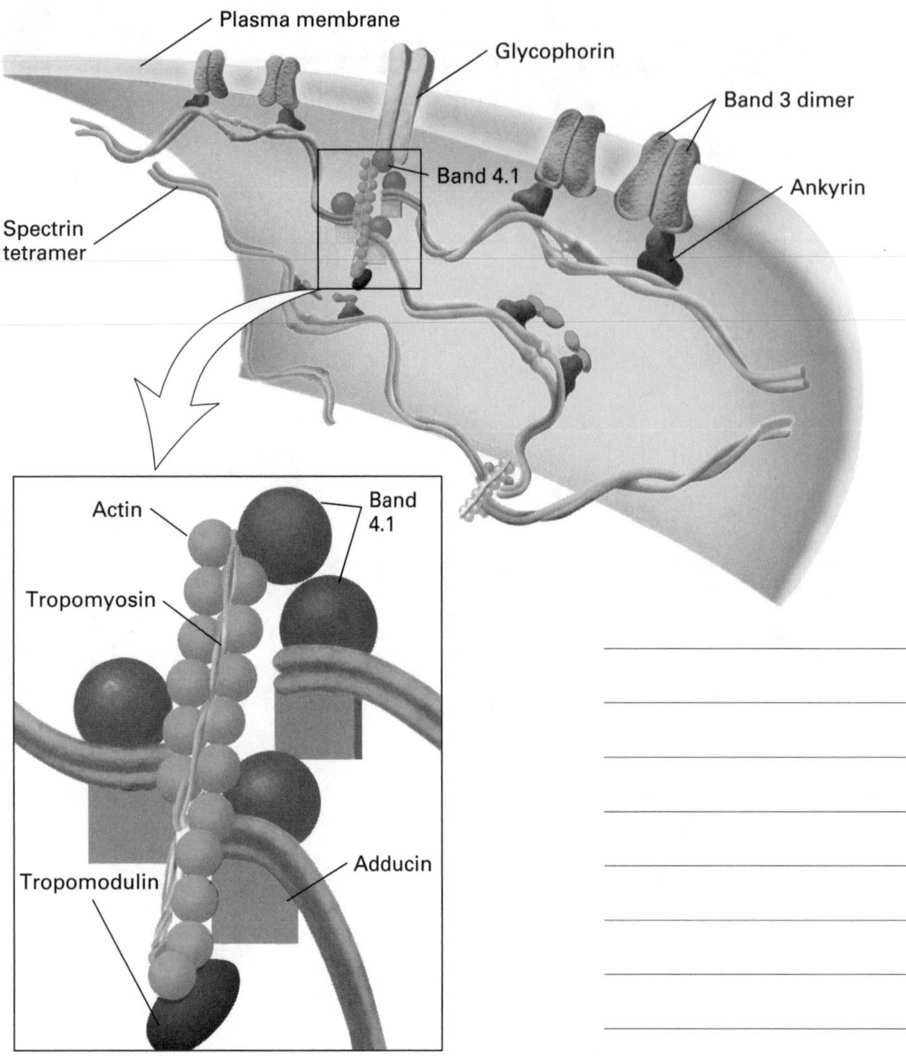

FIGURE 18-7, page 758
The organization of the major erythrocyte cytoskeletal proteins and their interactions with integral membrane proteins.

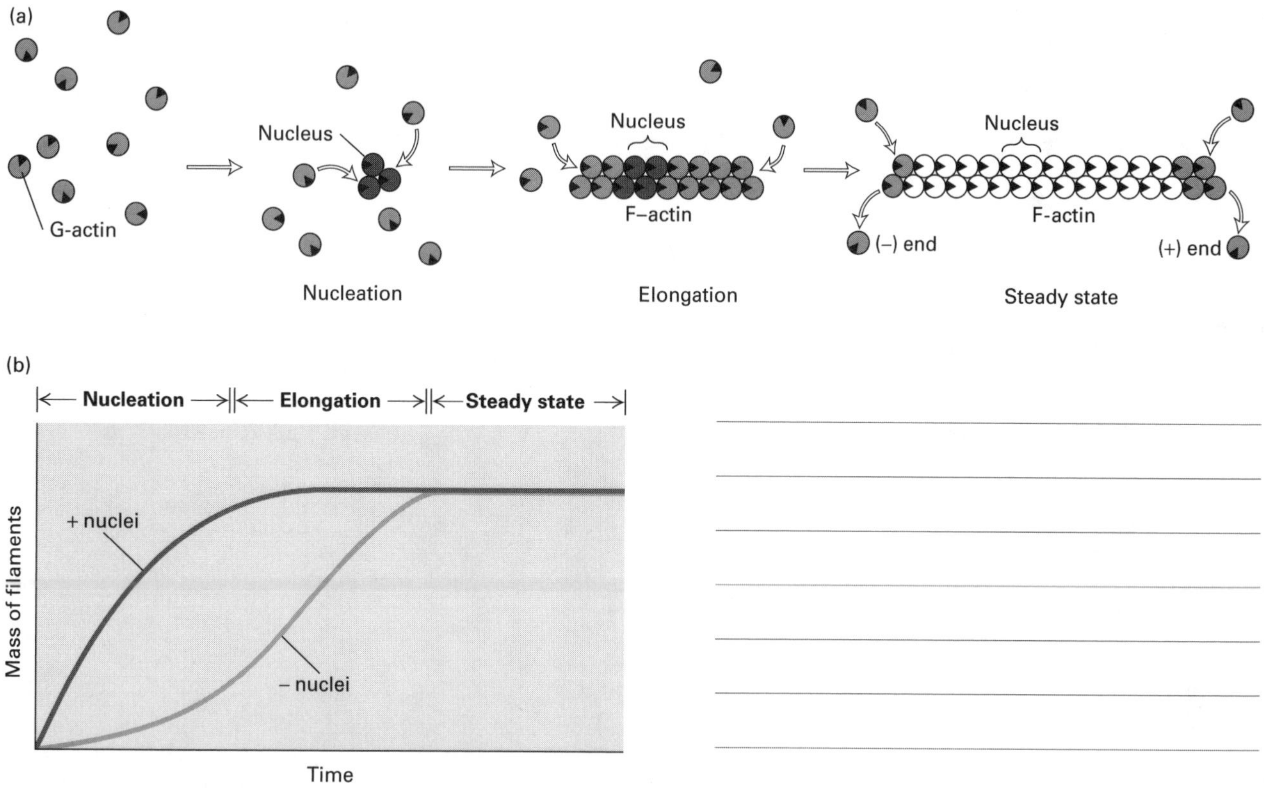

FIGURE 18-11, page 762
The three phases of G-actin polymerization in vitro.

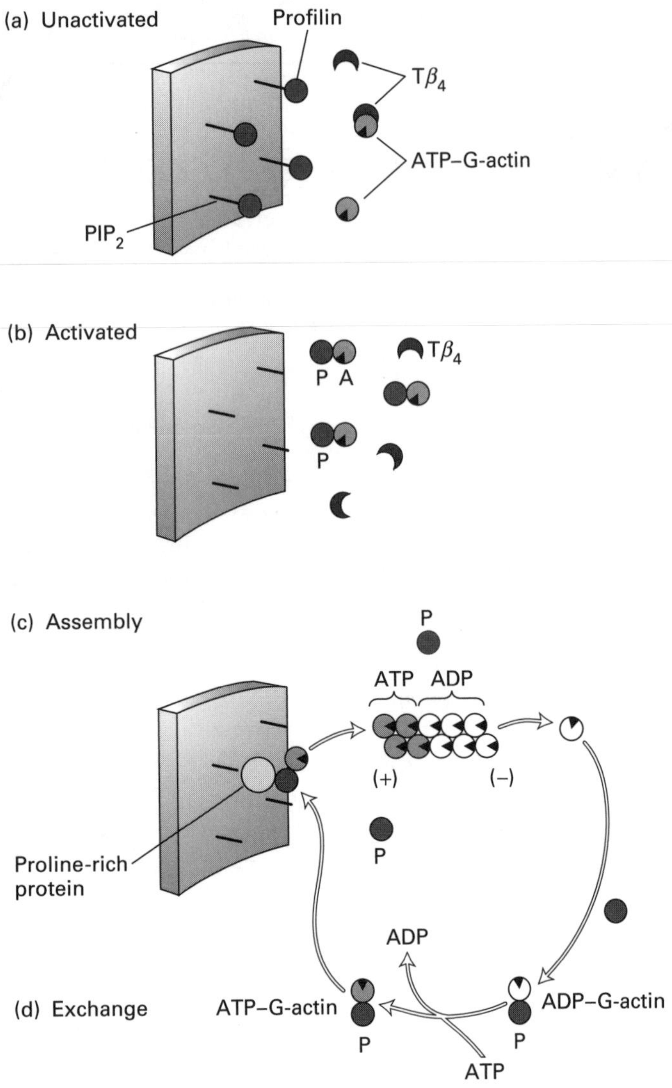

FIGURE 18-15, page 765
Model of the complementary roles of profilin and thymosin $\beta 4$ in regulating polymerization of G-actin.

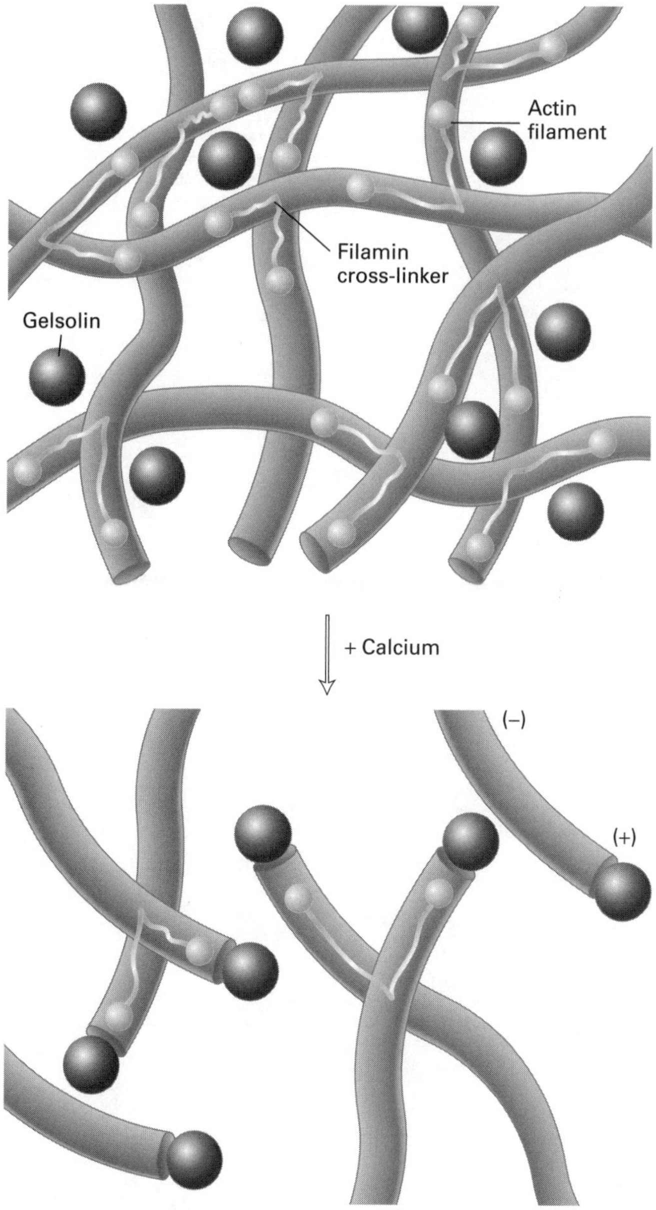

FIGURE 18-16, page 766
Action of gelsolin in severing actin filaments.

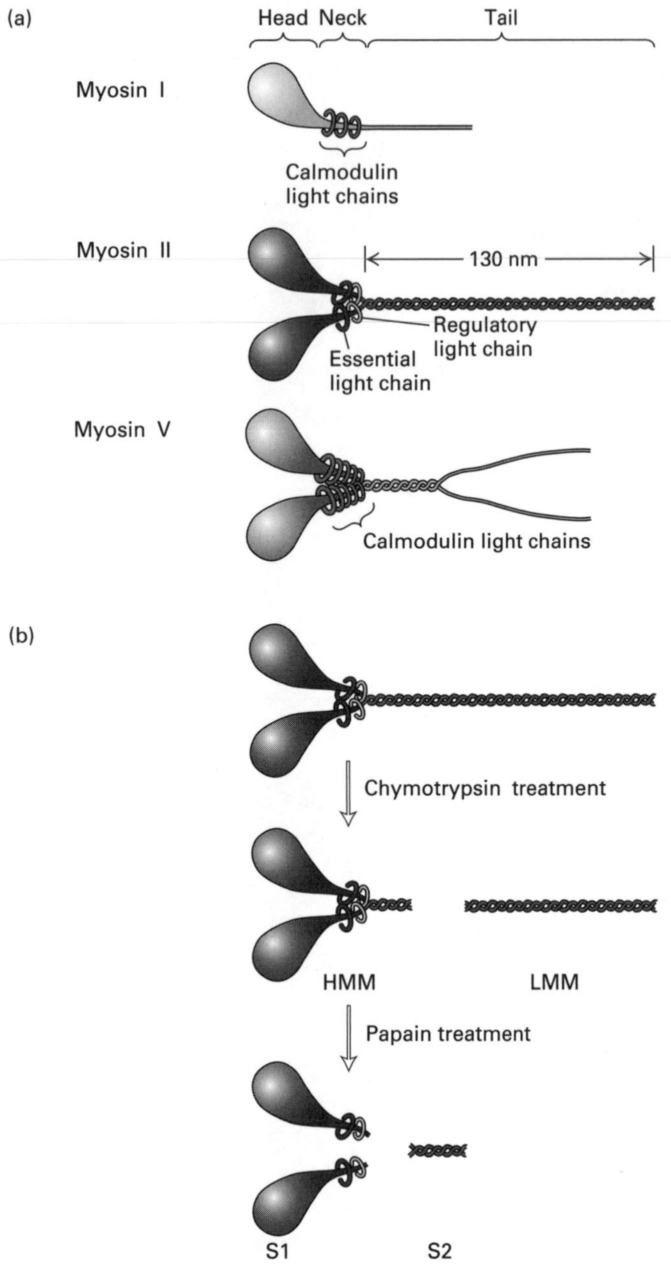

FIGURE 18-20, page 769
Structure of various myosin molecules.

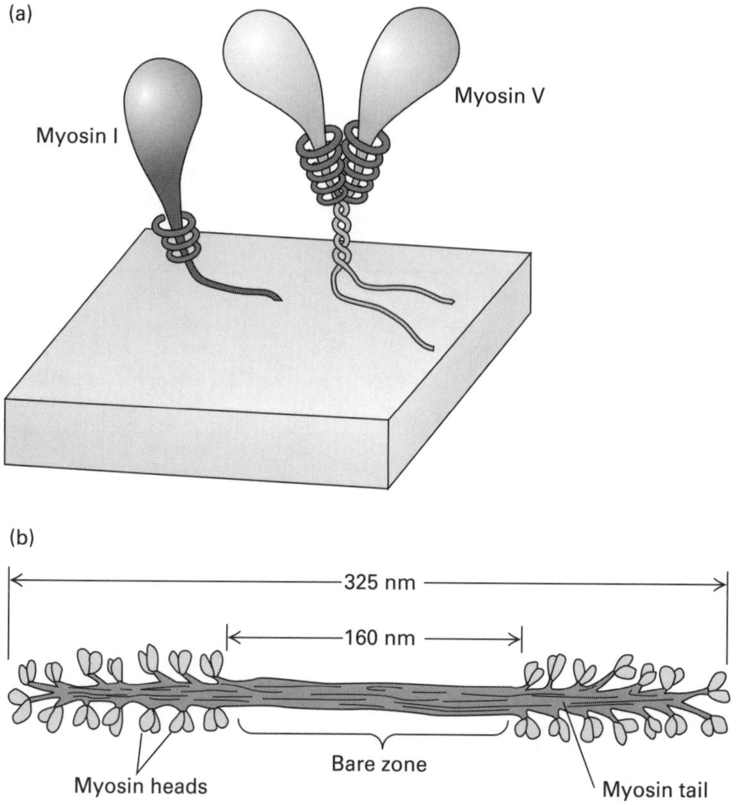

FIGURE 18-21, page 770
Functions of the myosin tail domain.

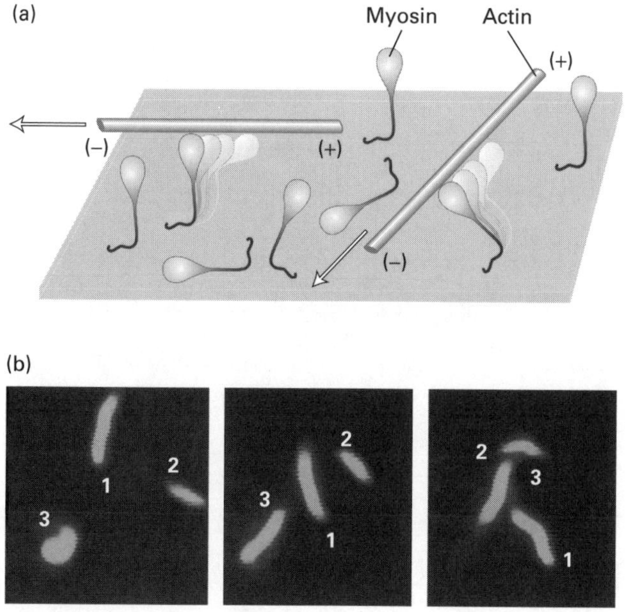

FIGURE 18-22, page 771
The sliding-filament assay.

MEDIA CONNECTIONS
In Vitro Motility Assay

MEDIA CONNECTIONS

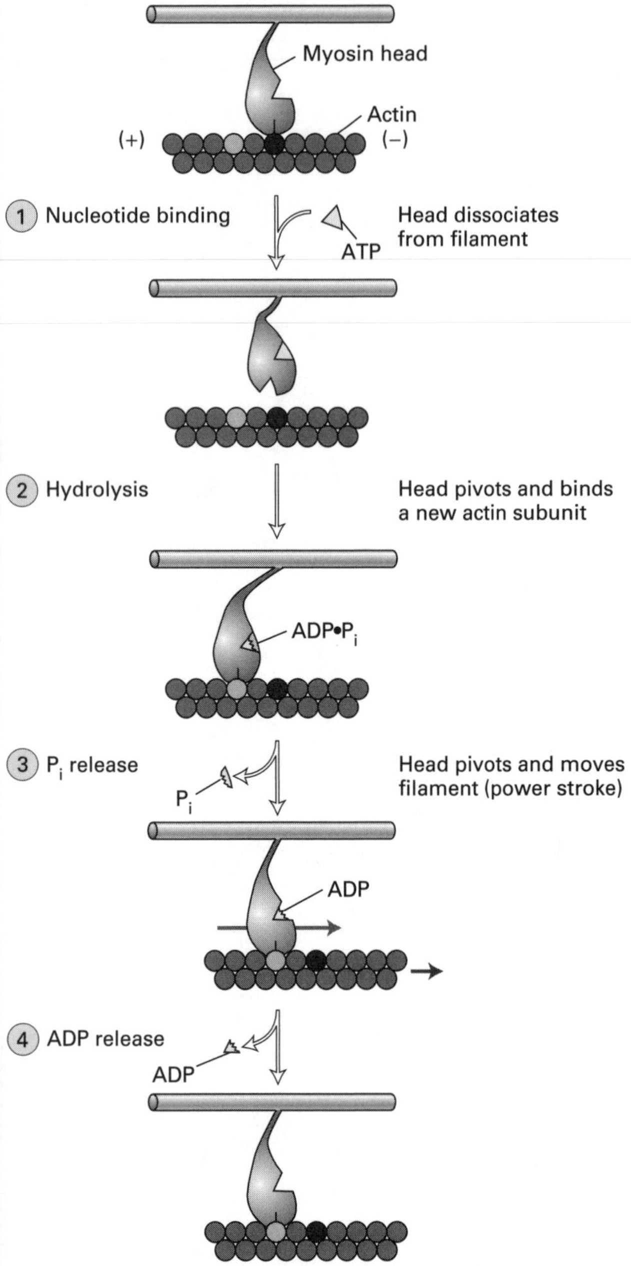

FIGURE 18-25, page 774
The coupling of ATP hydrolysis to movement of myosin along an actin filament.

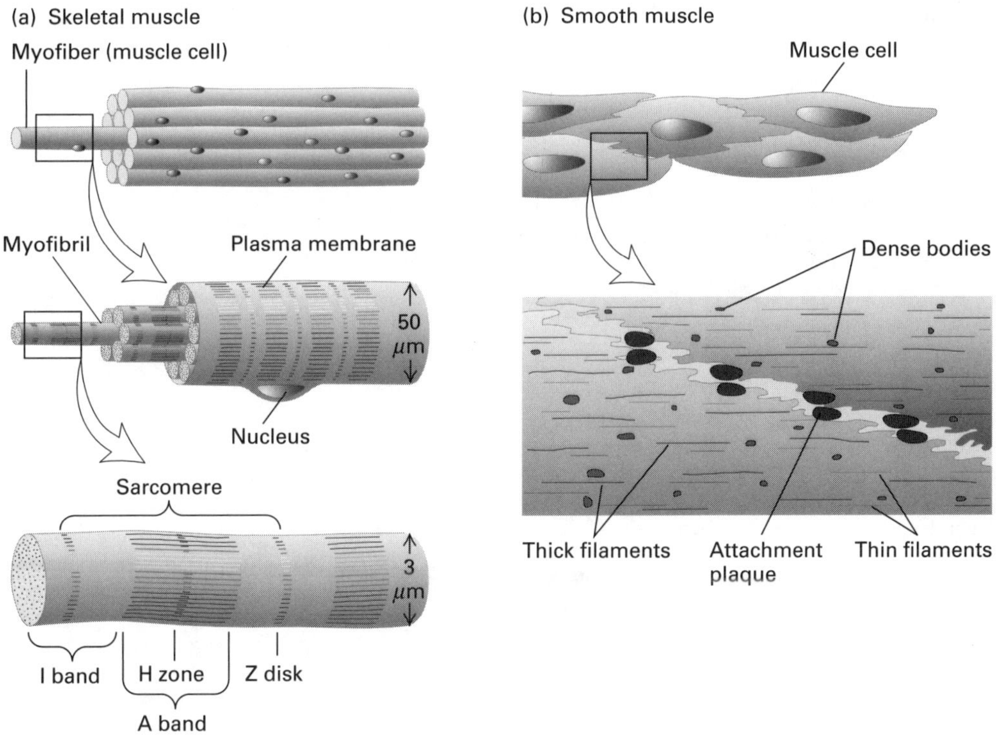

FIGURE 18-26, page 775
General structure of skeletal and smooth muscle.

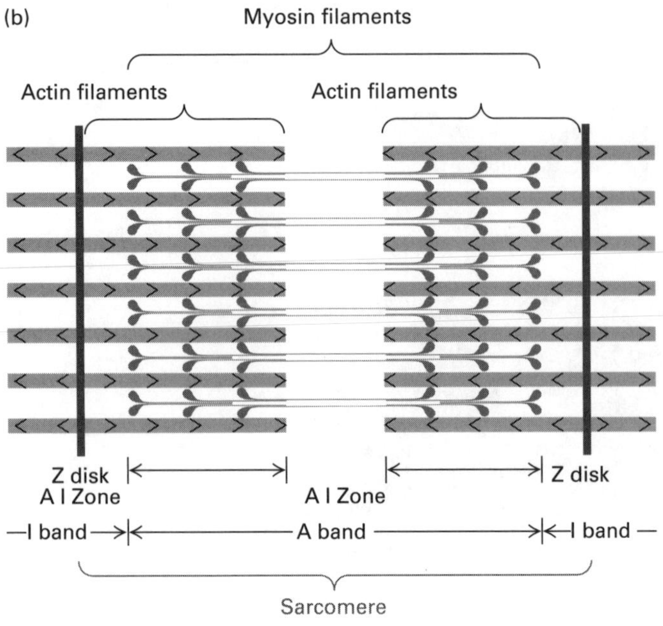

FIGURE 18-27 (b), page 776
Structure of the sarcomere.

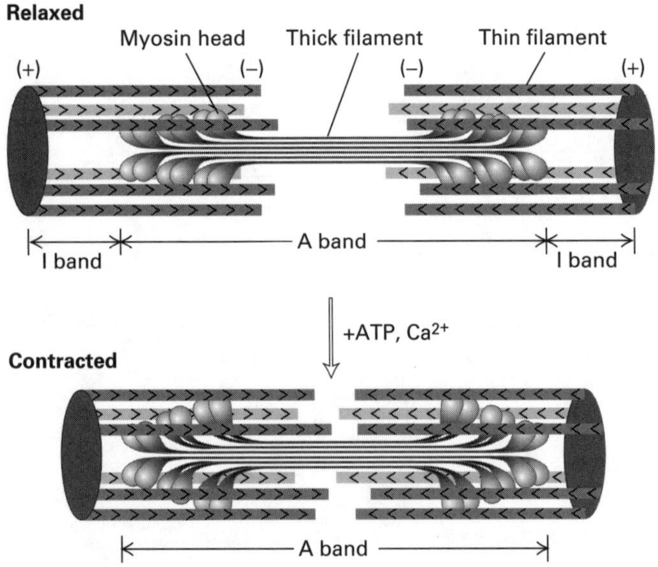

FIGURE 18-29, page 777
The sliding-filament model of contraction in striated muscle.

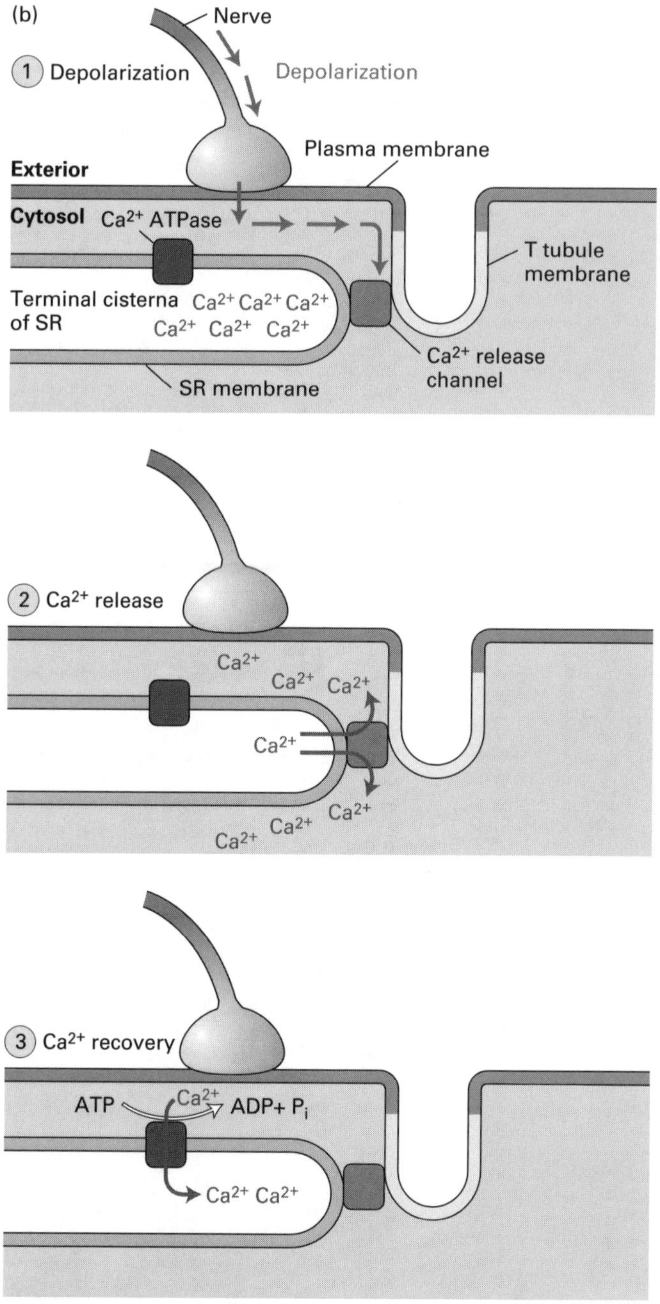

FIGURE 18-31 (b), page 779
The sarcoplasmic reticulum (SR) regulates the cytosolic Ca^{2+} level in skeletal muscle.

MEDIA CONNECTIONS
Cell Motility

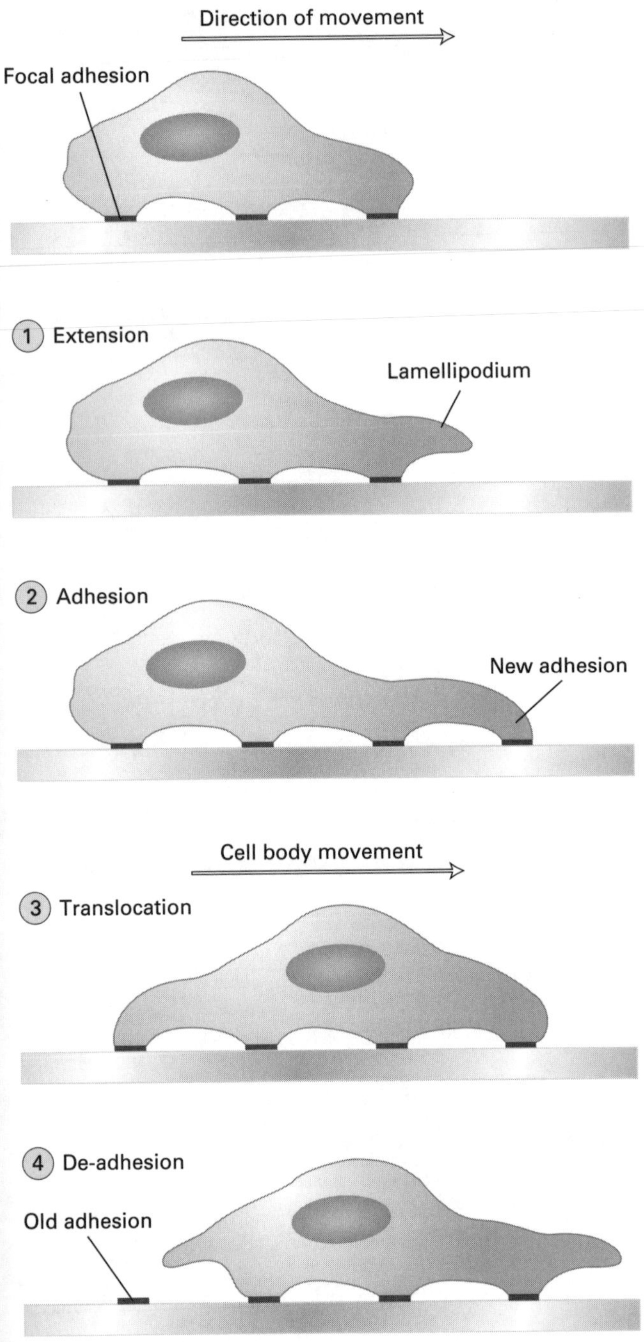

FIGURE 18-41, page 788
Steps in keratinocyte movement.

Additional Notes

Cell Motility and Shape II: Microtubules and Intermediate Filaments

19

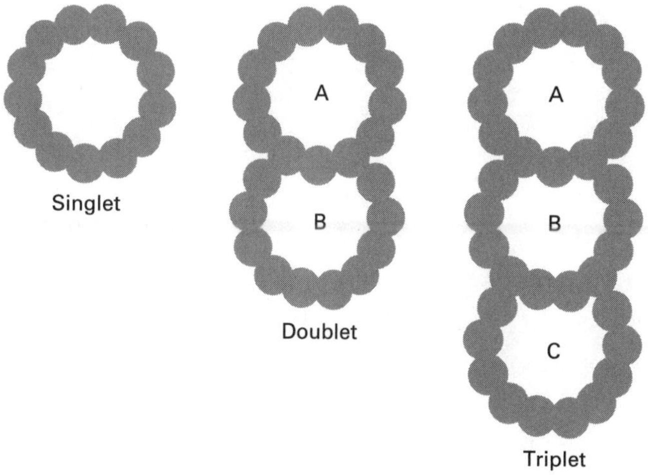

FIGURE 19-3, page 797
Arrangement of protofilaments in singlet, doublet, and triplet microtubules.

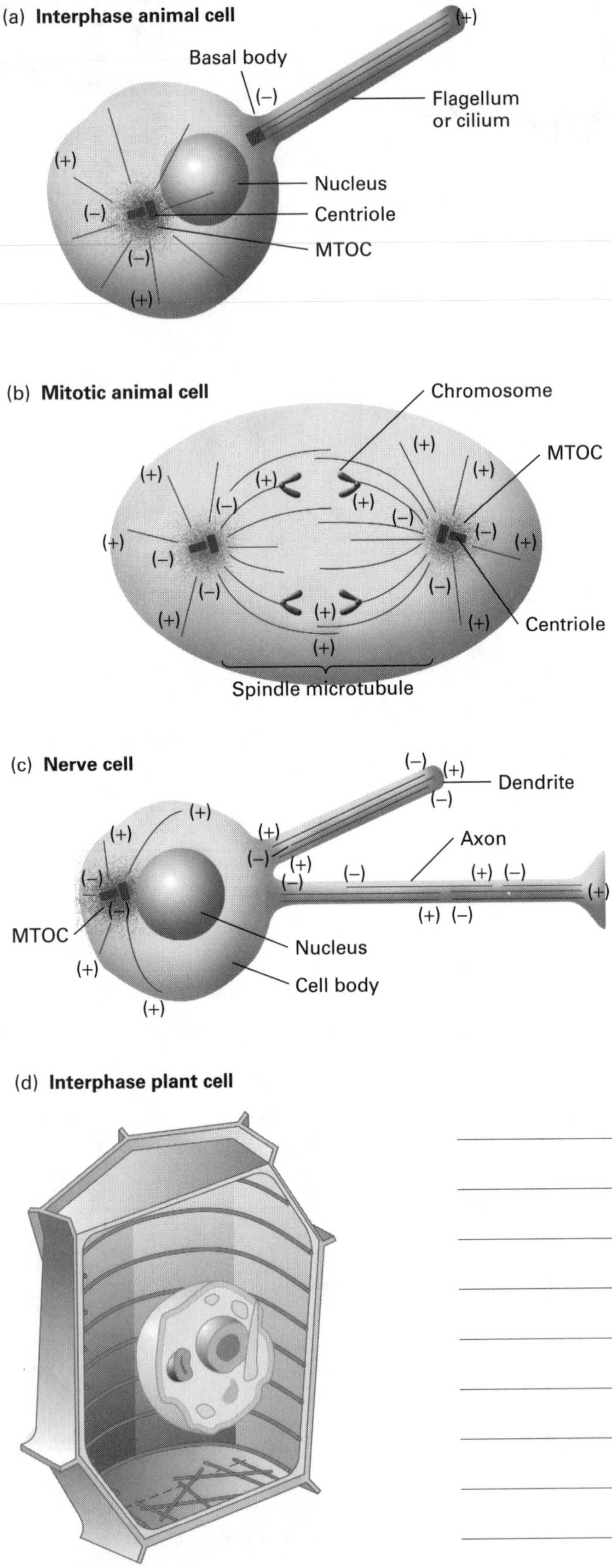

FIGURE 19-7, page 801
Orientation of cellular microtubules.

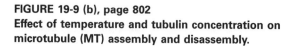

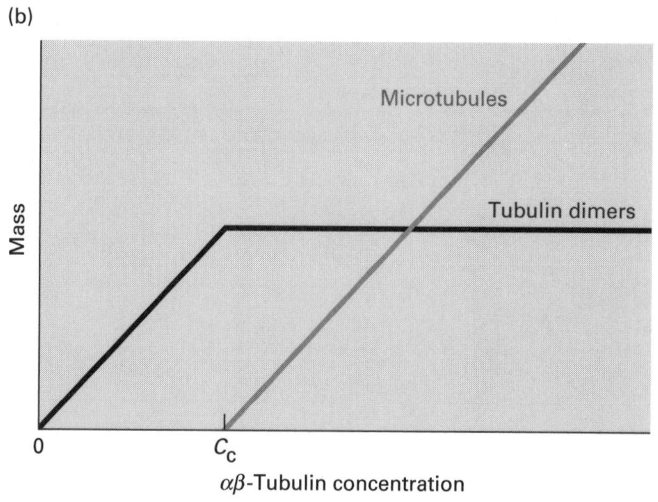

FIGURE 19-9 (b), page 802
Effect of temperature and tubulin concentration on microtubule (MT) assembly and disassembly.

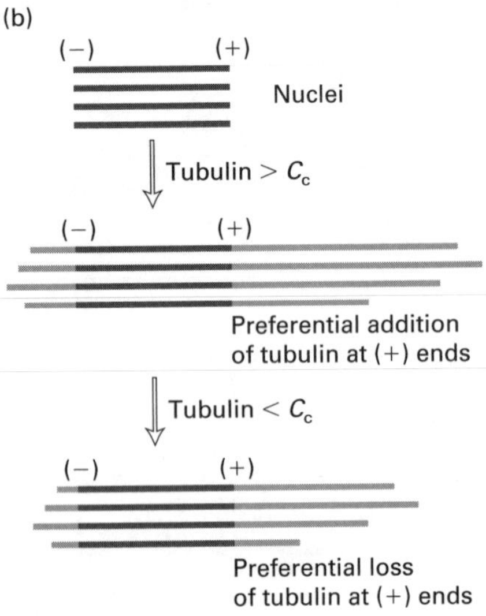

FIGURE 19-10 (b), page 803
Polarity of tubulin polymerization.

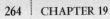

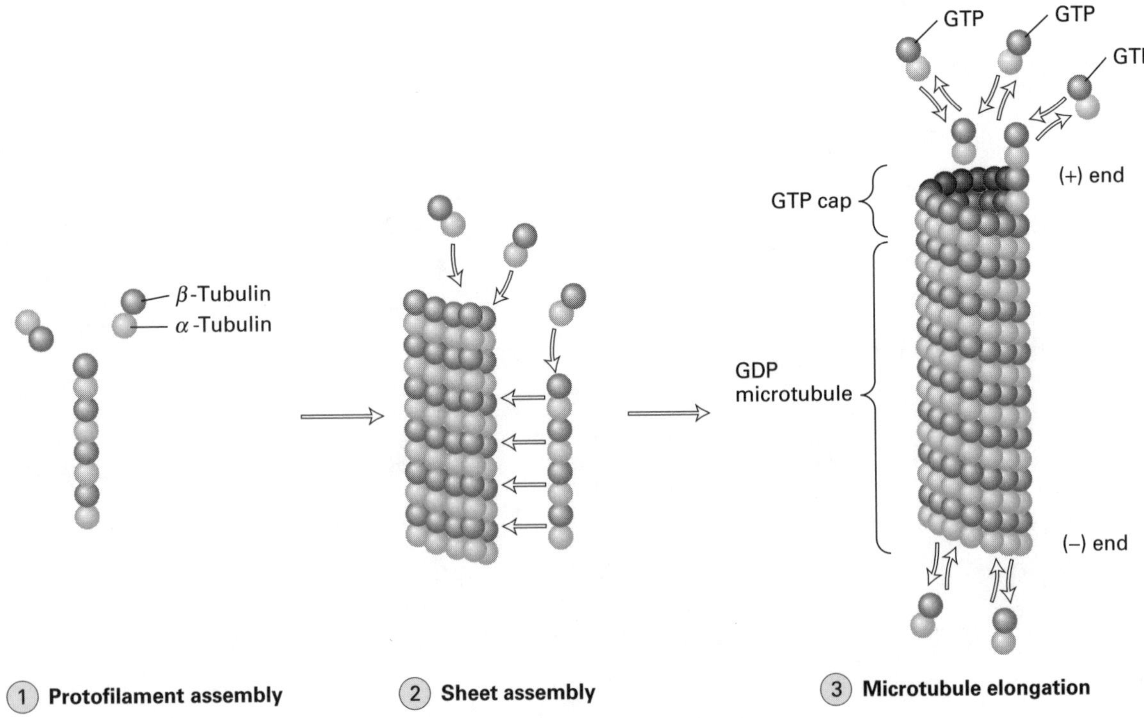

FIGURE 19-11, page 804
Assembly of microtubules.

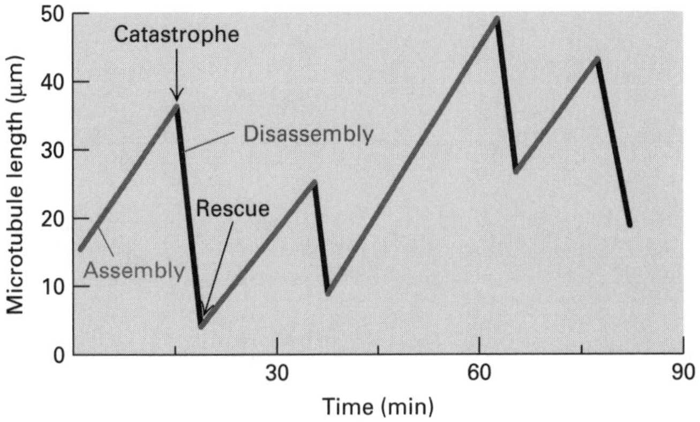

FIGURE 19-13, page 805
Dynamic instability of microtubules in vitro.

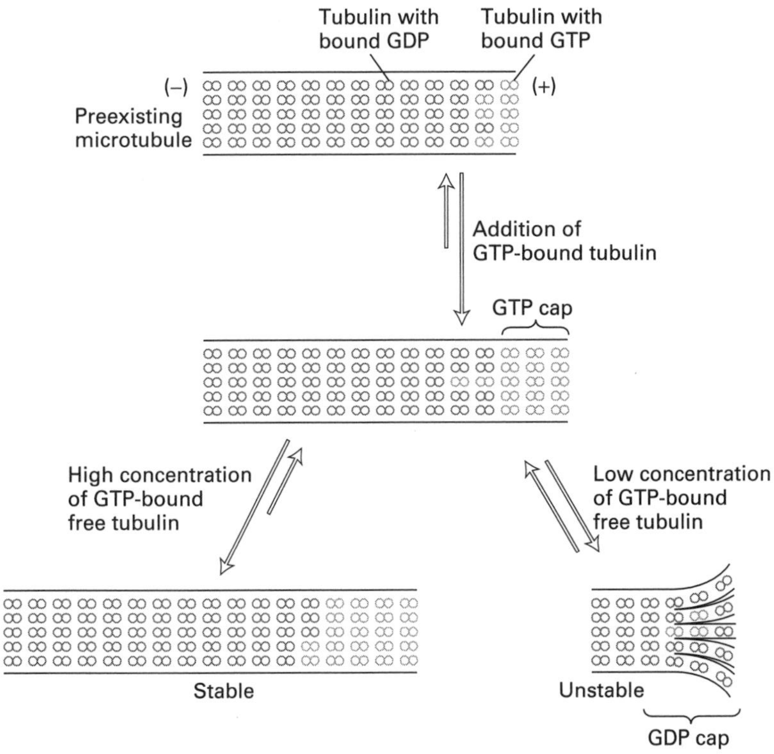

FIGURE 19-15, page 806
Dynamic instability model of microtubule growth and shrinkage.

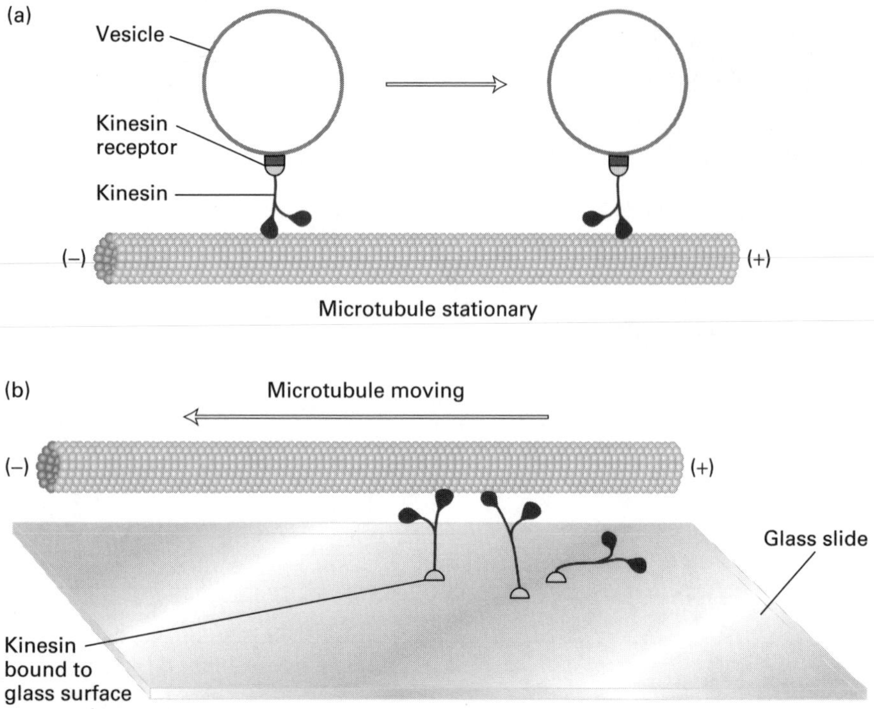

FIGURE 19-24, page 814
Model of kinesin-catalyzed anterograde transport.

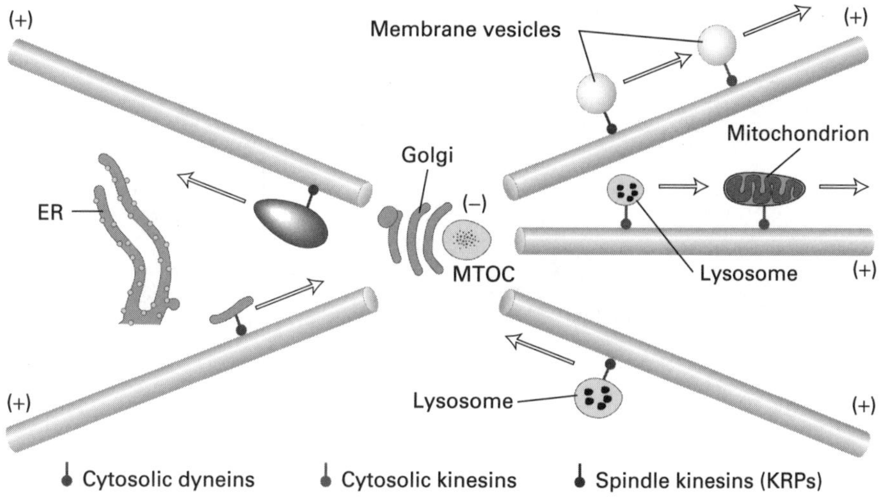

FIGURE 19-26, page 816
A general model for kinesin- and dynein-mediated transport in a typical cell.

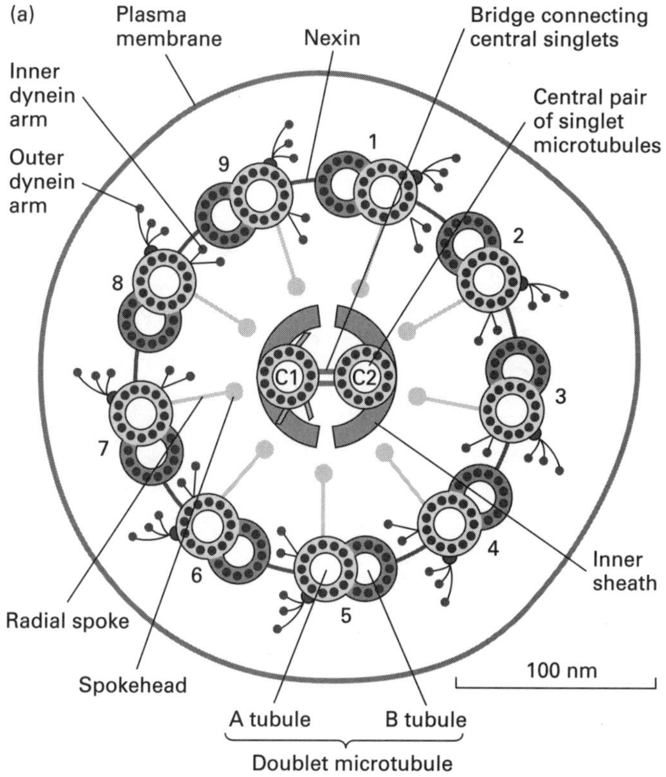

FIGURE 19-28 (a), page 819
Structure of ciliary and flagellar axonemes.

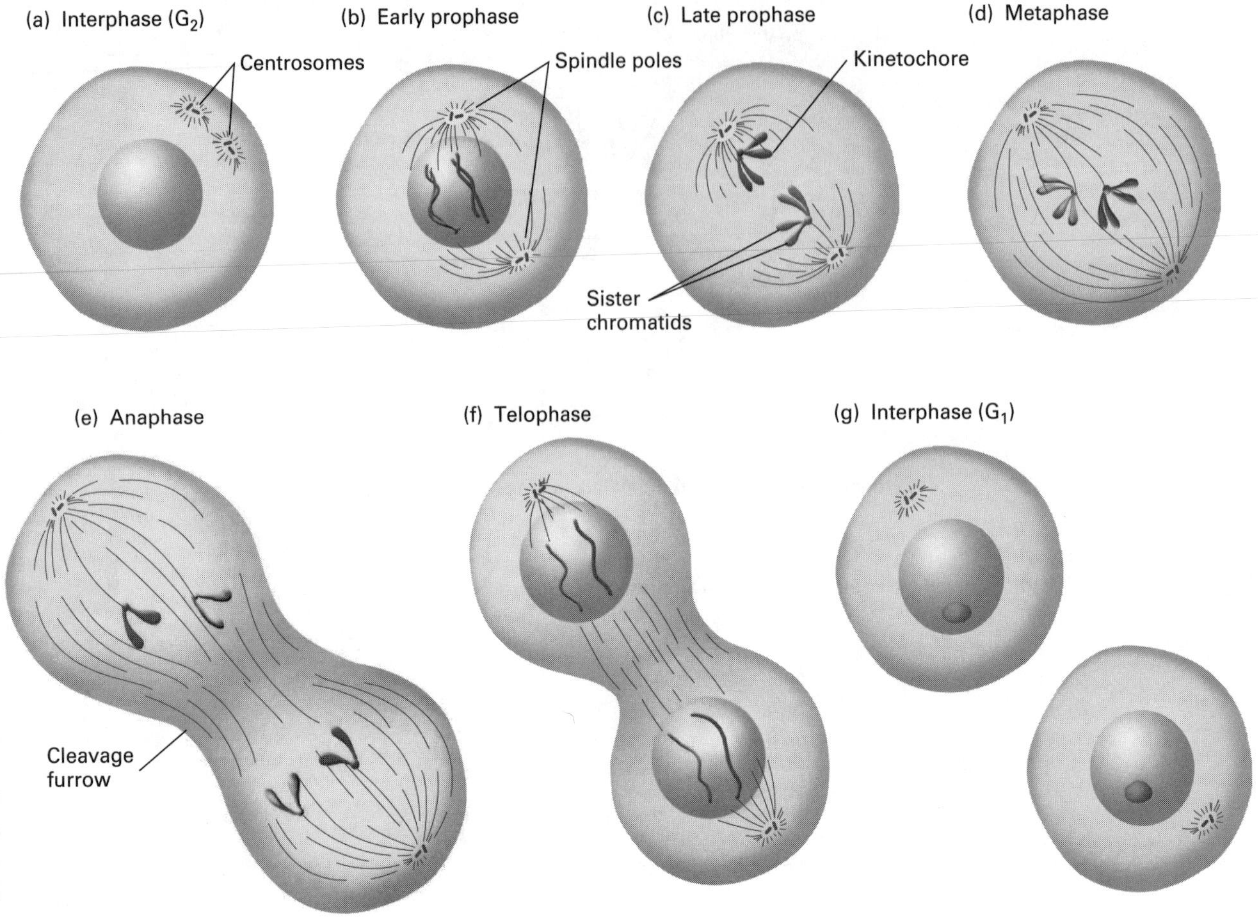

FIGURE 19-34, page 824
The stages of mitosis and cytokinesis in an animal cell.

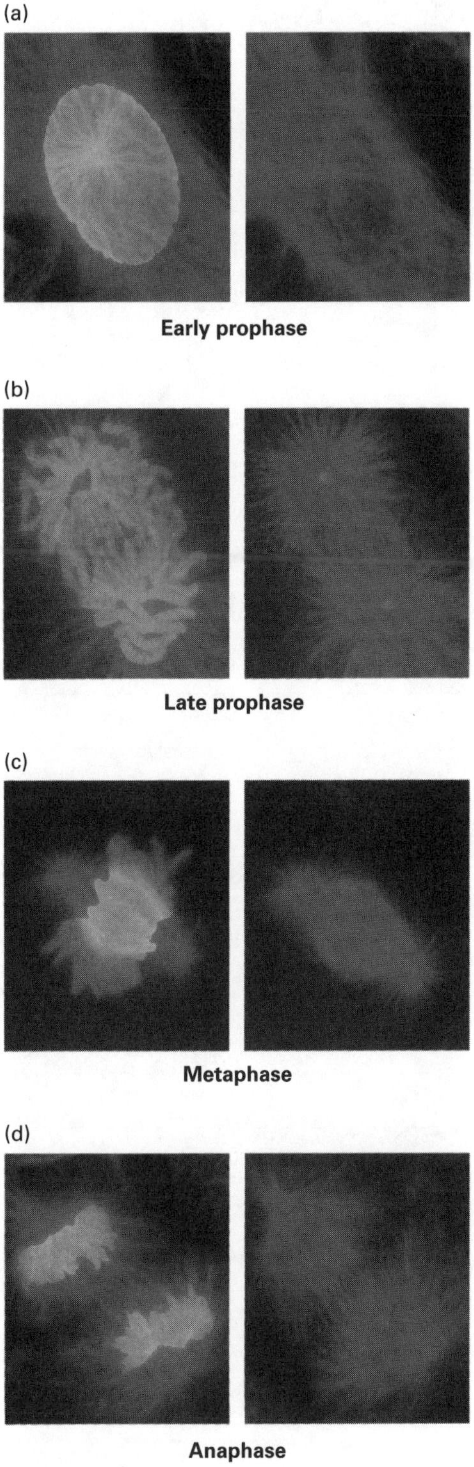

FIGURE 19-35, page 825
Fluorescence micrographs showing the organization of chromosomes and microtubules during four mitotic stages.

MEDIA CONNECTIONS
Microtubule Dynamics

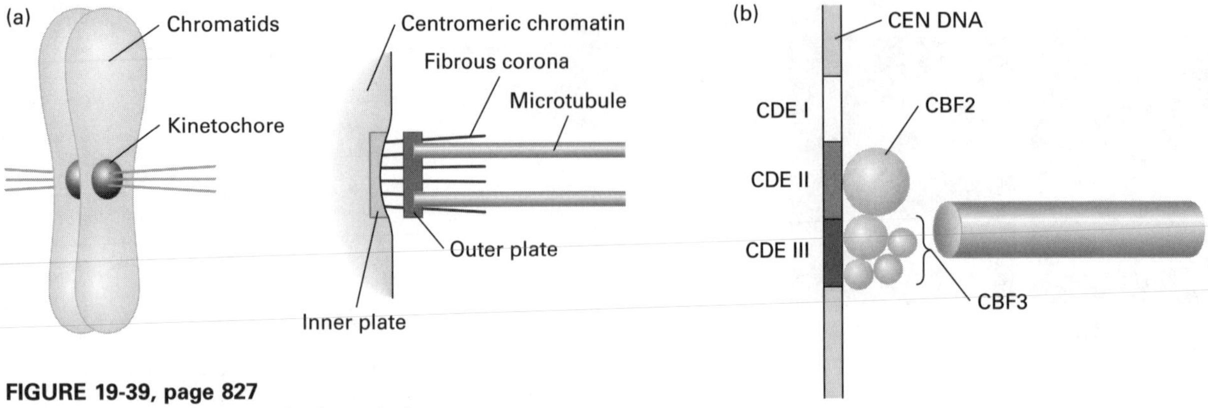

FIGURE 19-39, page 827
Centromeric attachment of microtubules.

MEDIA CONNECTIONS
Microtubule Dynamics

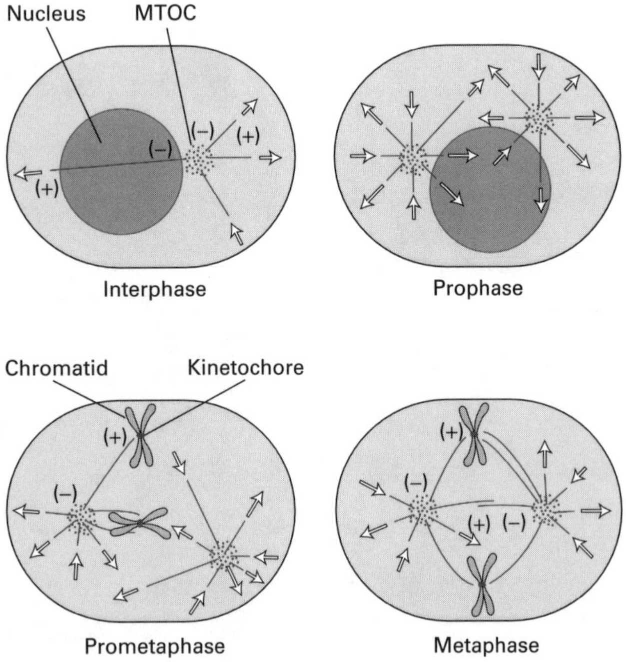

FIGURE 19-41, page 828
Microtube dynamics during mitosis.

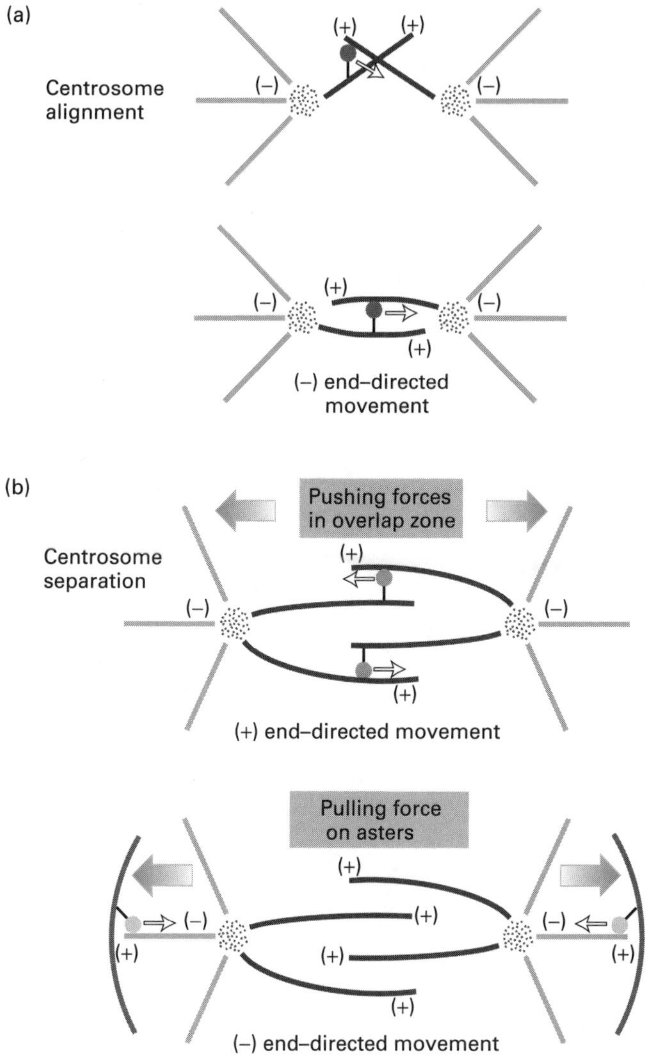

FIGURE 19-42, page 829
Model for participation of microtubule motor proteins in centrosome movements during mitosis.

MEDIA CONNECTIONS
Microtubule Dynamics

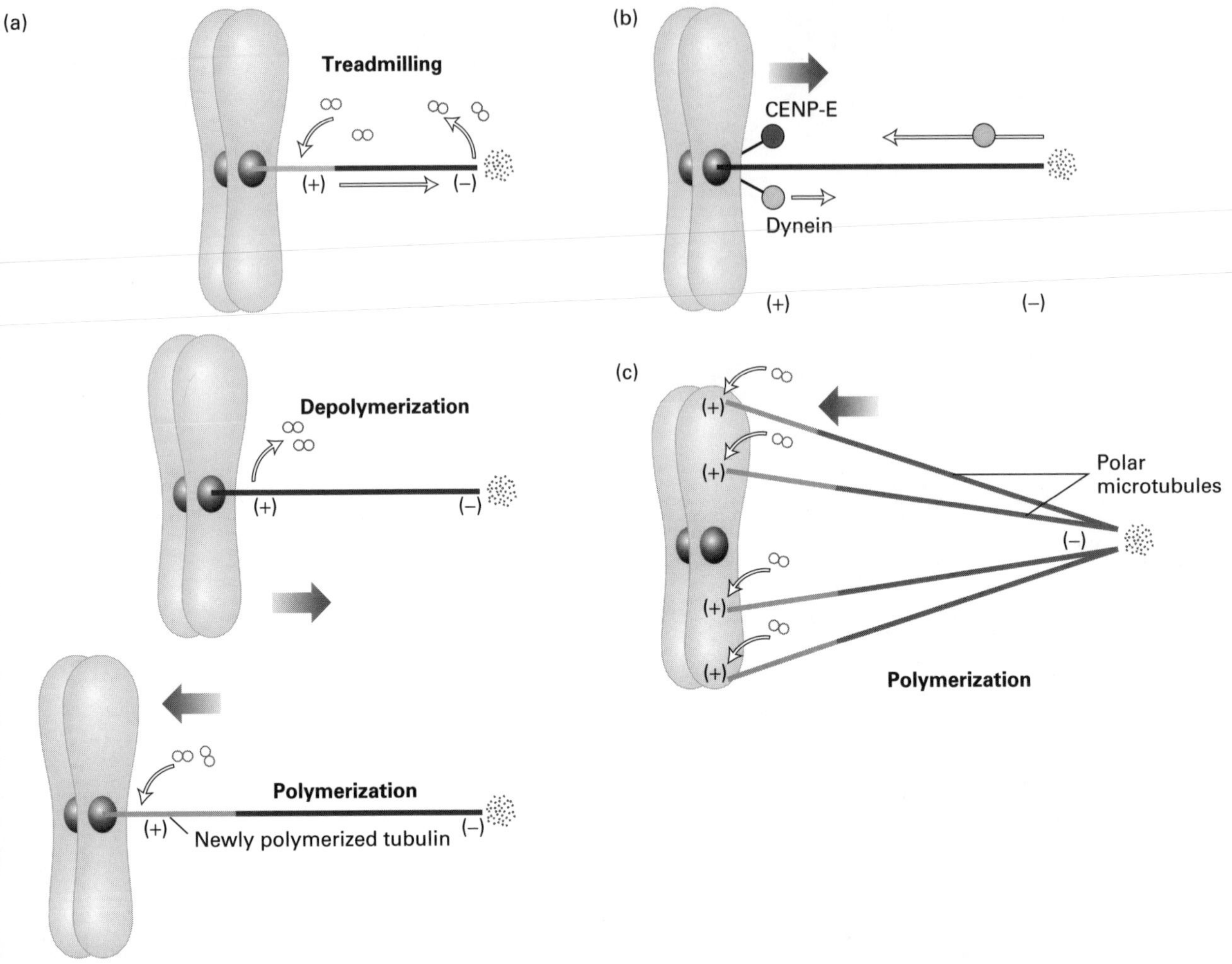

FIGURE 19-45, page 831
Proposed alternative mechanisms for chromosome congression.

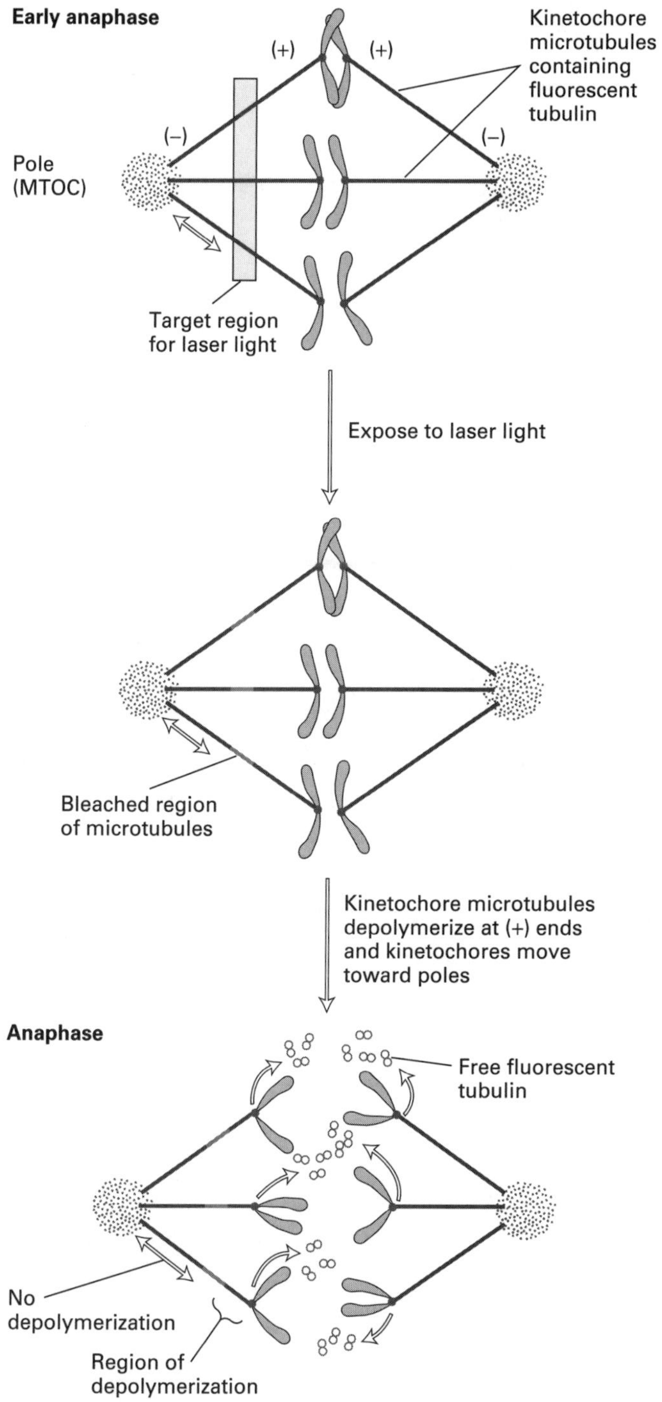

FIGURE 19-46, page 832
Experimental demonstration that during anaphase A chromosomes move poleward along stationary kinetochore microtubules, which coordinately disassemble from their kinetochore ends.

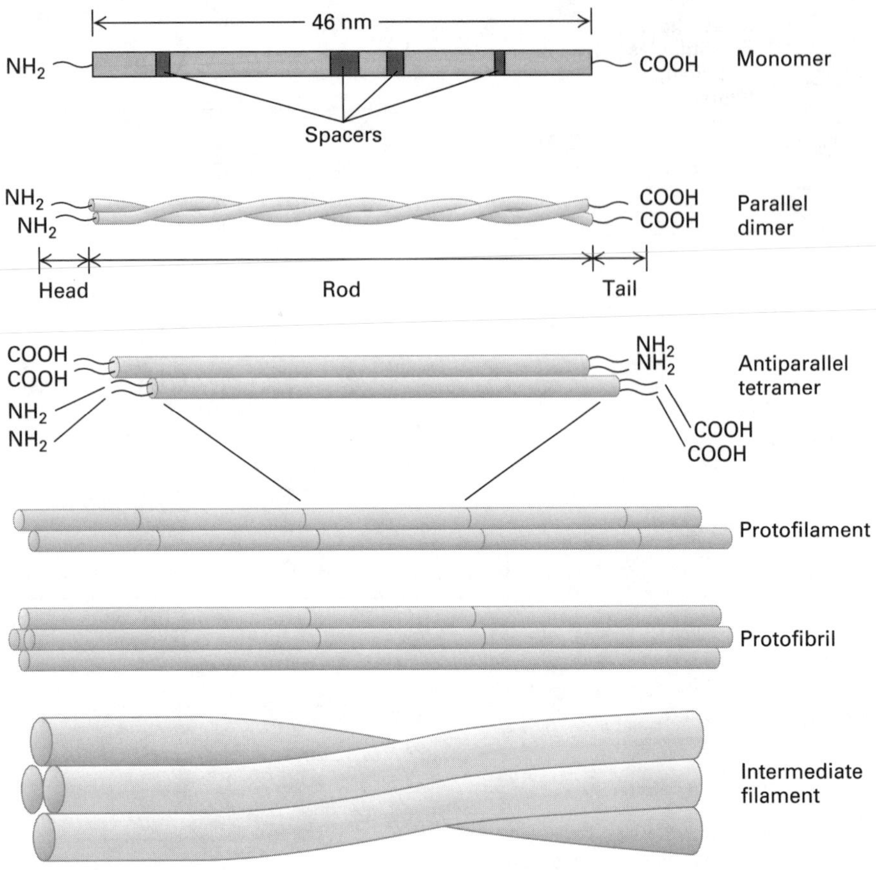

FIGURE 19-51, page 839
Levels of organization and assembly of intermediate filaments.

Additional Notes

Cell-to-Cell Signaling: Hormones and Receptors

20

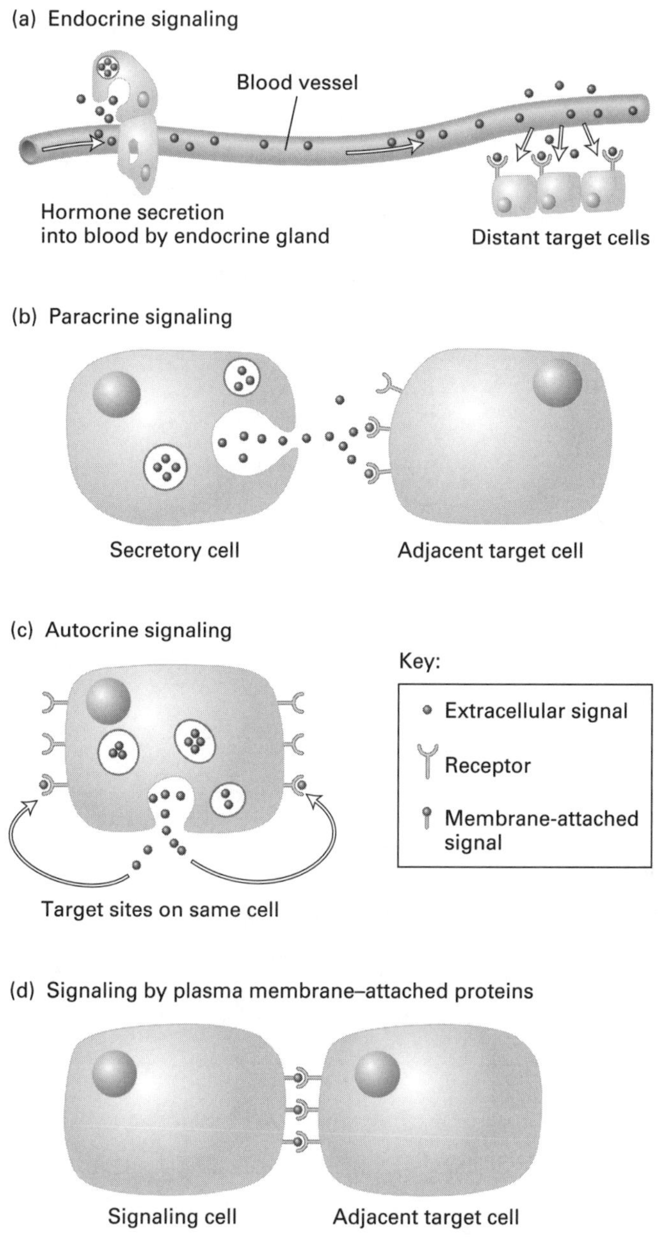

FIGURE 20-1, page 849
General schemes of intercellular signaling in animals.

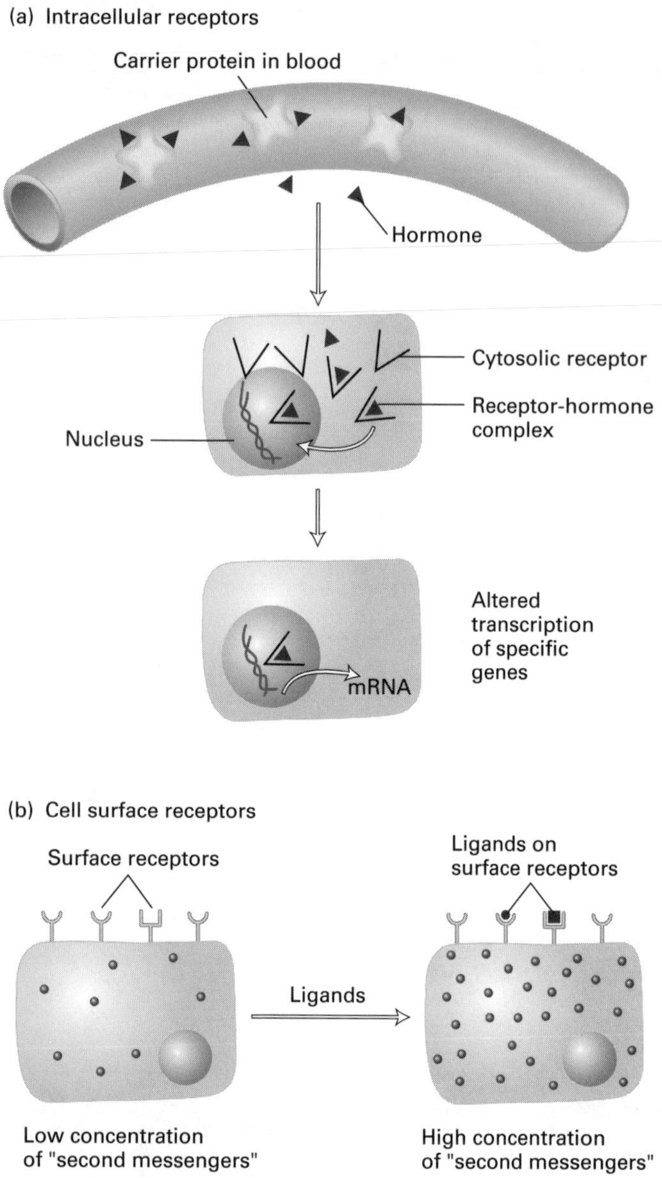

FIGURE 20-2, page 851
Some hormones bind to intracellular receptors; others, to cell-surface receptors.

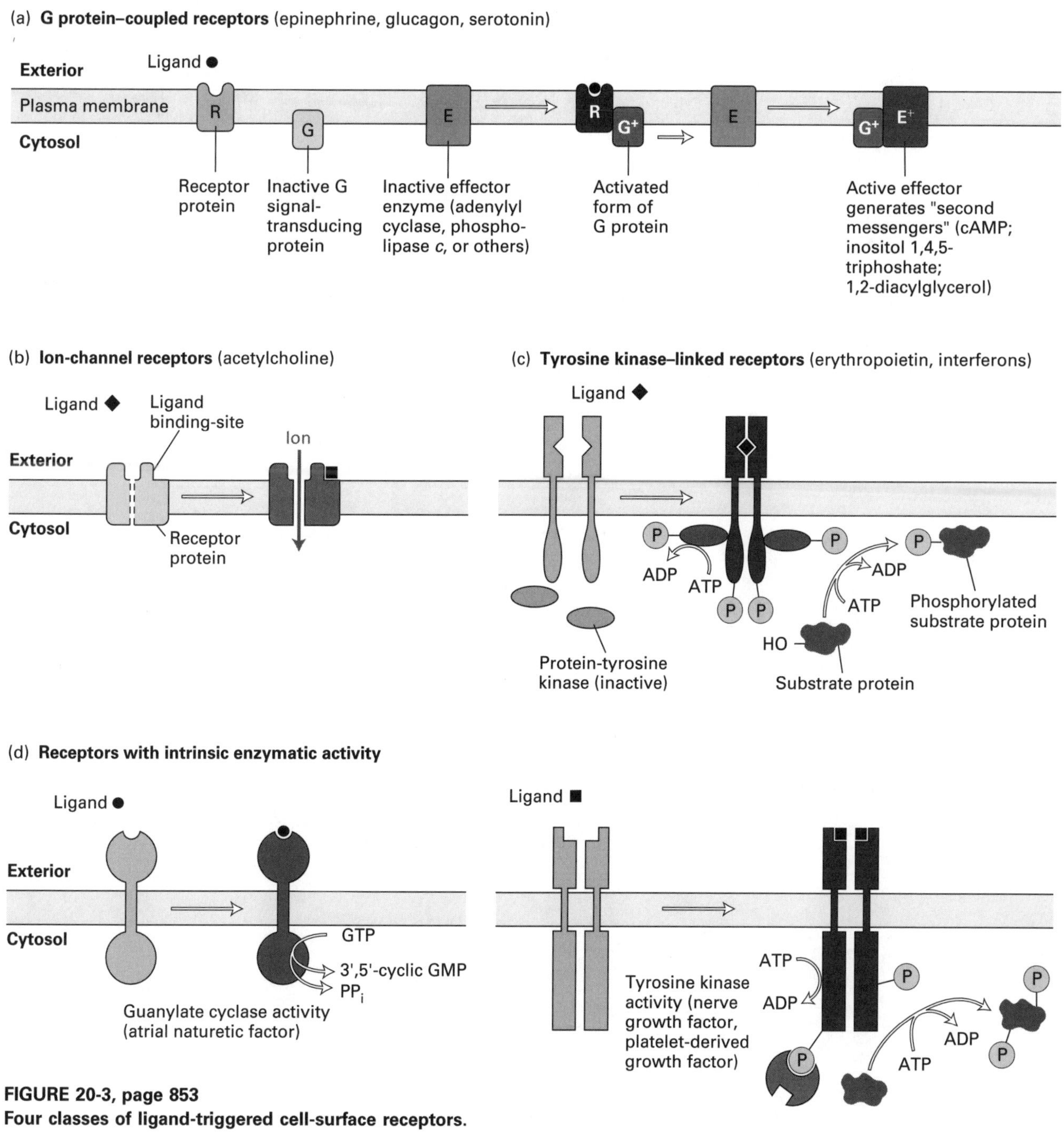

(a) G protein–coupled receptors (epinephrine, glucagon, serotonin)

Exterior Ligand ●

Plasma membrane

Cytosol

Receptor protein

Inactive G signal-transducing protein

Inactive effector enzyme (adenylyl cyclase, phospholipase *c*, or others)

Activated form of G protein

Active effector generates "second messengers" (cAMP; inositol 1,4,5-triphoshate; 1,2-diacylglycerol)

(b) Ion-channel receptors (acetylcholine)

Ligand ◆ Ligand binding-site

Exterior Ion

Cytosol Receptor protein

(c) Tyrosine kinase–linked receptors (erythropoietin, interferons)

Ligand ◆

ADP ATP ADP Phosphorylated substrate protein

ATP

HO

Protein-tyrosine kinase (inactive)

Substrate protein

(d) Receptors with intrinsic enzymatic activity

Ligand ●

Exterior

Cytosol

GTP

3',5'-cyclic GMP

PP_i

Guanylate cyclase activity (atrial naturetic factor)

Ligand ■

Exterior

Cytosol

ATP

ADP

Tyrosine kinase activity (nerve growth factor, platelet-derived growth factor)

ATP

ADP

FIGURE 20-3, page 853
Four classes of ligand-triggered cell-surface receptors.

MEDIA CONNECTIONS
Second Messengers in Signaling Pathways

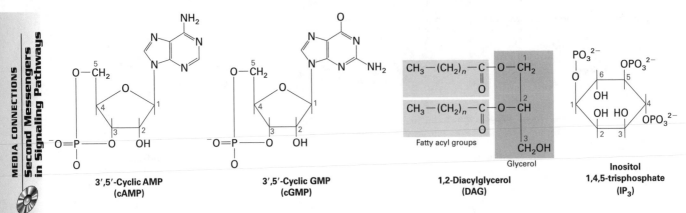

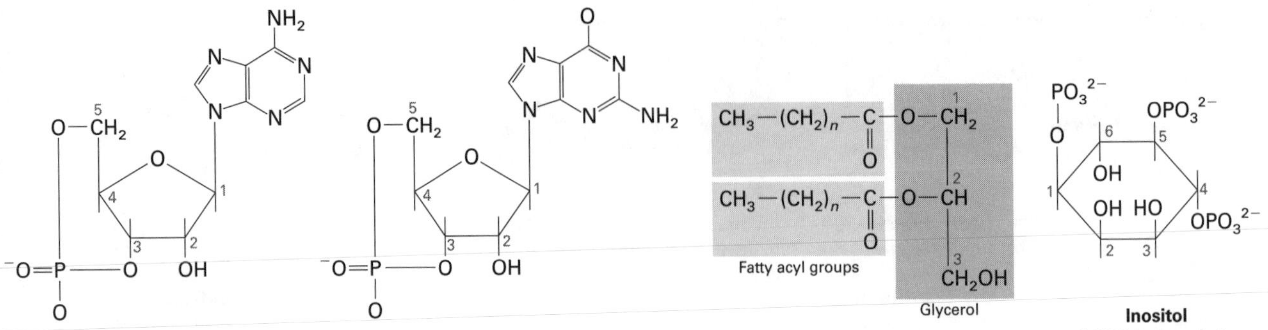

3′,5′-Cyclic AMP (cAMP)

3′,5′-Cyclic GMP (cGMP)

1,2-Diacylglycerol (DAG)

Inositol 1,4,5-trisphosphate (IP_3)

FIGURE 20-4, page 854
Structural formulas of four common intracellular second messengers.

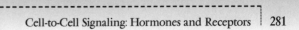

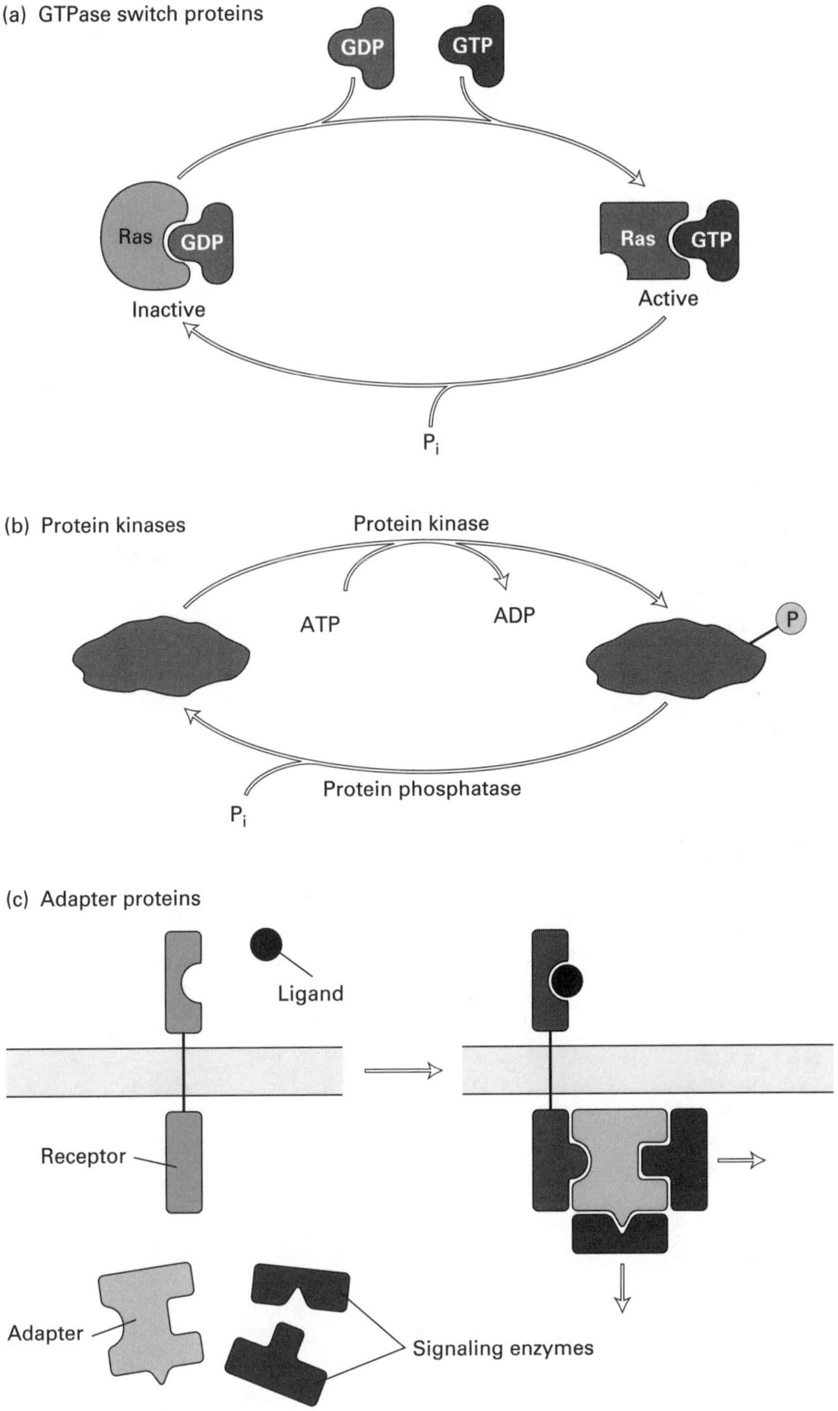

FIGURE 20-5, page 855
Common intracellular signaling proteins.

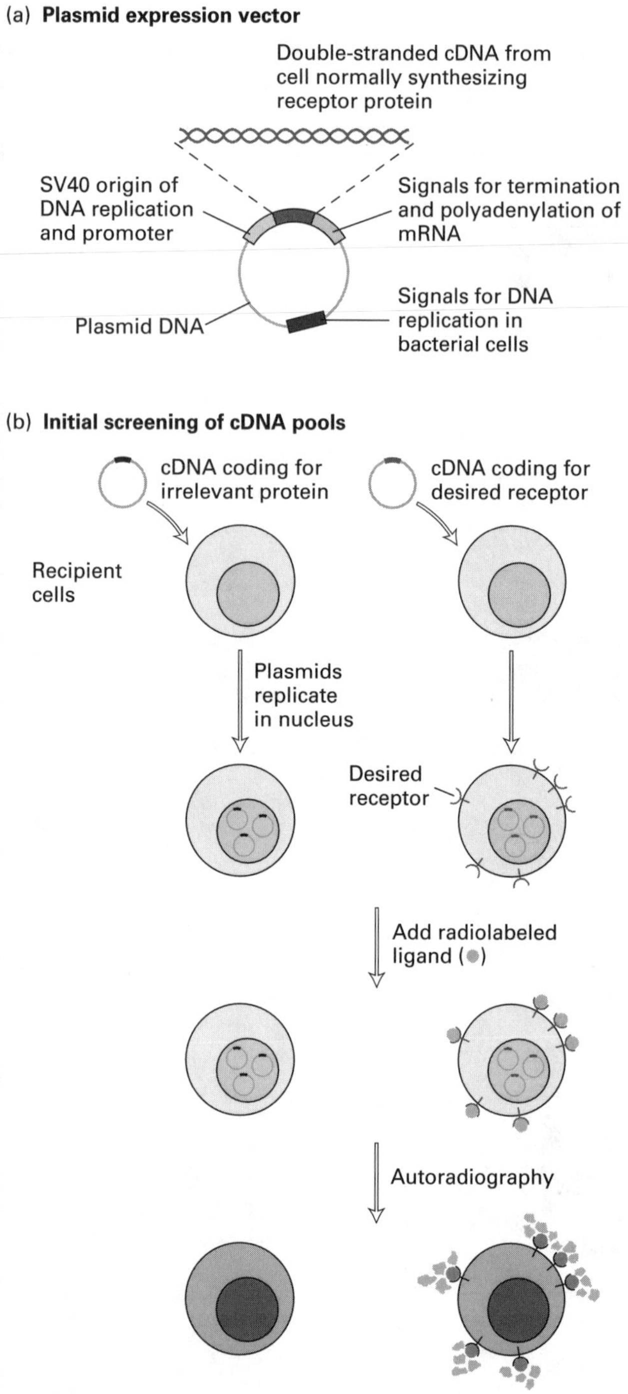

FIGURE 20-9, page 861
Identification and isolation of a cDNA encoding a desired cell-surface receptor by plasmid expression cloning.

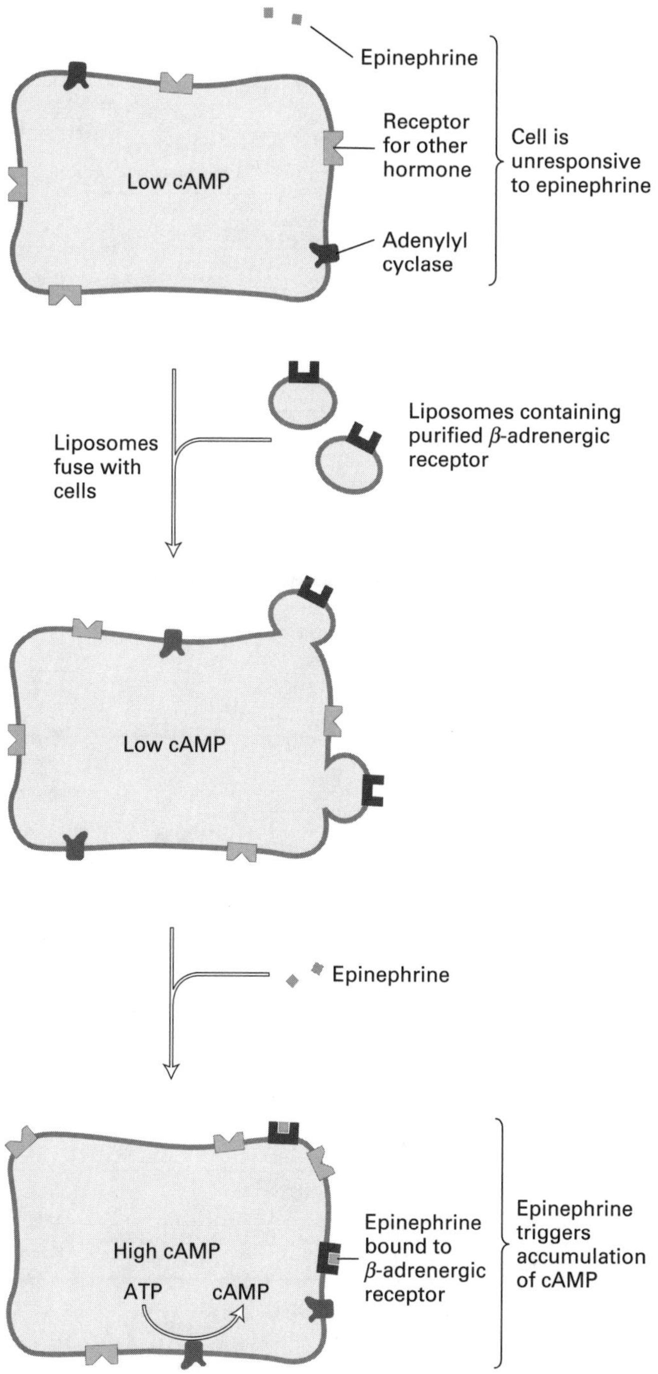

FIGURE 20-12, page 864
Experimental demonstration that β-adrenergic receptors mediate the induction of epinephrine-initiated cAMP synthesis.

MEDIA CONNECTIONS Extracellular Signaling

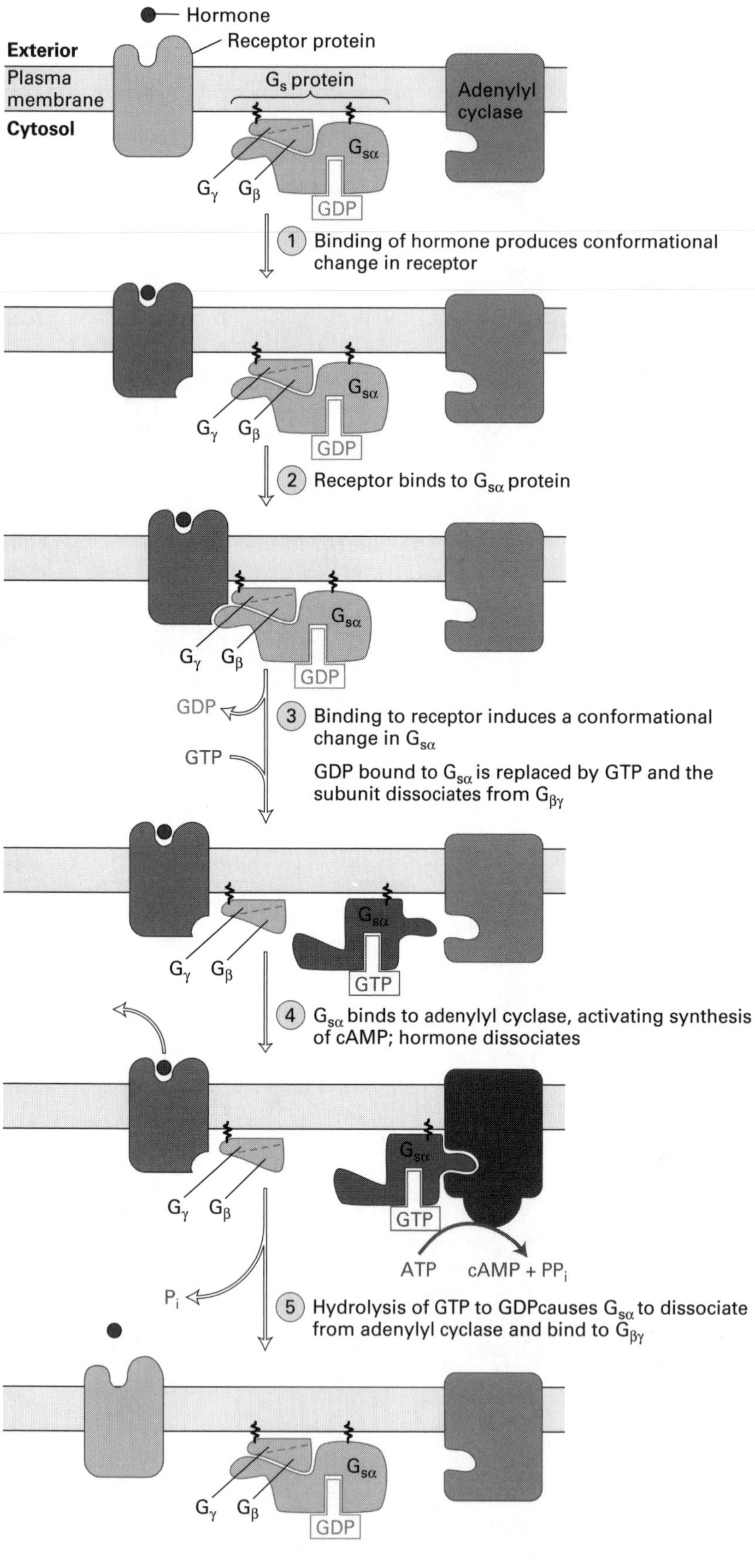
Hormone
Receptor protein
Exterior
Plasma membrane
Cytosol
G_s protein
Adenylyl cyclase
$G_{s\alpha}$
G_γ G_β
GDP
1 Binding of hormone produces conformational change in receptor
2 Receptor binds to $G_{s\alpha}$ protein
GDP
GTP
3 Binding to receptor induces a conformational change in $G_{s\alpha}$
GDP bound to $G_{s\alpha}$ is replaced by GTP and the subunit dissociates from $G_{\beta\gamma}$
GTP
4 $G_{s\alpha}$ binds to adenylyl cyclase, activating synthesis of cAMP; hormone dissociates
ATP cAMP + PP_i
P_i
5 Hydrolysis of GTP to GDPcauses $G_{s\alpha}$ to dissociate from adenylyl cyclase and bind to $G_{\beta\gamma}$

FIGURE 20-16, page 867
Activation of adenylyl cyclase following binding of an appropriate hormone (e.g., epinephrine, glucagon) to a G_s protein–coupled receptor.

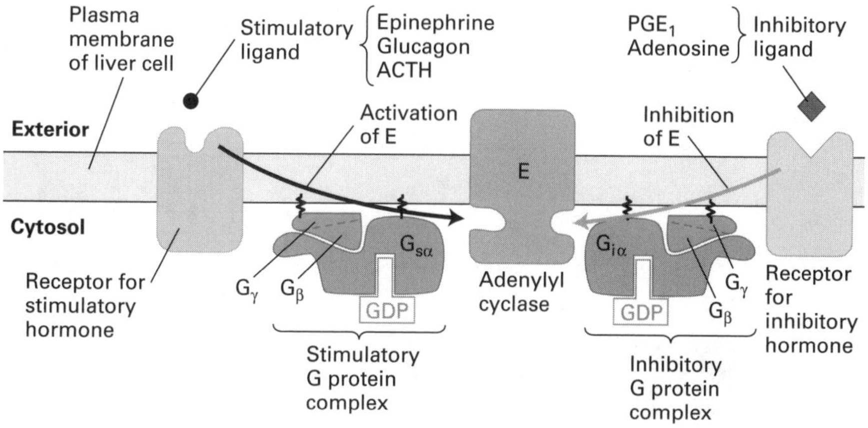

FIGURE 20-18, page 869
Hormone-induced activation and inhibition of adenylyl cyclase is mediated by $G_{s\alpha}$ (gray arrow) and $G_{i\alpha}$ (black arrow), respectively.

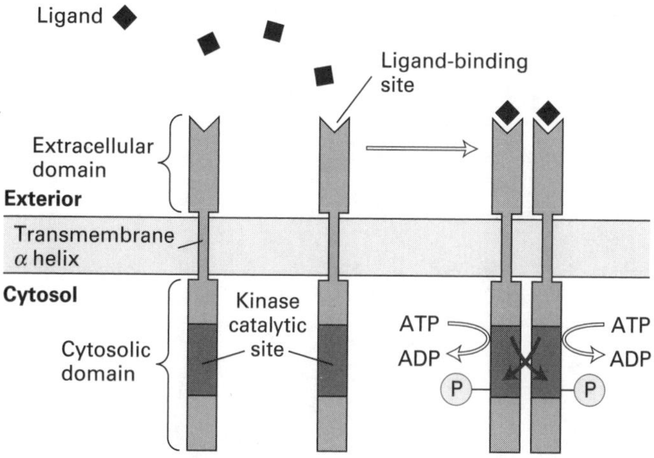

FIGURE 20-21, page 872
General structure and activation of receptor tyrosine kinases (RTKs).

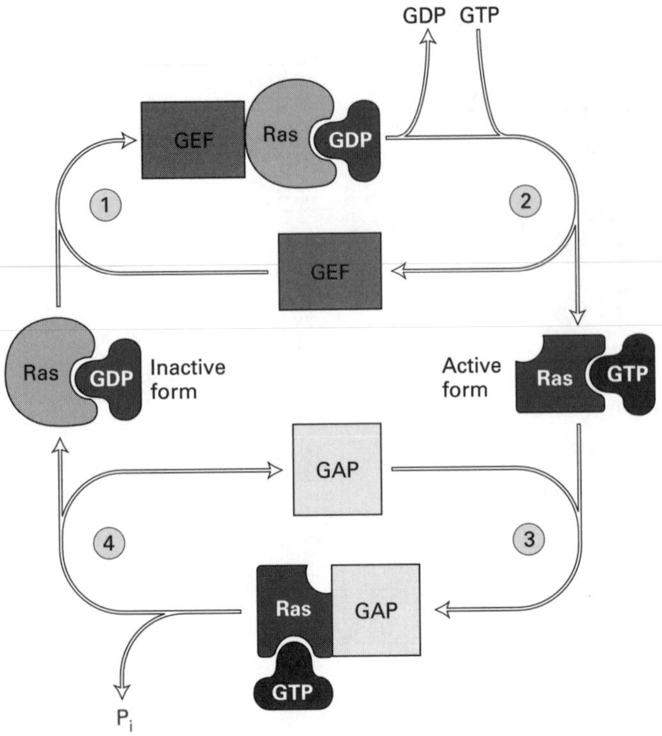

FIGURE 20-22, page 873
Cycling of the Ras protein between the inactive form with bound GDP and the active form with bound GTP occurs in four steps.

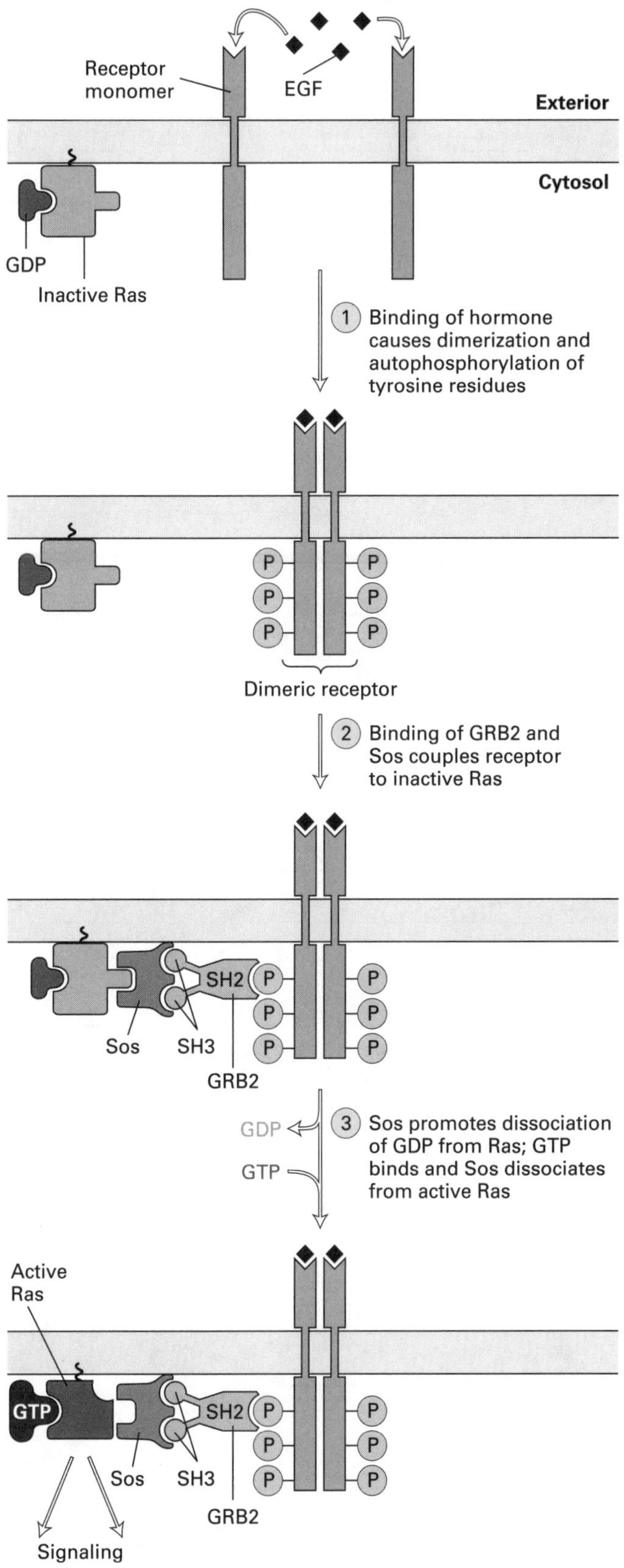
Receptor monomer
EGF
Exterior
Cytosol
GDP
Inactive Ras
1 Binding of hormone causes dimerization and autophosphorylation of tyrosine residues
P
Dimeric receptor
2 Binding of GRB2 and Sos couples receptor to inactive Ras
SH2
Sos
SH3
GRB2
GDP
GTP
3 Sos promotes dissociation of GDP from Ras; GTP binds and Sos dissociates from active Ras
Active Ras
GTP
Signaling

FIGURE 20-23, page 874
Activation of Ras following binding of a hormone (e.g., EGF) to an RTK.

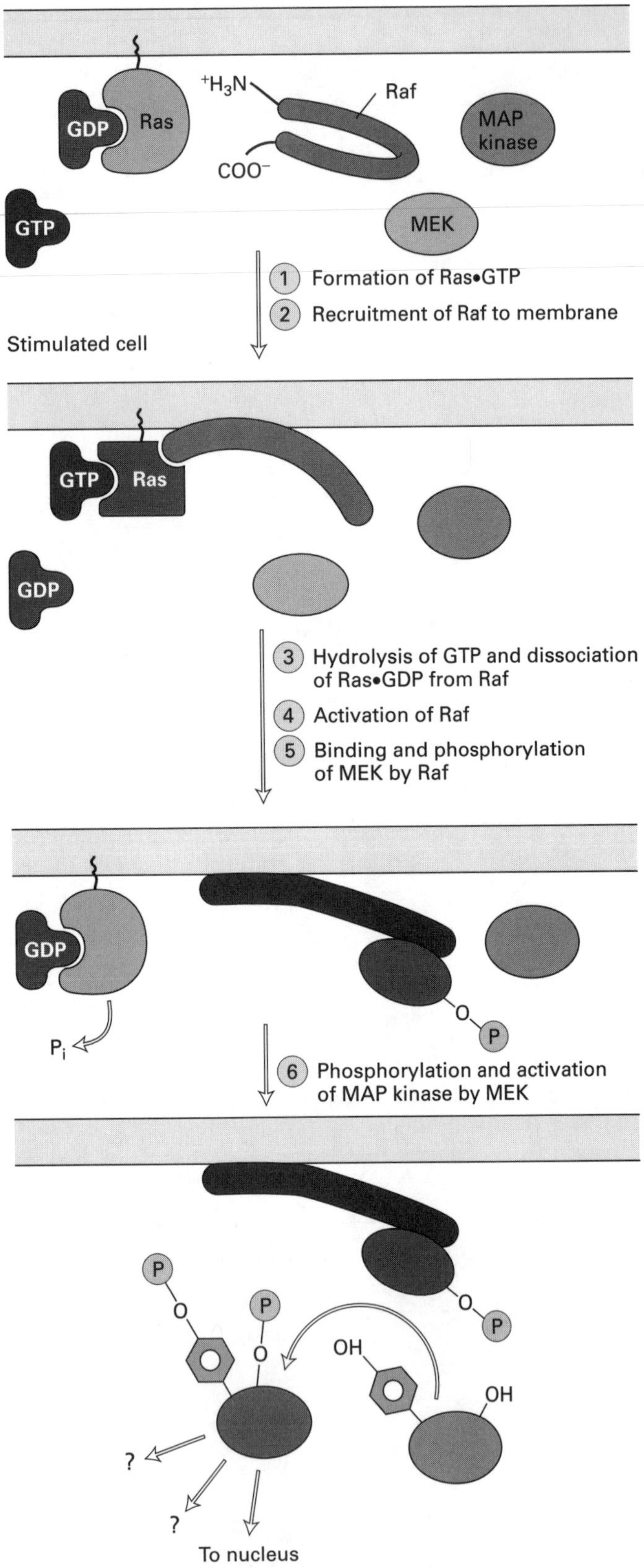

FIGURE 20-28, page 879
Kinase cascade that transmits signals downstream from activated Ras protein.

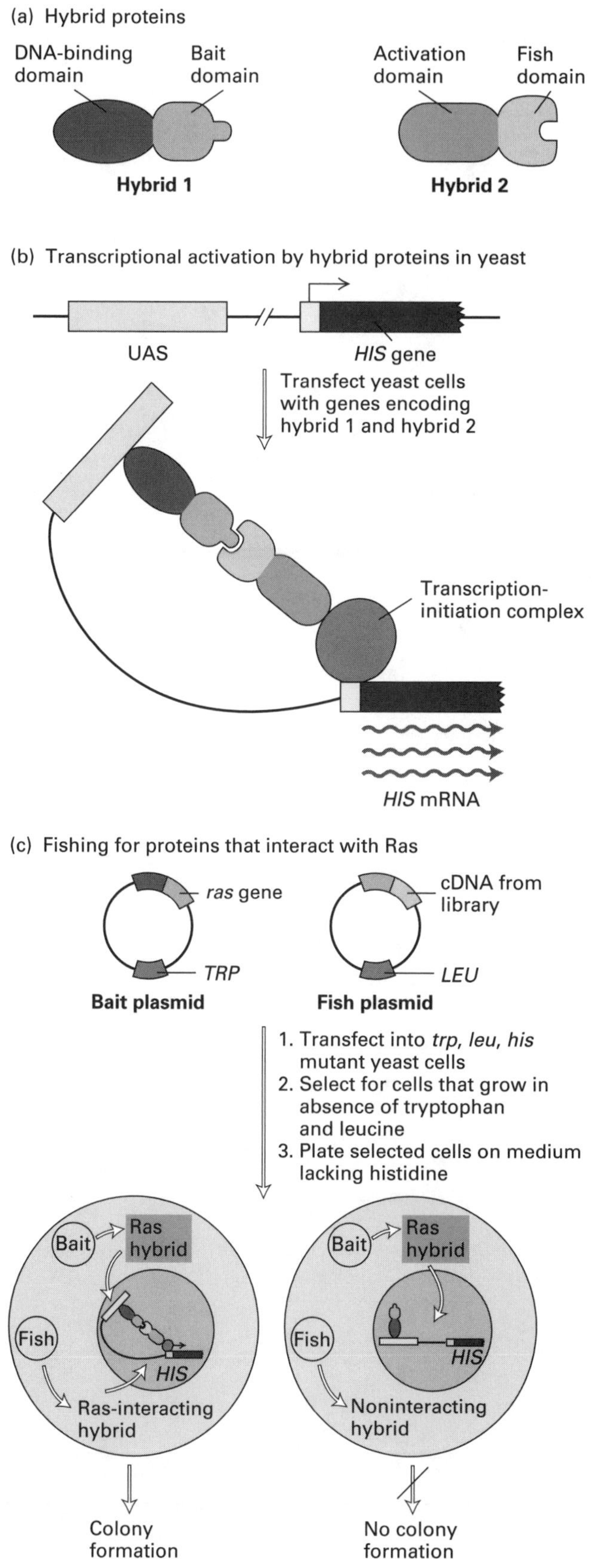

FIGURE 20-29, page 880
Yeast two-hybrid system for detecting proteins that interact.

MEDIA CONNECTIONS
Yeast Two-Hybrid System

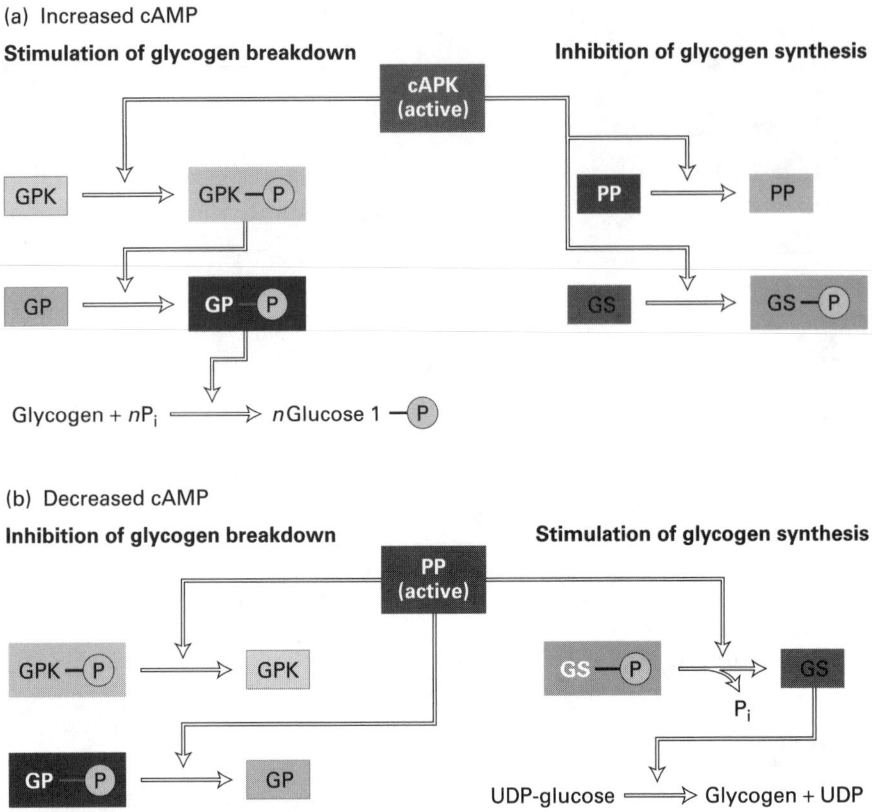

FIGURE 20-35, page 886
Regulation of glycogen breakdown and synthesis by cAMP in liver and muscle cells.

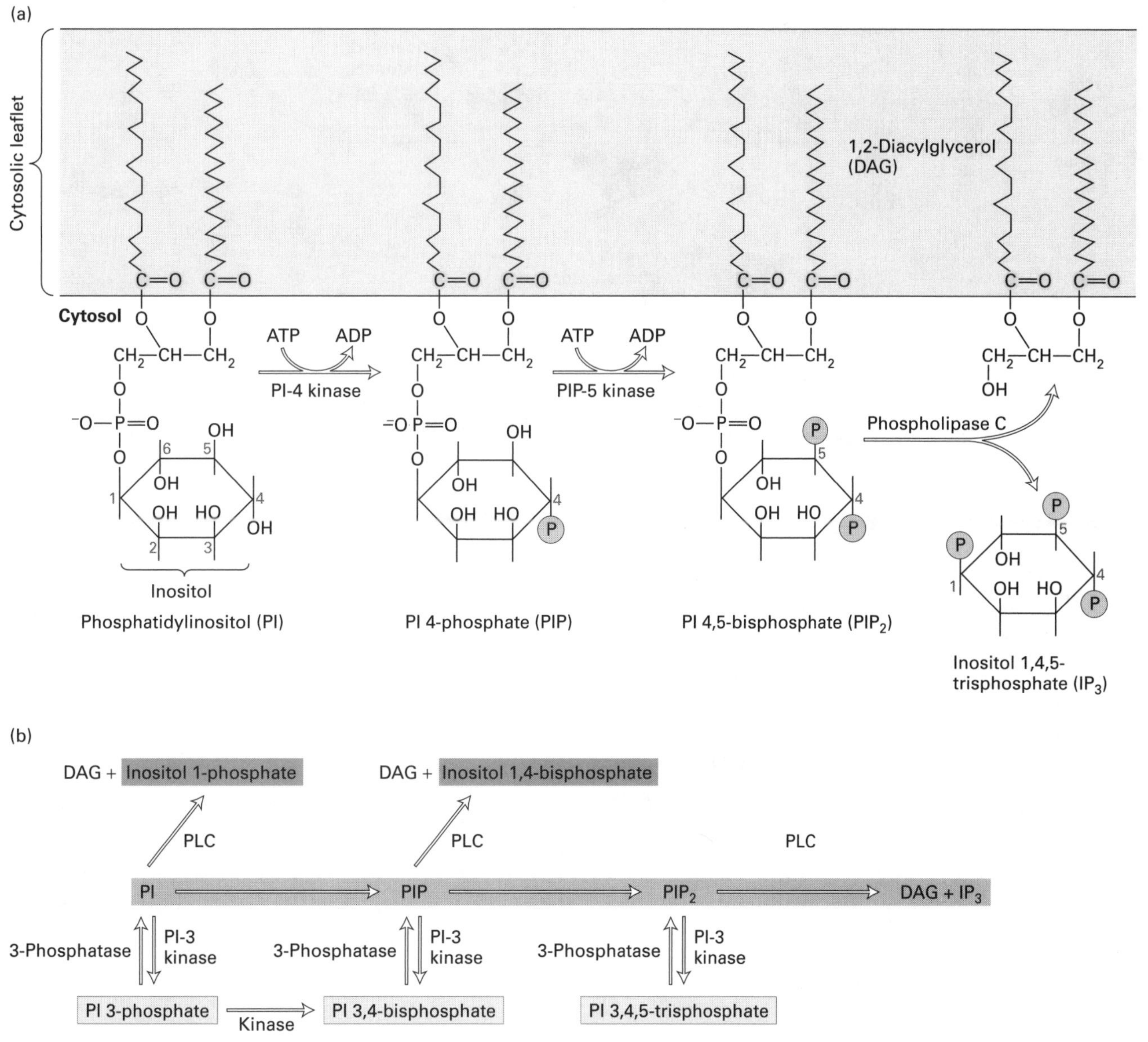

FIGURE 20-38, page 889
Several second messengers are derived from phosphatidylinositol (PI).

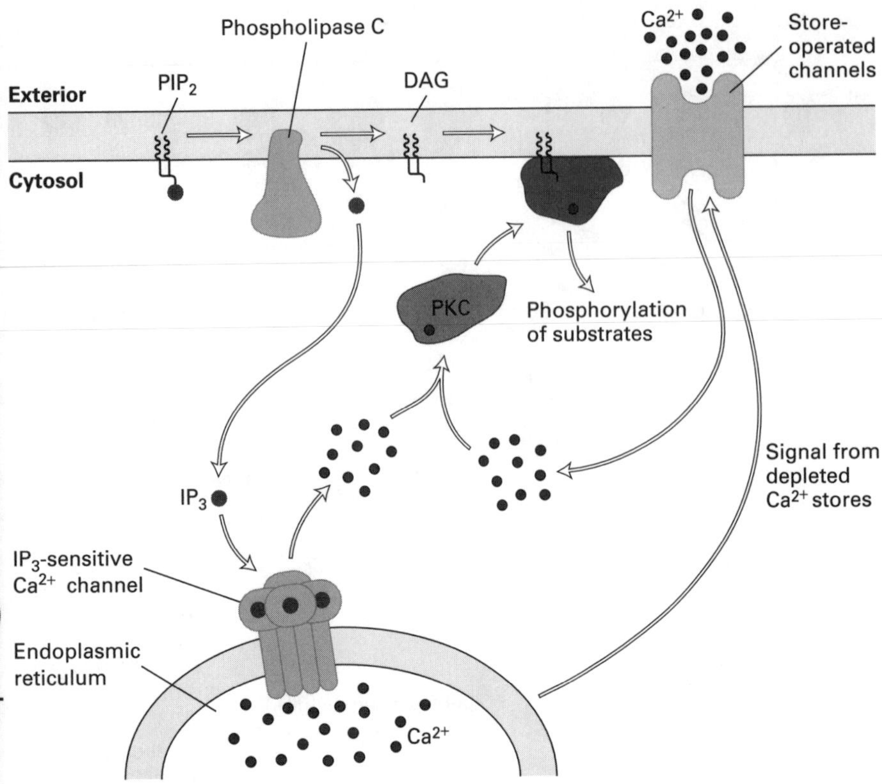

FIGURE 20-39, page 890
Elevation of cytosolic Ca^{2+} via the inositol-lipid signaling pathway.

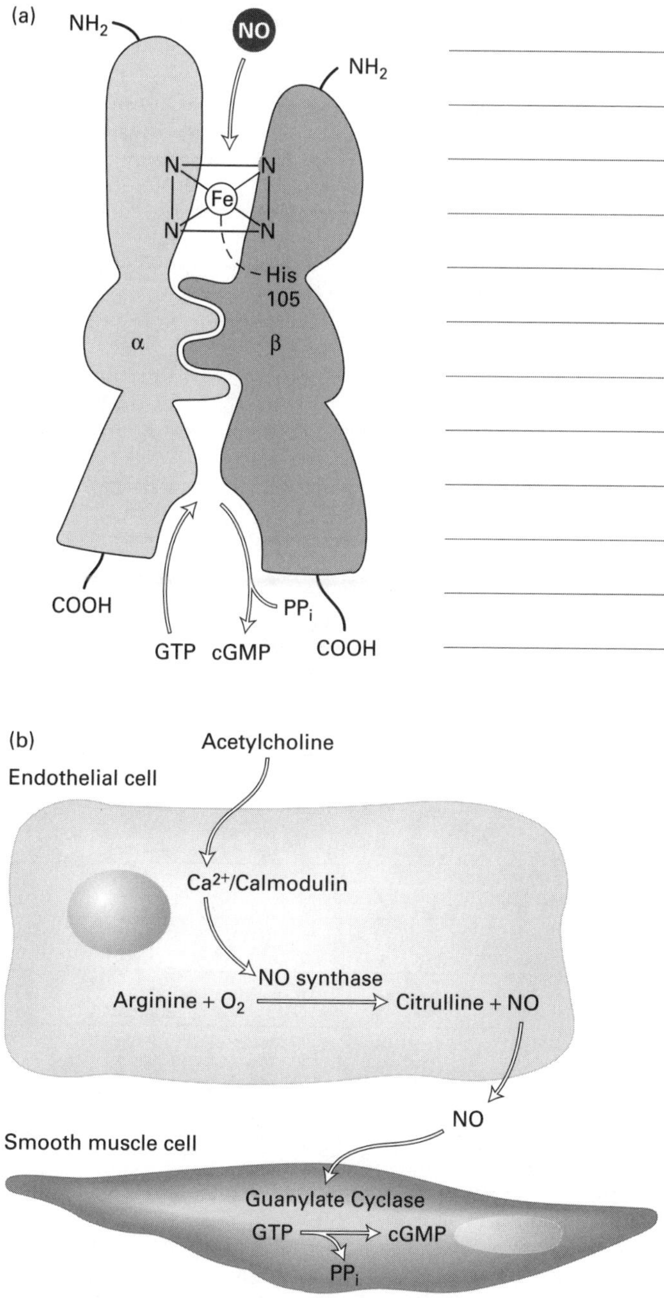

FIGURE 20-42, page 893
cGMP mediates local signaling by nitric oxide.

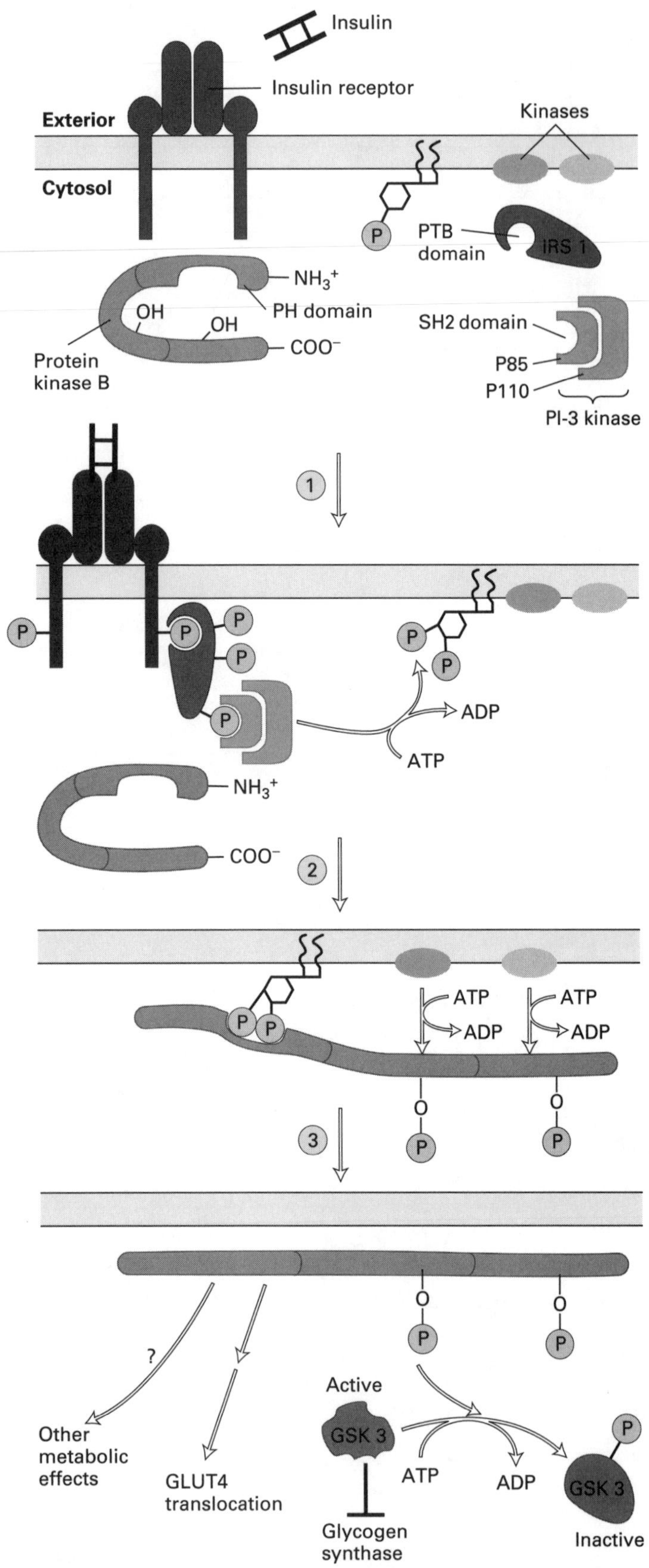

FIGURE 20-45, page 899
Activation of protein kinase B by the Ras-independent insulin signaling pathway.

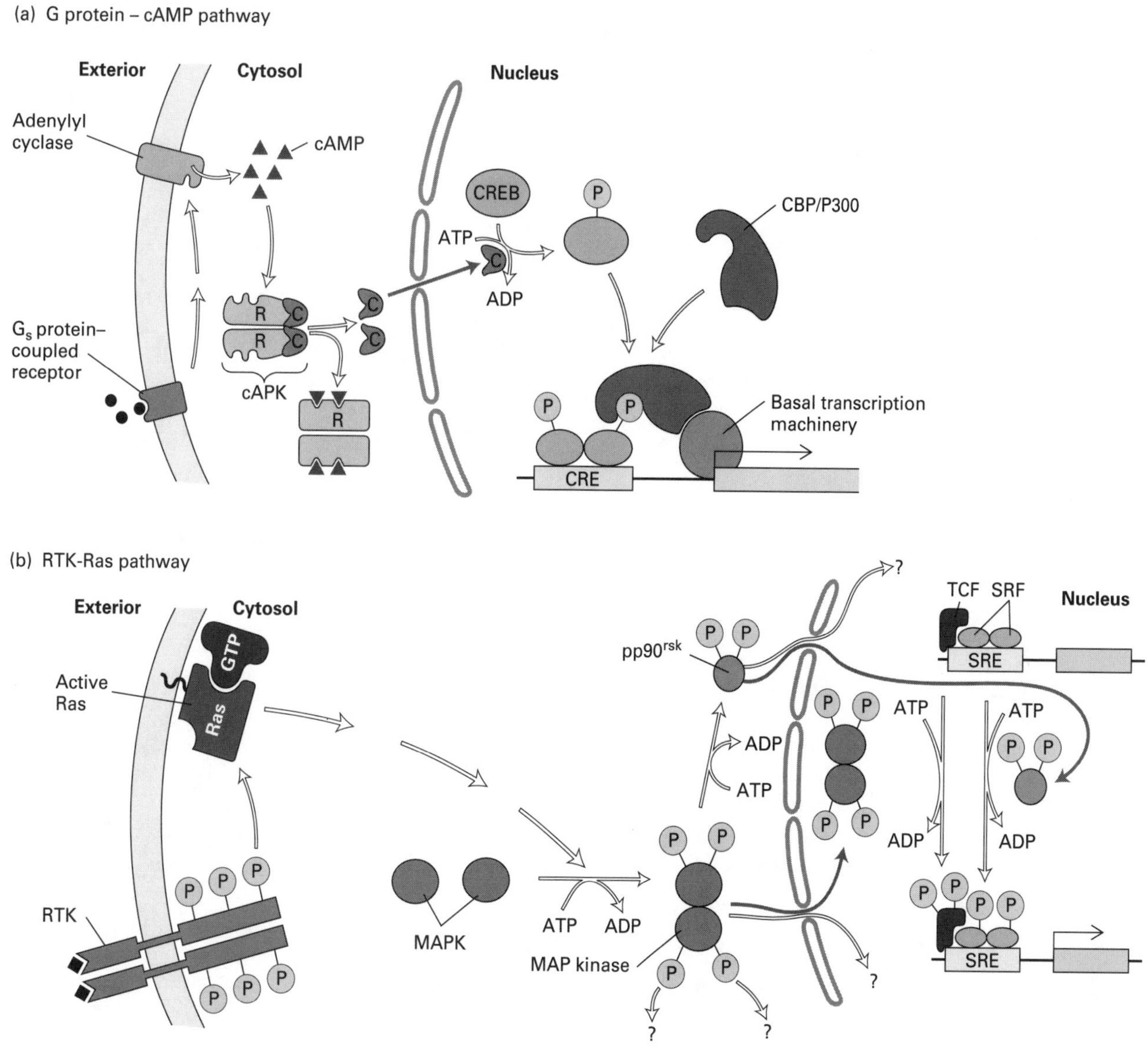

FIGURE 20-48, page 903
Signaling pathways leading to activation of transcription factors and modulation of gene expression following ligand binding to certain G_s protein–linked receptors (a) and receptor tyrosine kinases (b).

Additional Notes

Nerve Cells

21

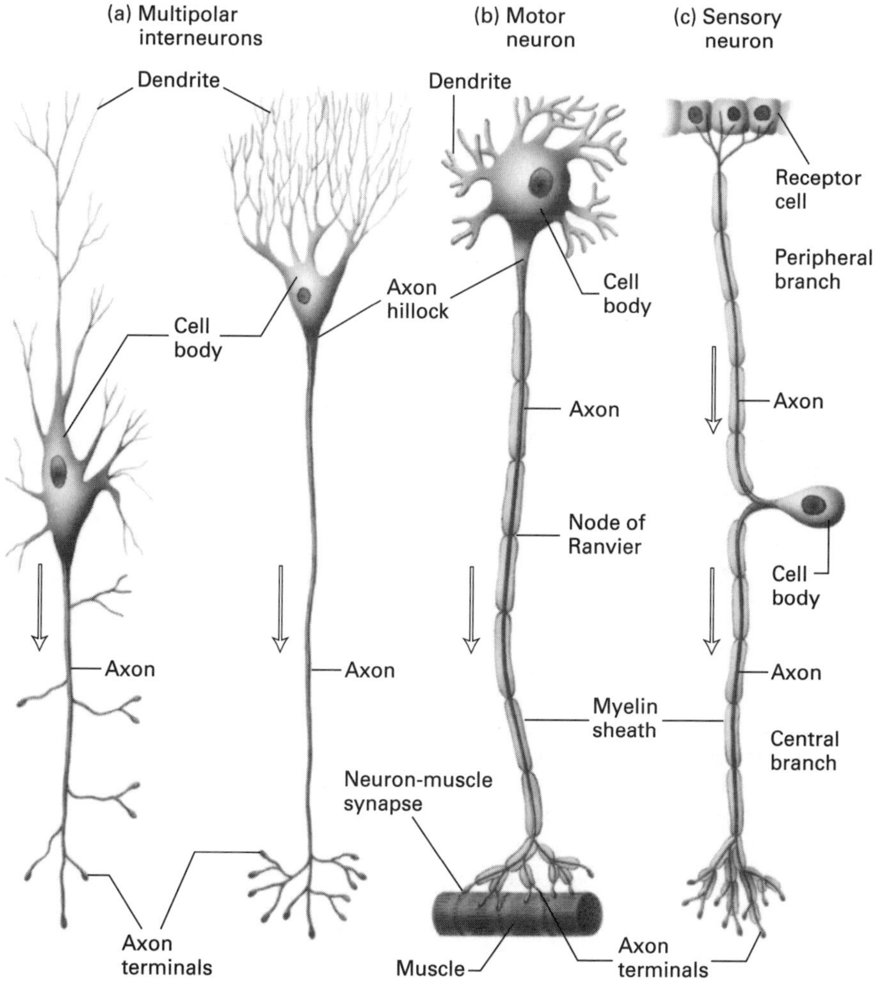

FIGURE 21-1, page 913
Structure of typical mammalian neurons.

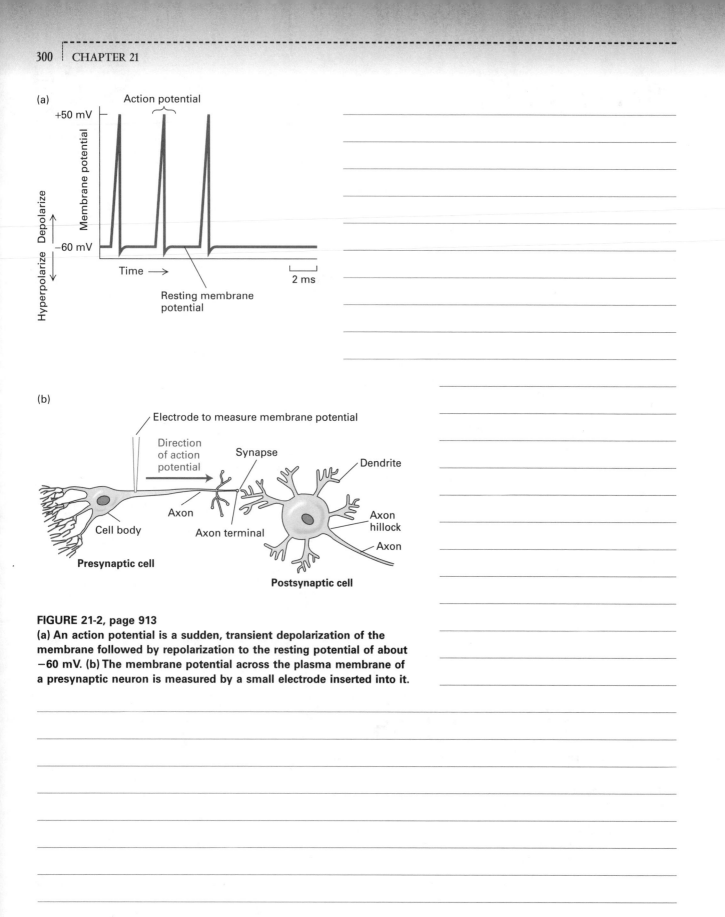

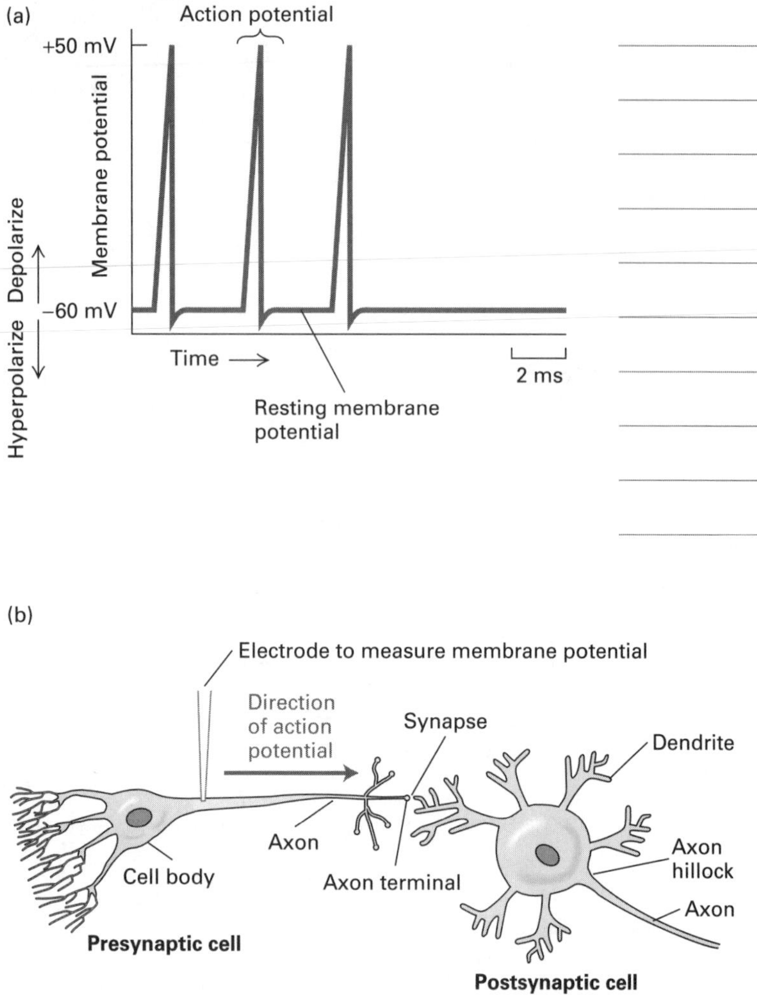

FIGURE 21-2, page 913
(a) An action potential is a sudden, transient depolarization of the membrane followed by repolarization to the resting potential of about −60 mV. (b) The membrane potential across the plasma membrane of a presynaptic neuron is measured by a small electrode inserted into it.

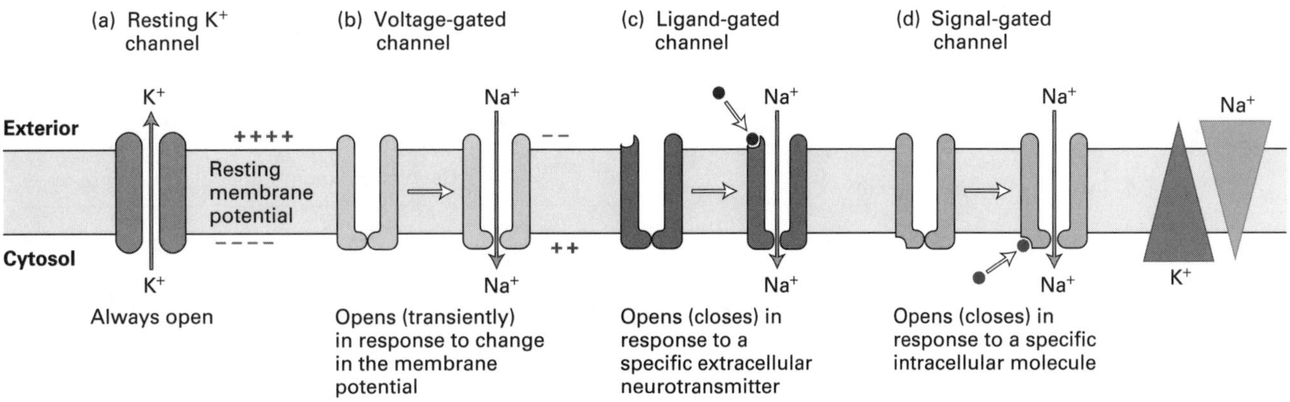

FIGURE 21-8, page 917
Ion channels in neuronal plasma membranes.

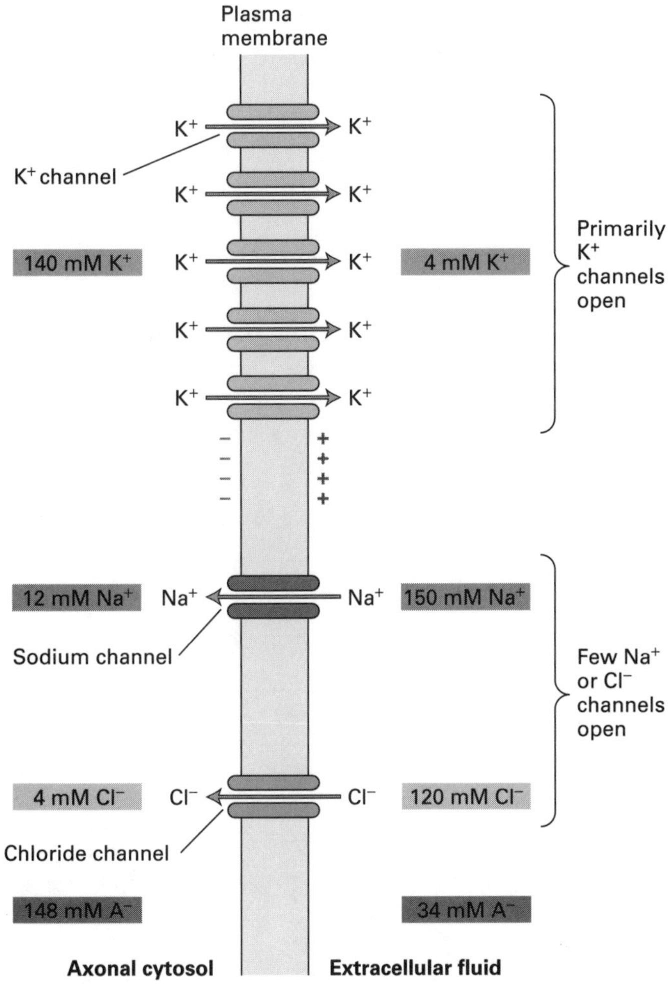

FIGURE 21-9, page 918
Origin of the resting potential in a typical vertebrate neuron.

MEDIA CONNECTIONS
Biological Energy Interconversions

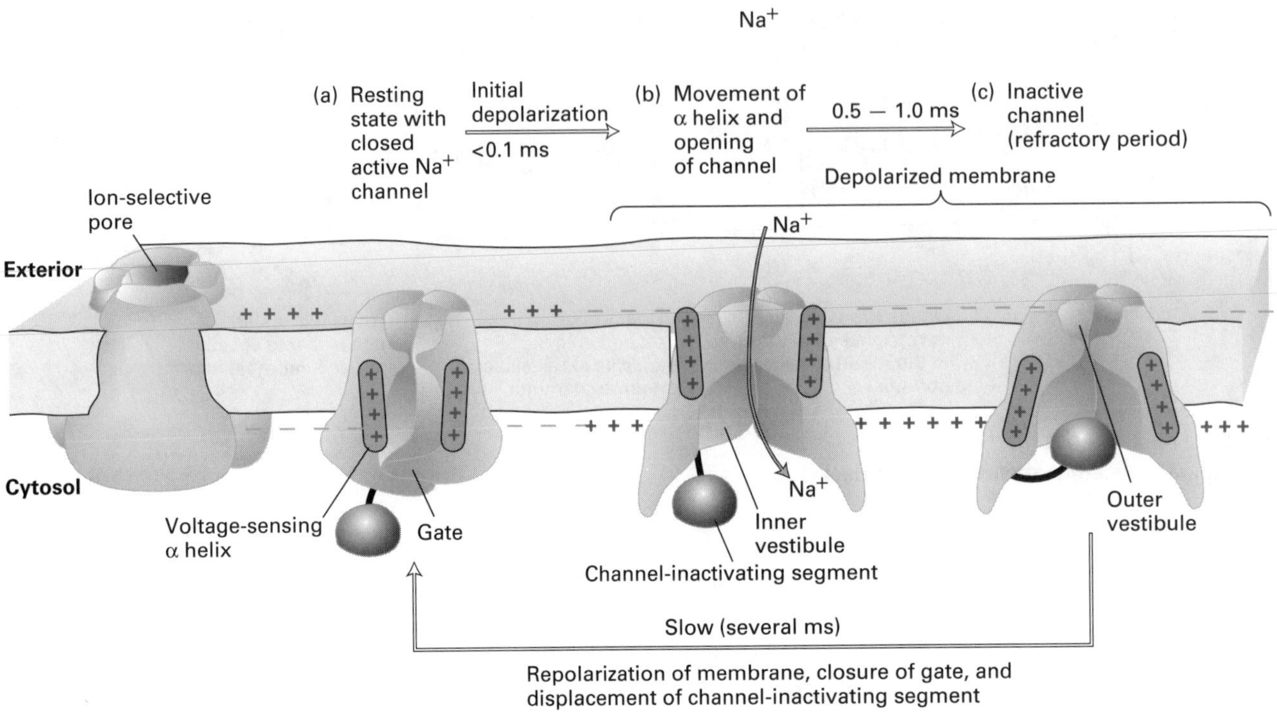

FIGURE 21-13, page 922
Structure and function of the voltage-gated Na^+ channel.

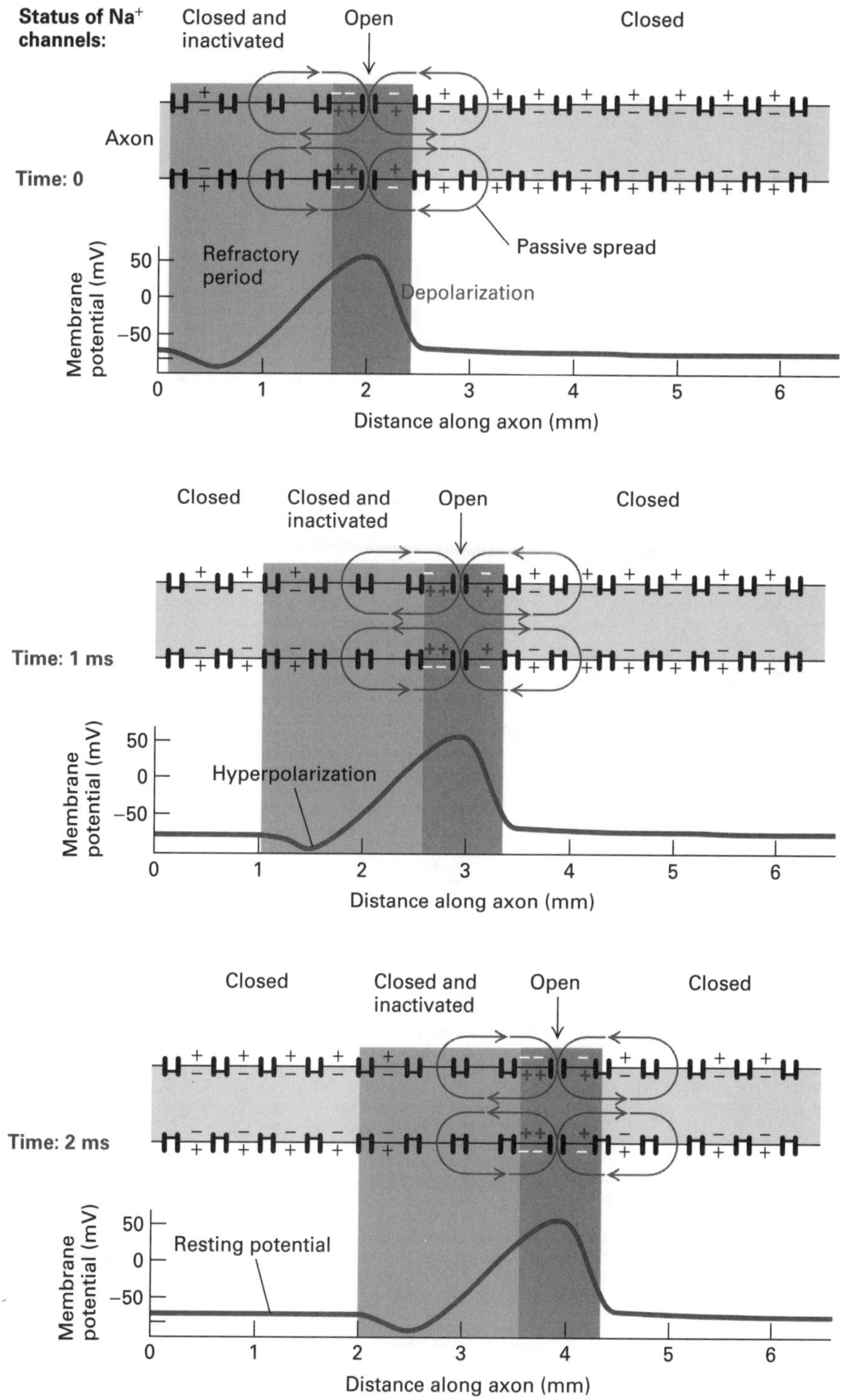

FIGURE 21-14, page 924
Unidirectional conduction of an action potential due to transient inactivation of voltage-gated Na^+ channels.

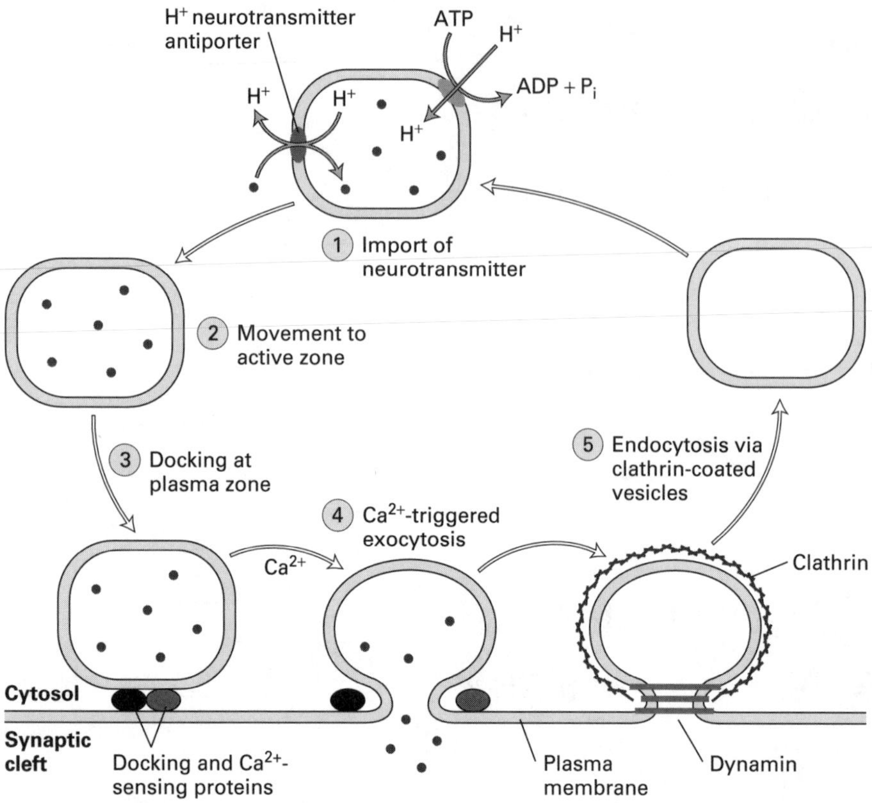

FIGURE 21-29, page 937
Release of neurotransmitters and the recycling of synaptic vesicles.

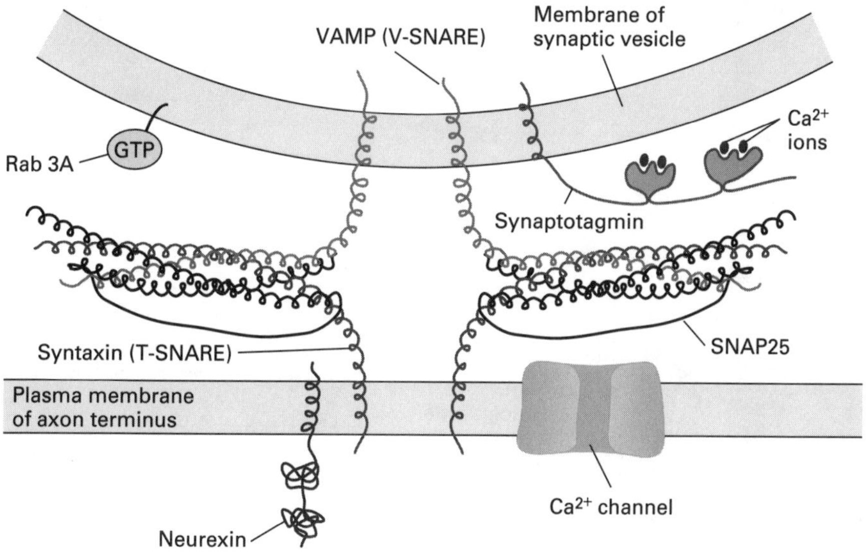

FIGURE 21-31, page 938
Synaptic-vesicle and plasma-membrane proteins important for vesicle docking and fusion.

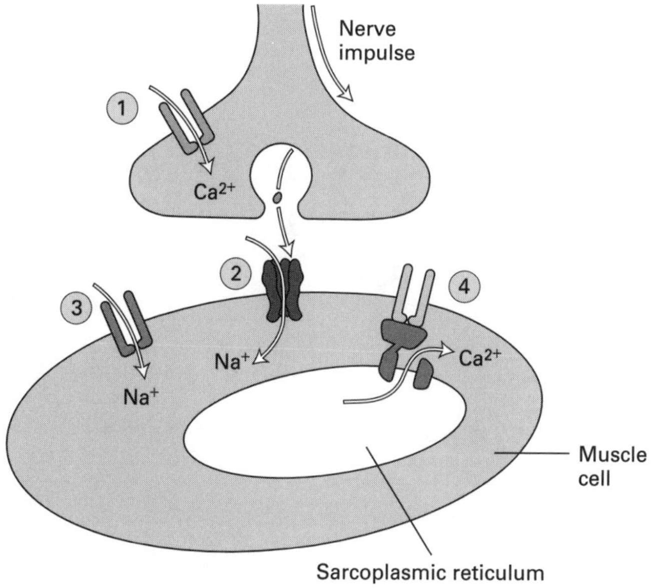

FIGURE 21-37, page 945
Sequential activation of gated ion channels at a neuromuscular junction.

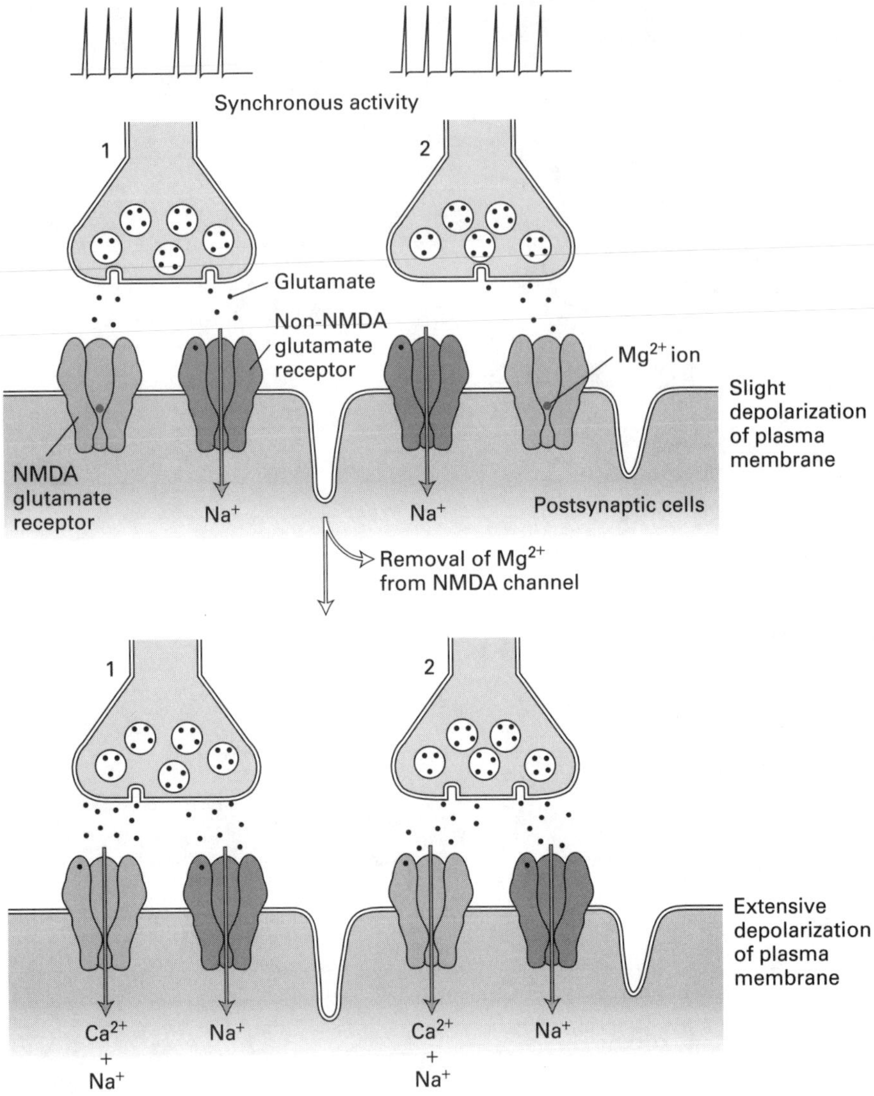

FIGURE 21-40, page 948
Different properties of two types of glutamate receptors found in the hippocampus region in the brain.

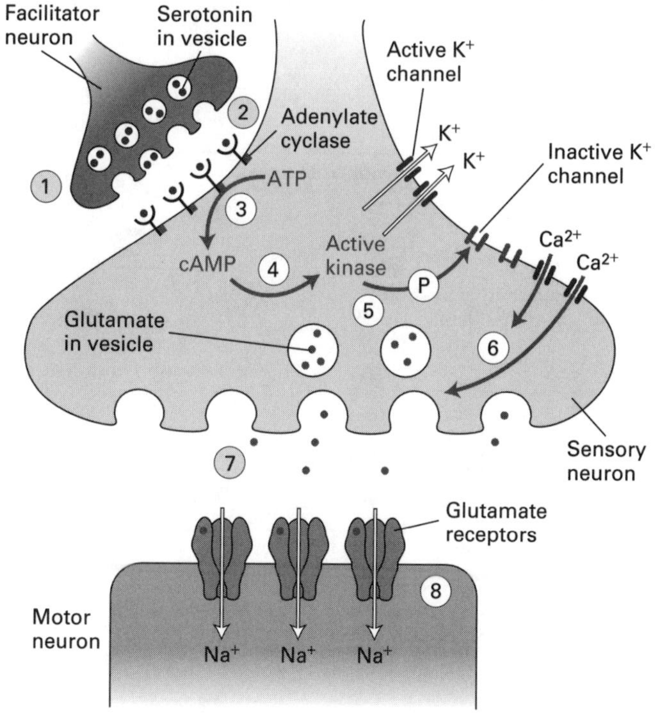

FIGURE 21-42, page 950
Action of a serotonin modulatory synapse in the sea slug *Aplysia punctata*.

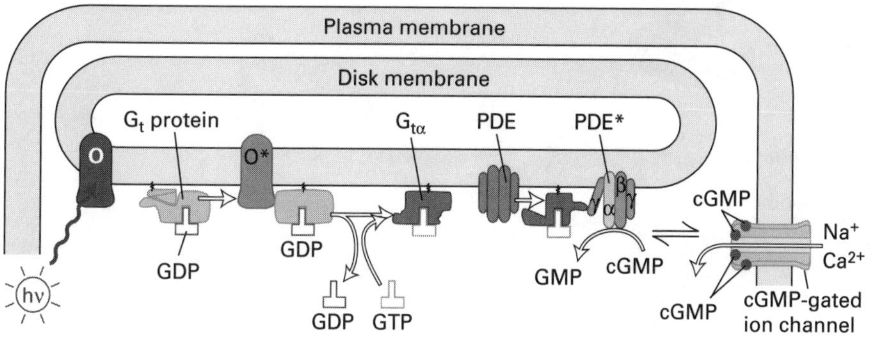

FIGURE 21-47, page 956
Coupling of light absorption by rhodopsin to activation of cGMP phosphodiesterase in rod cells.

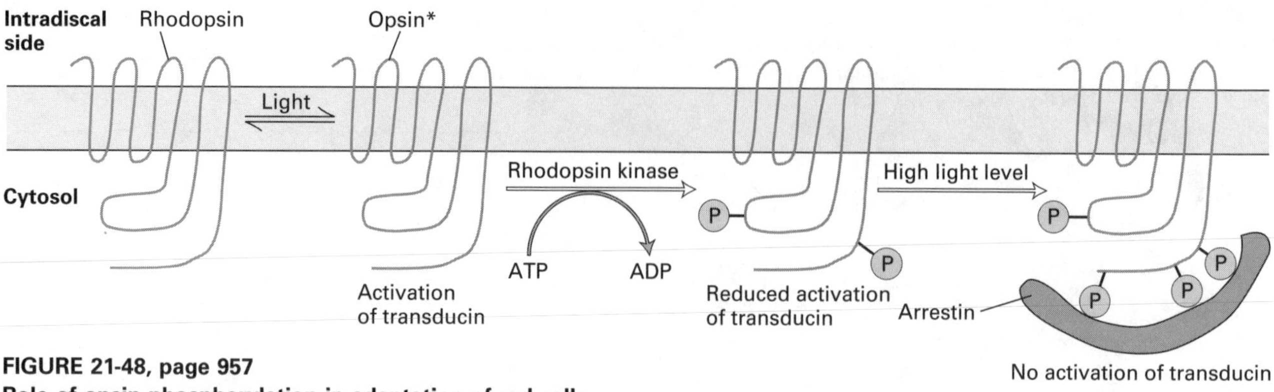

FIGURE 21-48, page 957
Role of opsin phosphorylation in adaptation of rod cells to changes in ambient light levels.

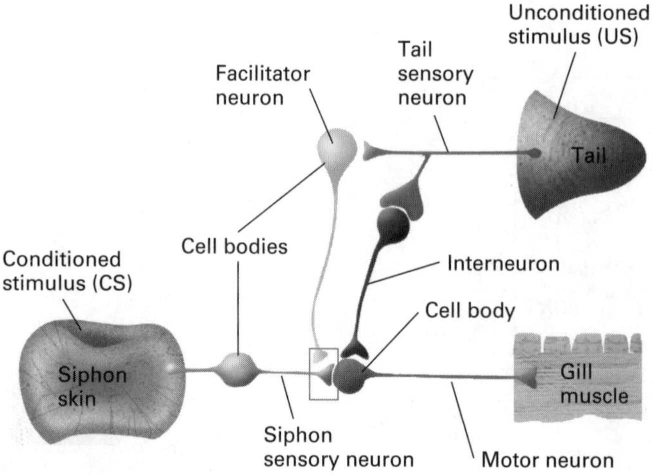

FIGURE 21-52, page 960
Neural circuits in the gill-withdrawal reflex of the sea slug *Aplysia*.

Additional Notes

Integrating Cells into Tissues

22

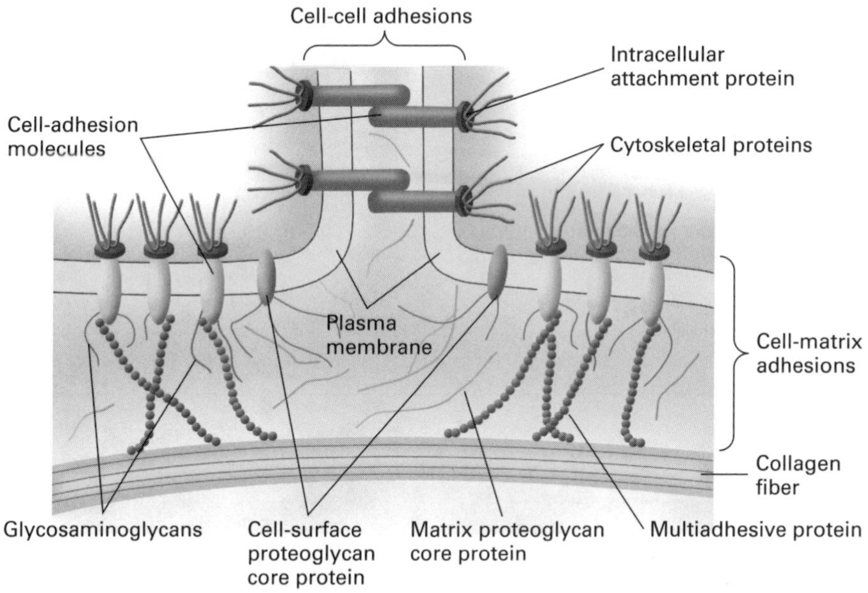

FIGURE 22-1, page 969
Schematic overview of the types of molecules that bind cells to each other and to the extracellular matrix.

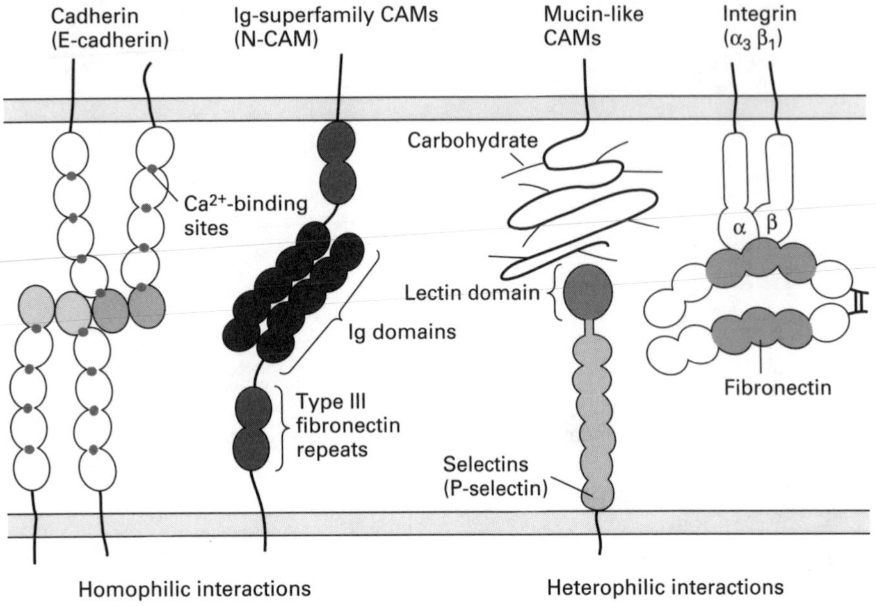

FIGURE 22-2, page 970
Major families of cell-adhesion molecules (CAMs).

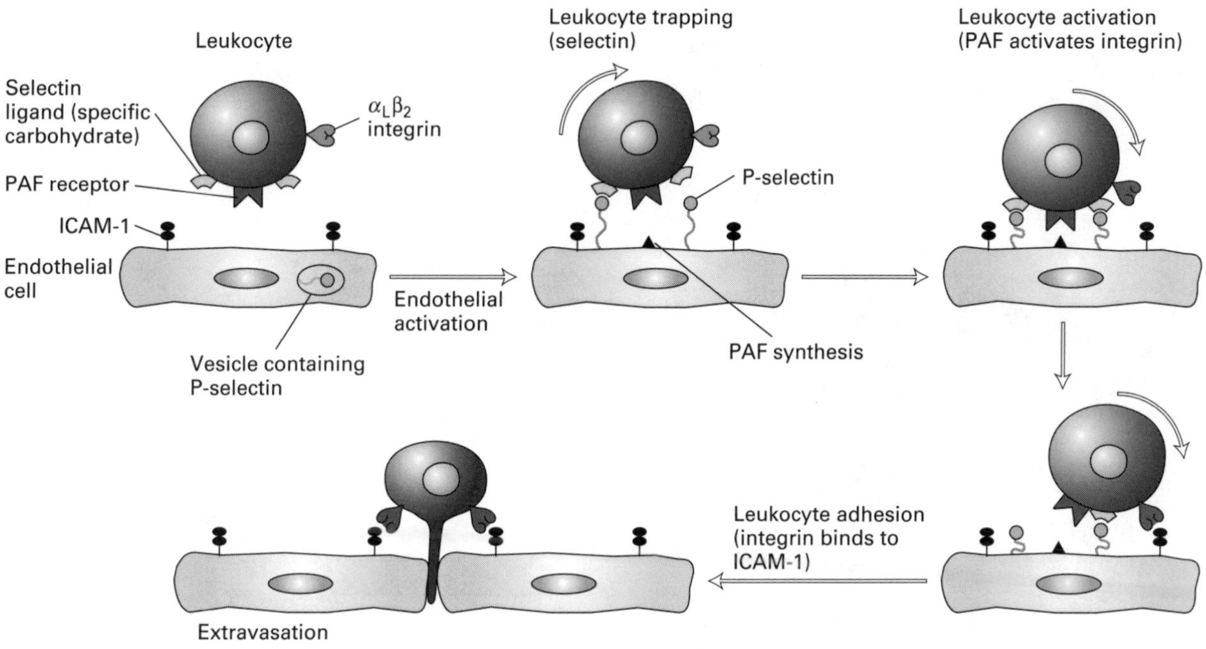

FIGURE 22-4, page 973
Interactions between cell-adhesion molecules during the initial binding and tight binding of T cells, a kind of leukocyte, to activation endothelial cells.

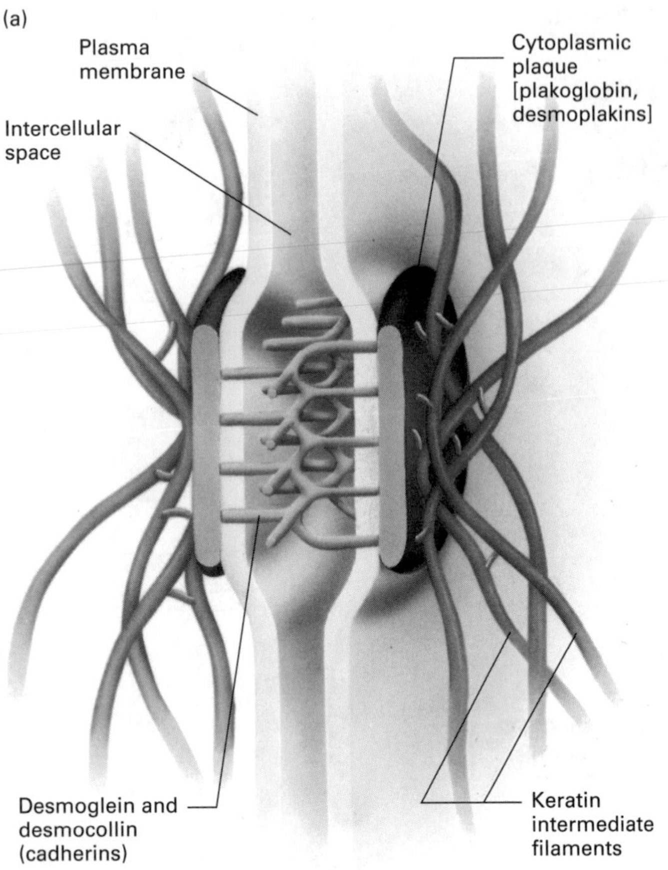

FIGURE 22-6 (a), page 974
Desmosomes.

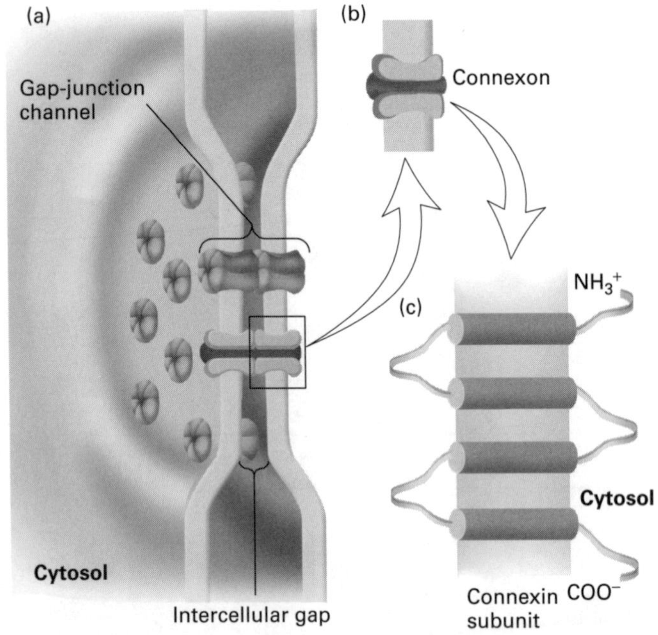

FIGURE 22-8, page 976
Structure of gap junctions.

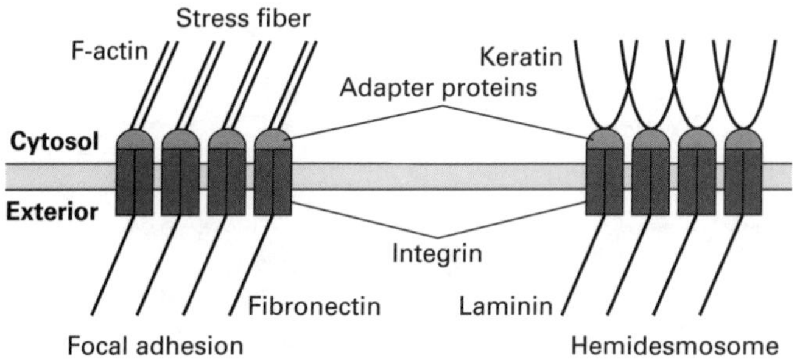

FIGURE 22-9, page 978
Adhesion molecules in junctions involved in cell-matrix adhesion.

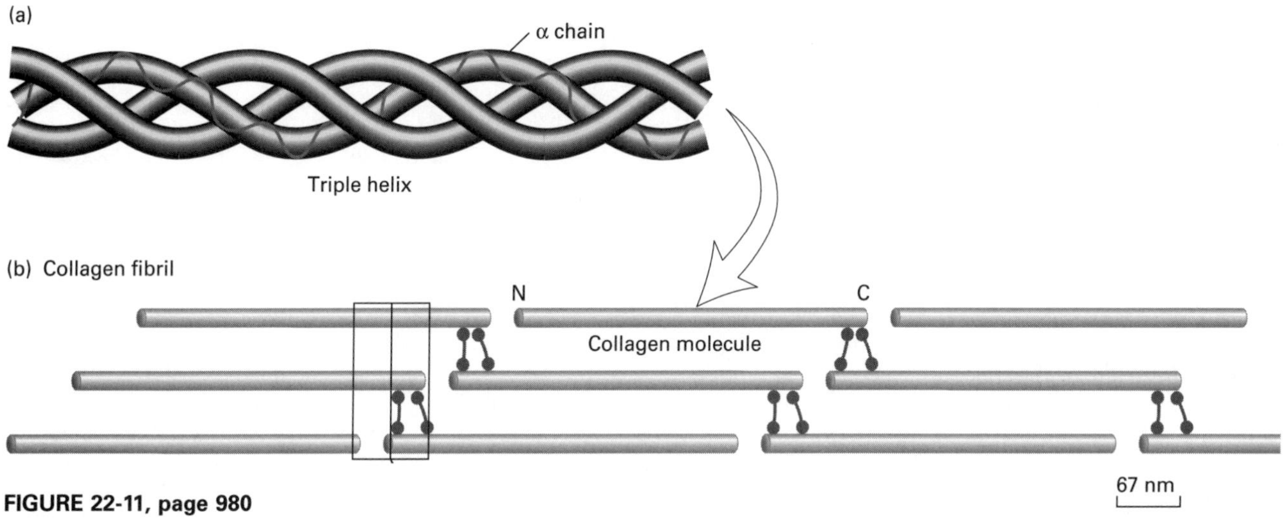

FIGURE 22-11, page 980
The structure of collagen.

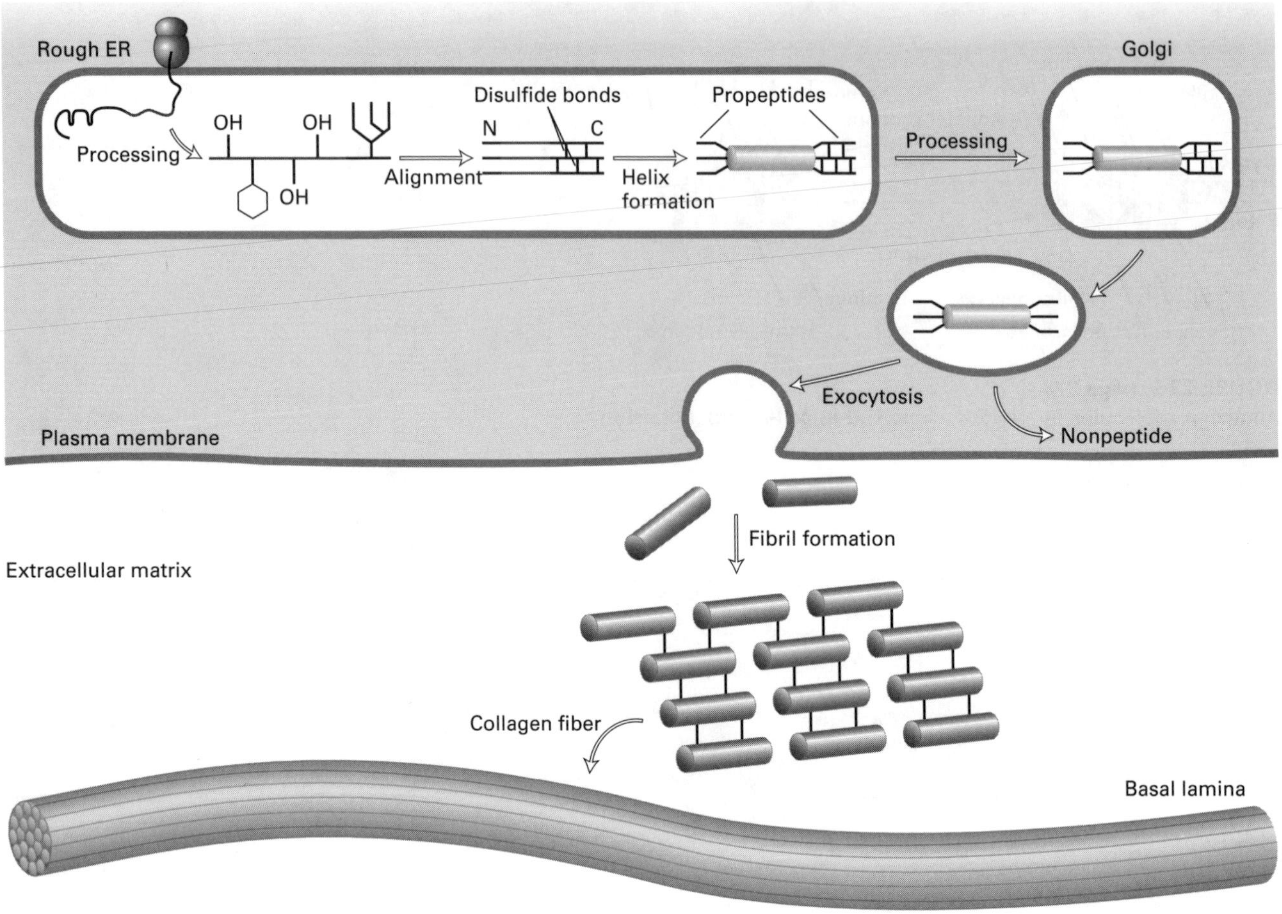

FIGURE 22-14, page 983
Major events in the biosynthesis of fibrous collagens.

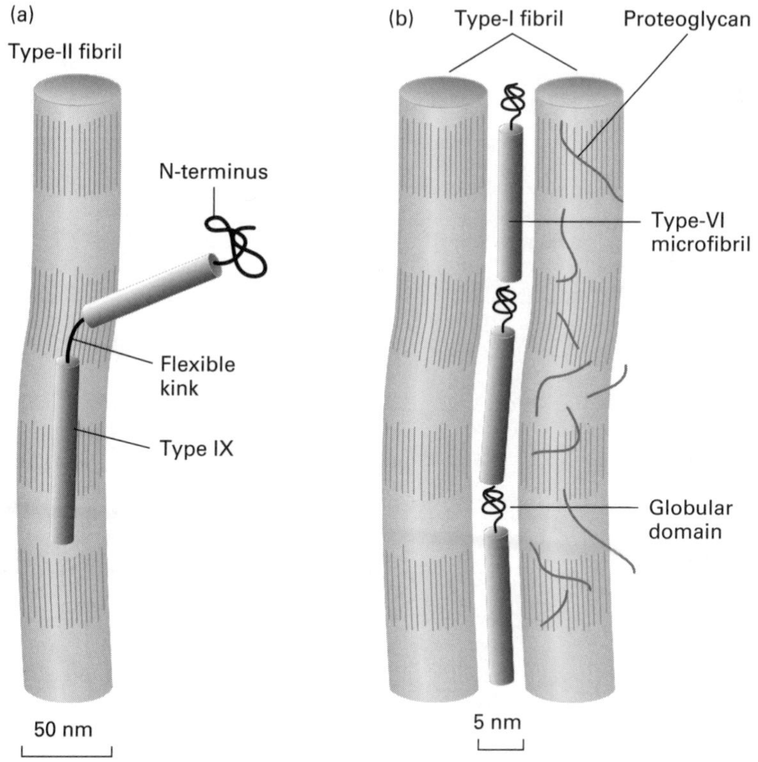

FIGURE 22-16, page 984
Interactions of fibrous and nonfibrous collagens.

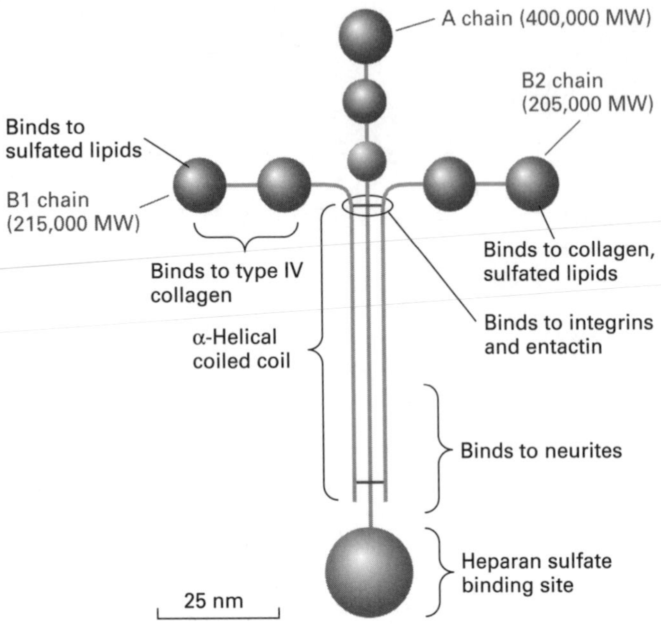

FIGURE 22-19, page 986
Structure of laminin, a large heterotrimeric multiadhesive matrix protein found in all basal laminae.

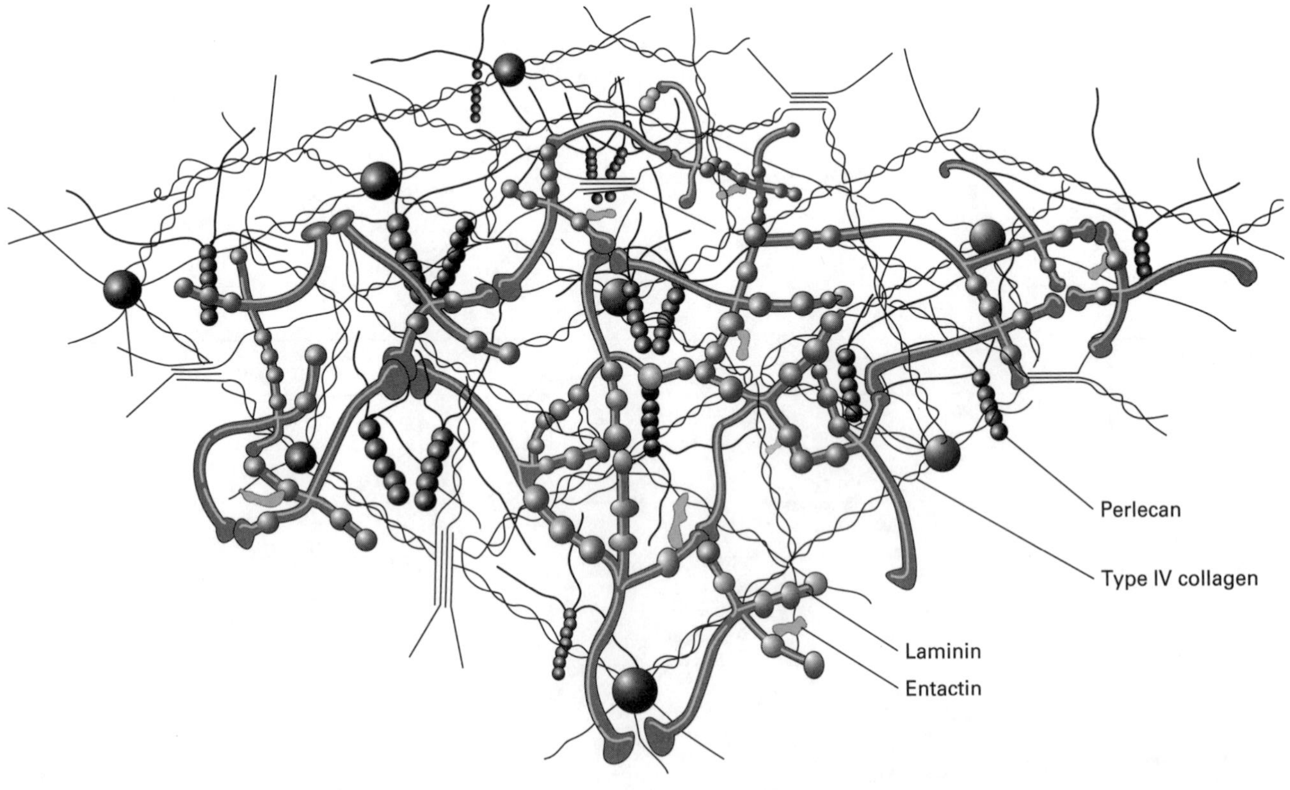

FIGURE 22-20, page 987
Model of the basal lamina. [Adapted from B. Alberts et al., 1994, *Molecular Biology of the Cell*, 3d ed, Garland, p. 991.]

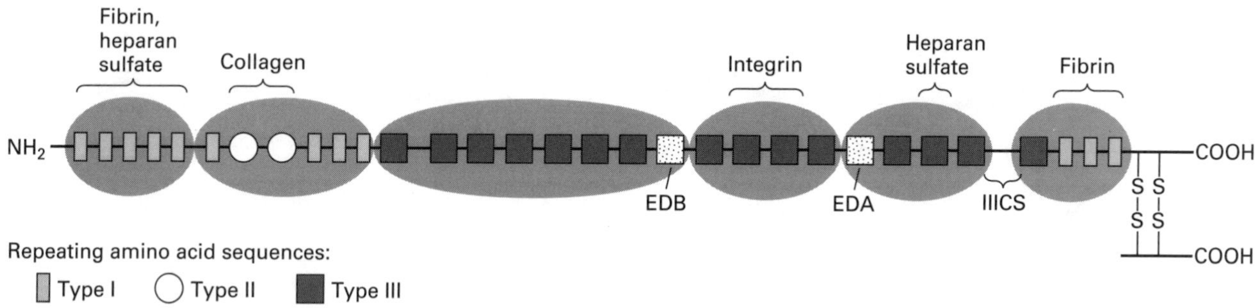

FIGURE 22-22, page 988
Structure of fibronectin chains.

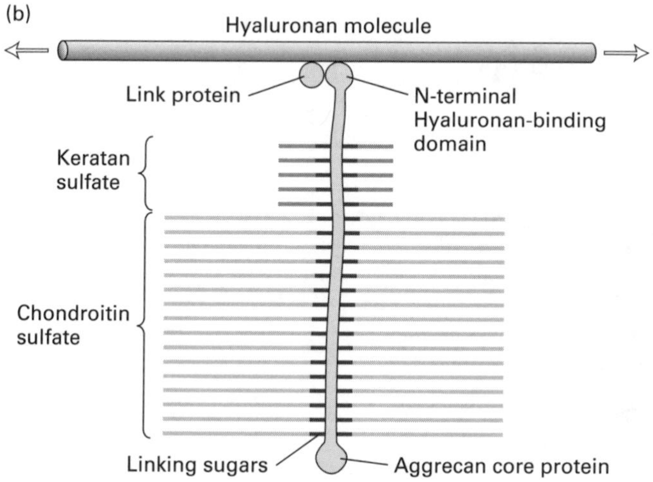

FIGURE 22-26 (b), page 991
Structure of cartilage proteoglycan aggregate.

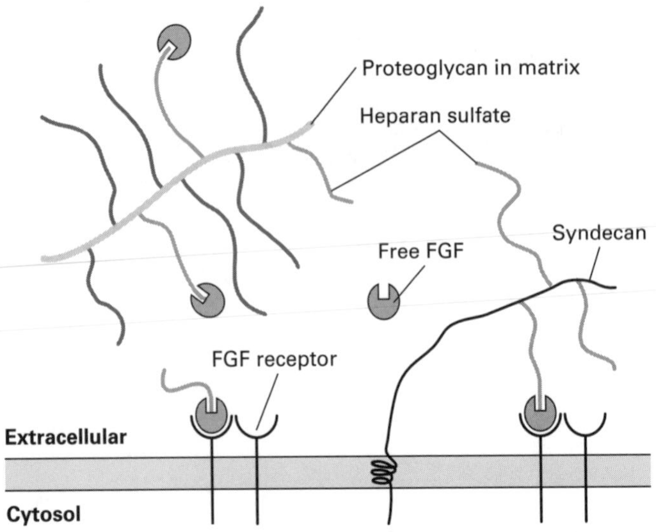

FIGURE 22-28, page 992
Modulation of activity of fibroblast growth factor (FGF) by heparan sulfate proteoglycans.

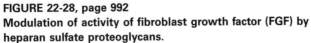

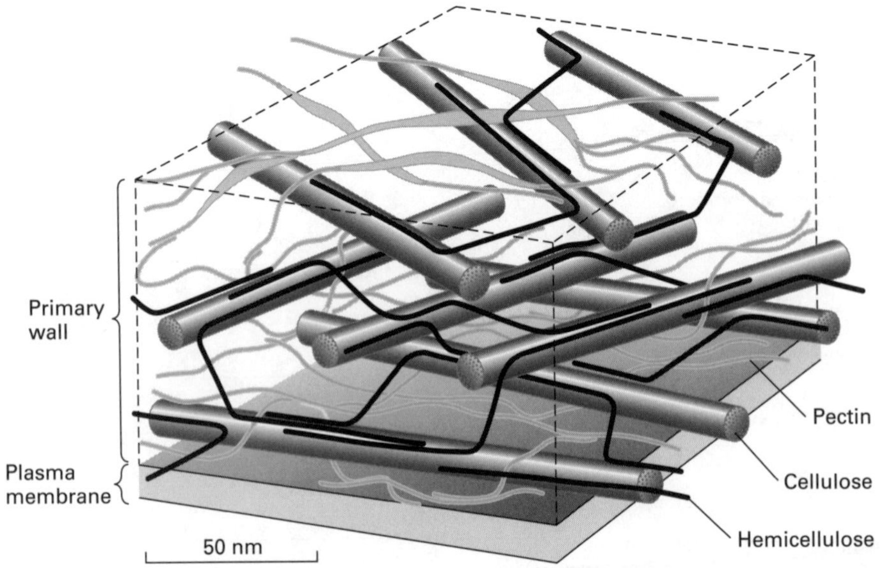

FIGURE 22-29, page 994
Schematic representation of the cell wall of an onion.

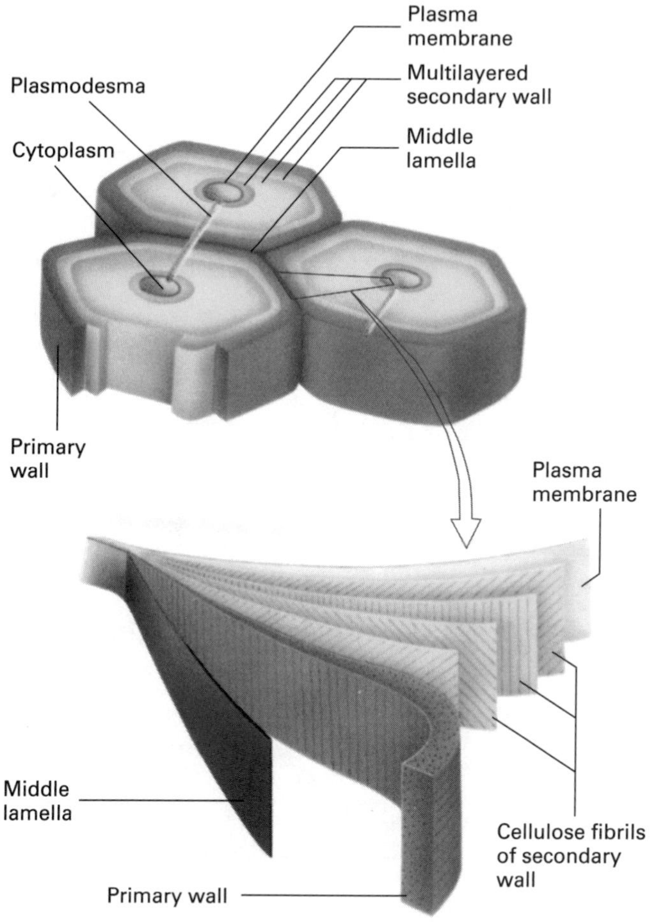

FIGURE 22-32, page 995
The structure of the secondary cell wall, built up of a series of layers of cellulose.

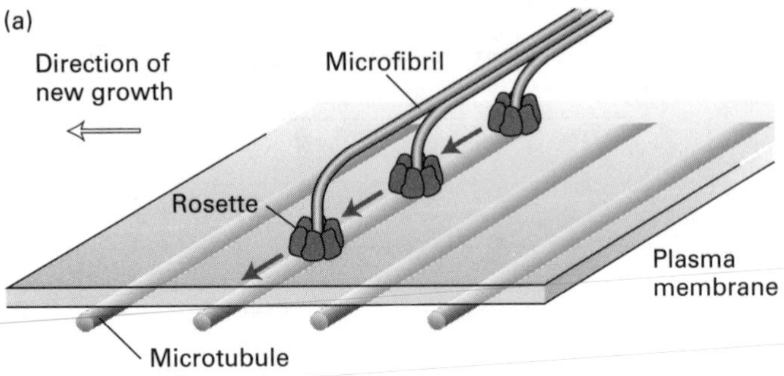

FIGURE 22-35 (a), page 997
Microtubules and cellulose synthesis in an elongating root tip cell.

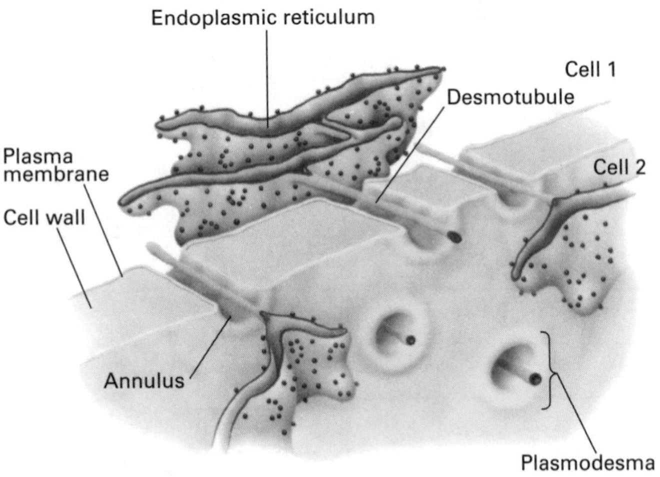

FIGURE 22-36, page 998
The structure of plasmodesmata.

Additional Notes

Cell Interactions in Development

23

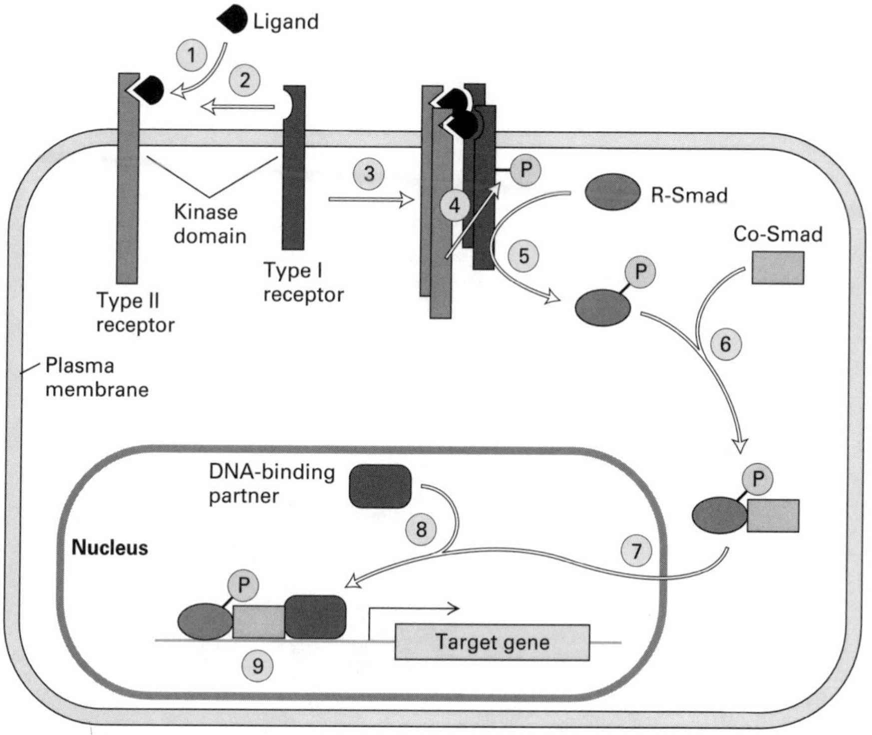

FIGURE 23-3, page 1006
TGFβ signaling pathway.

MEDIA CONNECTIONS
TGF β Signaling Pathway

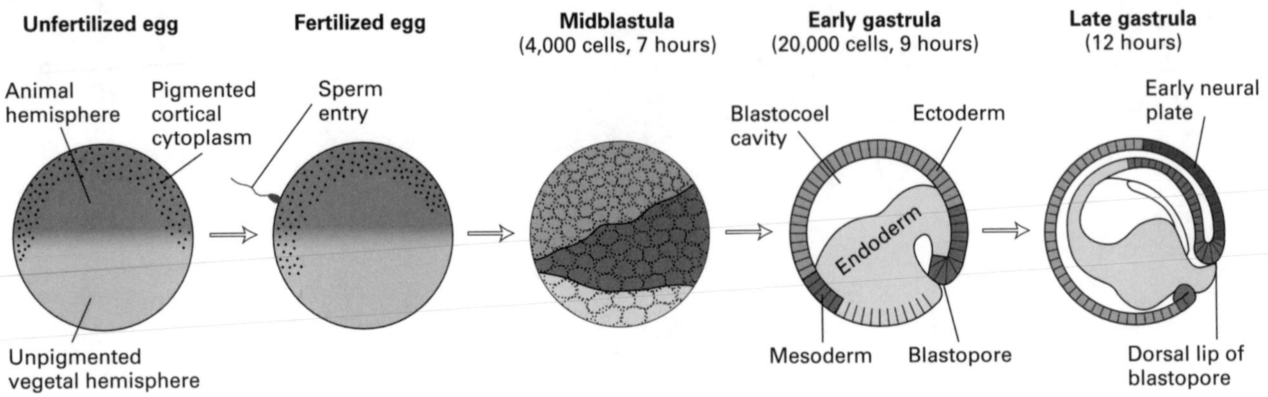

FIGURE 23-5, page 1009
Early embryogenesis of the frog *Xenopus laevis*.

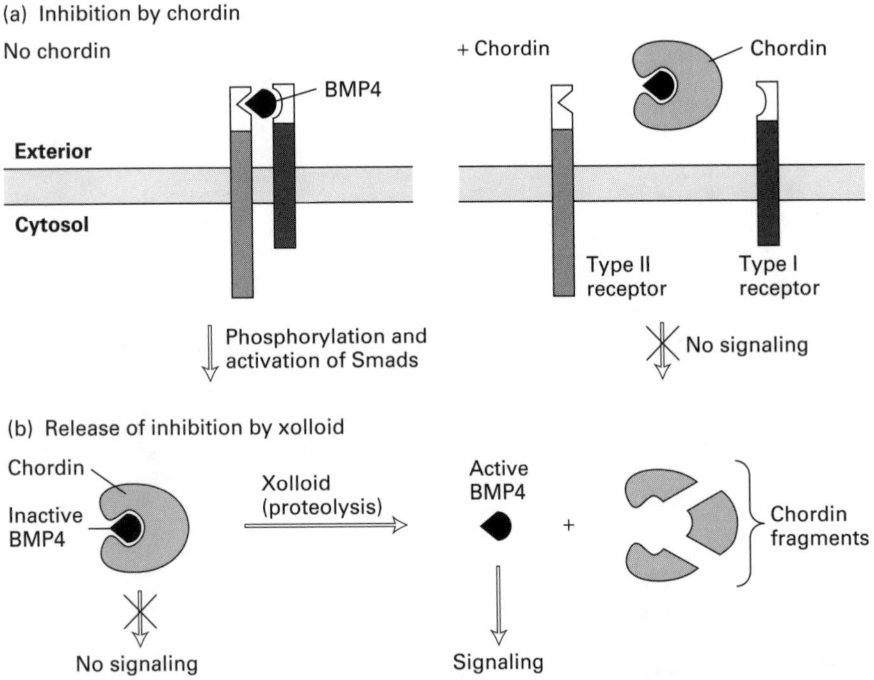

FIGURE 23-10, page 1011
Modulation of BMP4 signaling in *Xenopus* by chordin and Xolloid.

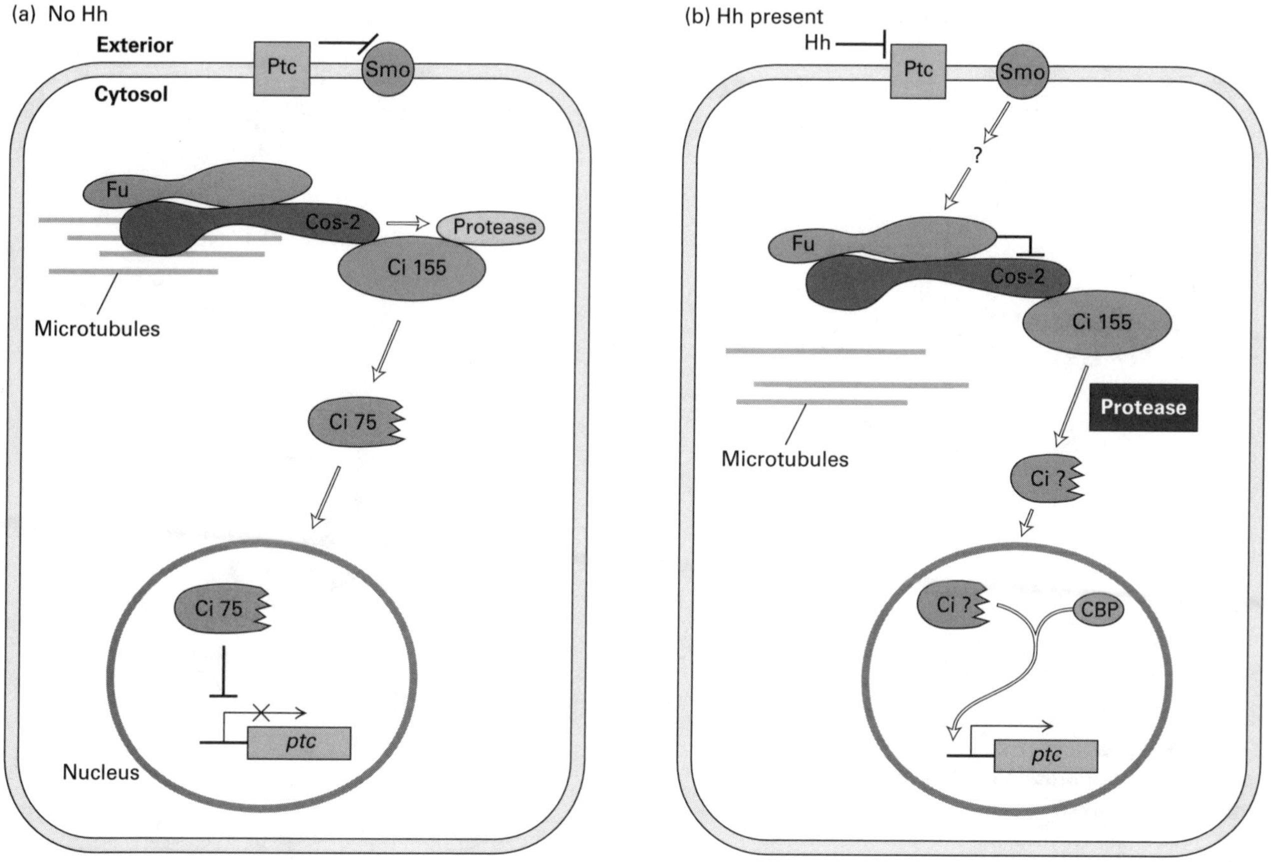

FIGURE 23-13, page 1015
A model of the Hedgehog (Hh) signaling pathway.

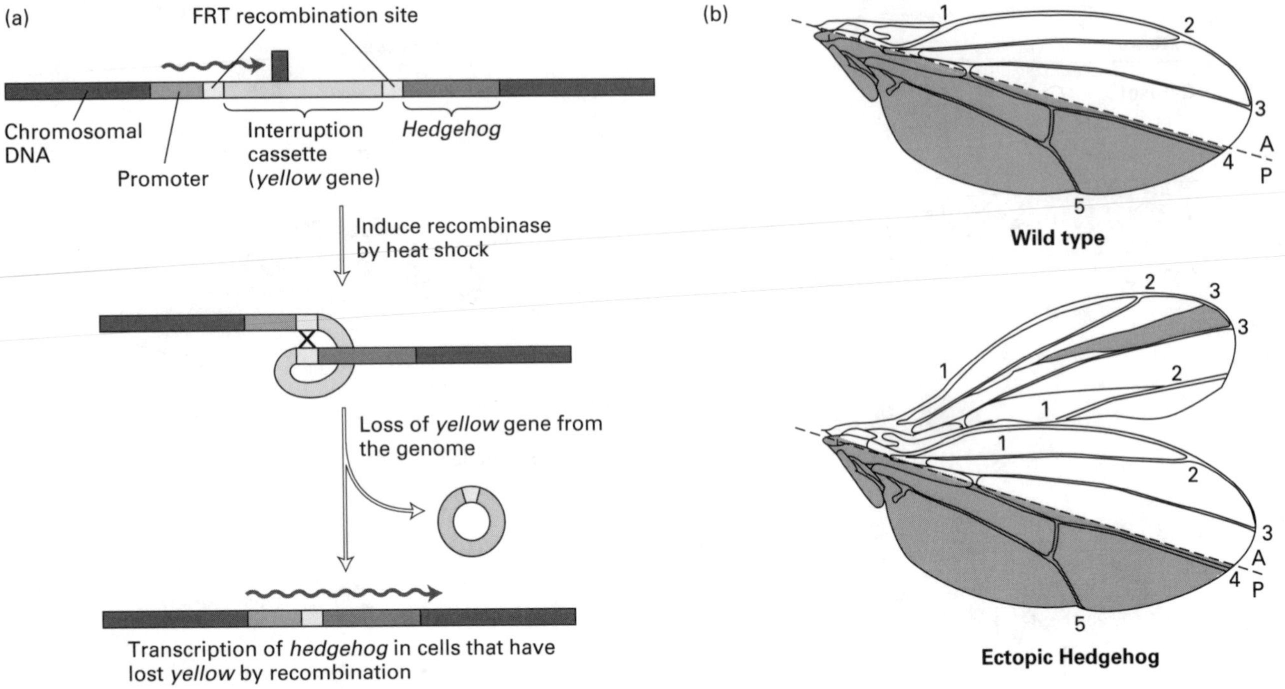

FIGURE 23-15, page 1017
Effect of ectopic expression of Hedgehog on *Drosophila* wing development.

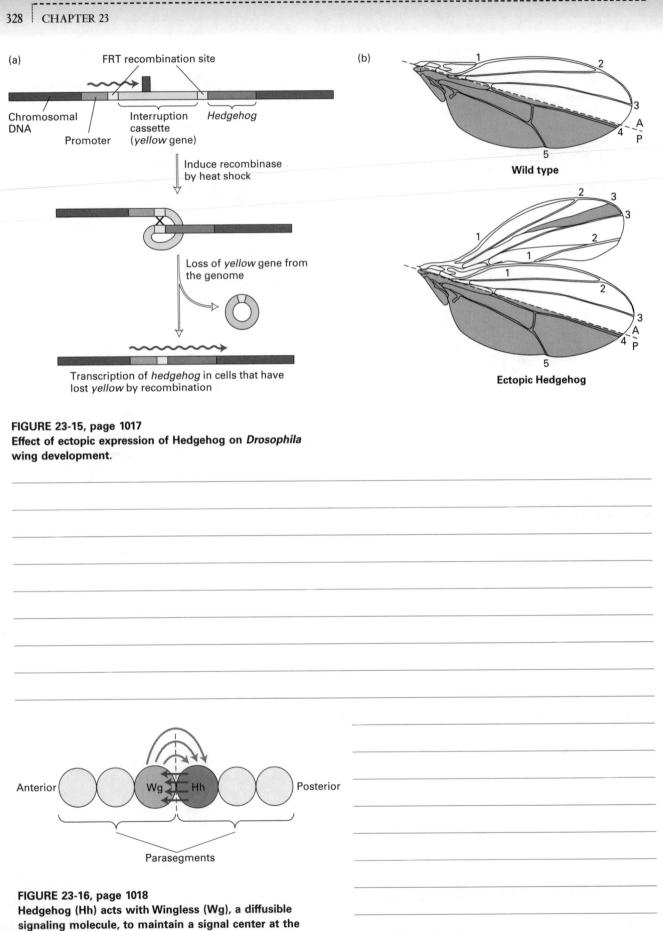

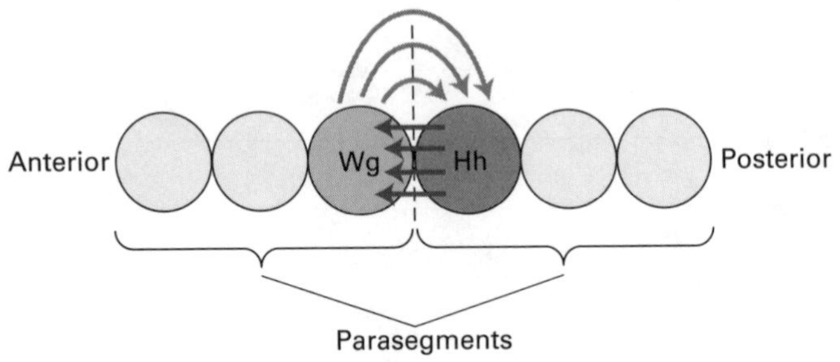

FIGURE 23-16, page 1018
Hedgehog (Hh) acts with Wingless (Wg), a diffusible signaling molecule, to maintain a signal center at the parasegment boundary in the *Drosophila* embryo.

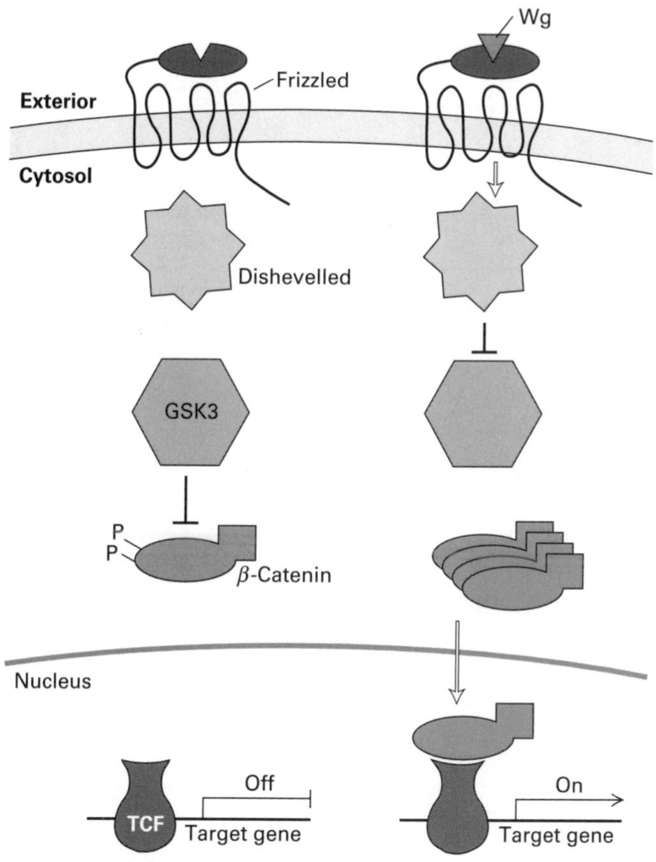

FIGURE 23-17, page 1018
The Wingless (Wg) signaling pathway.

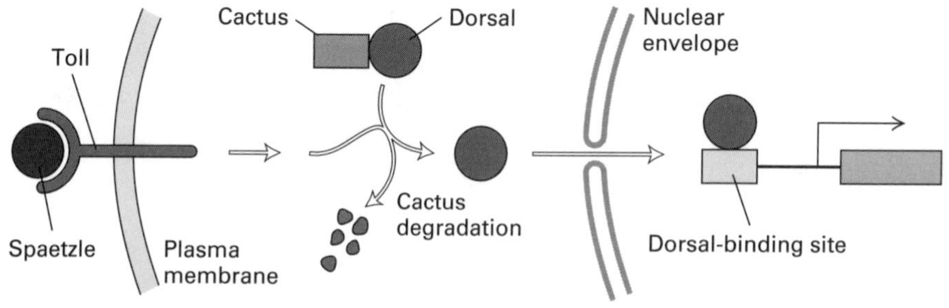

FIGURE 23-19, page 1020
Toll-mediated nuclear localization of Dorsal.

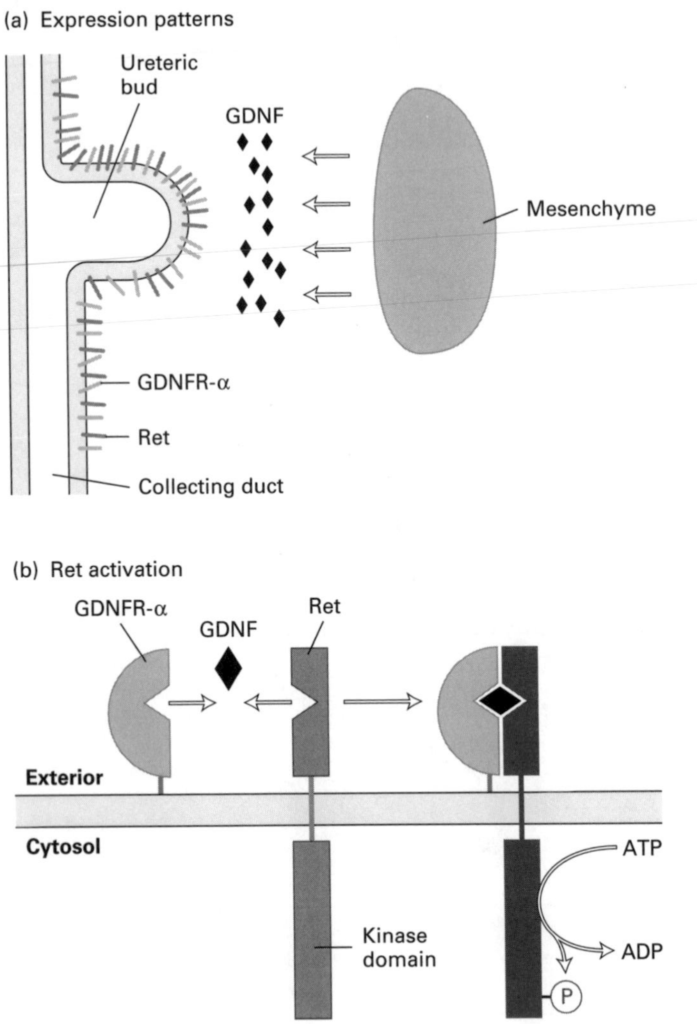

FIGURE 23-23, page 1024
Mechanism of mesenchymal inductive effect on the ureteric bud.

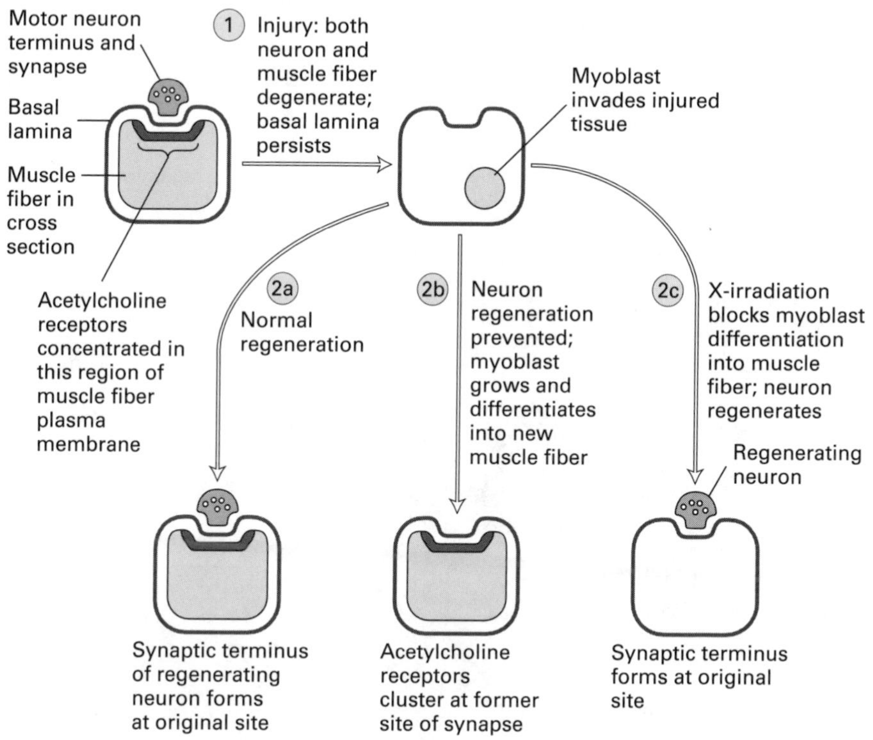

FIGURE 23-42, page 1042
A specialized region of the basal lamina determines the site of the neuromuscular junction.

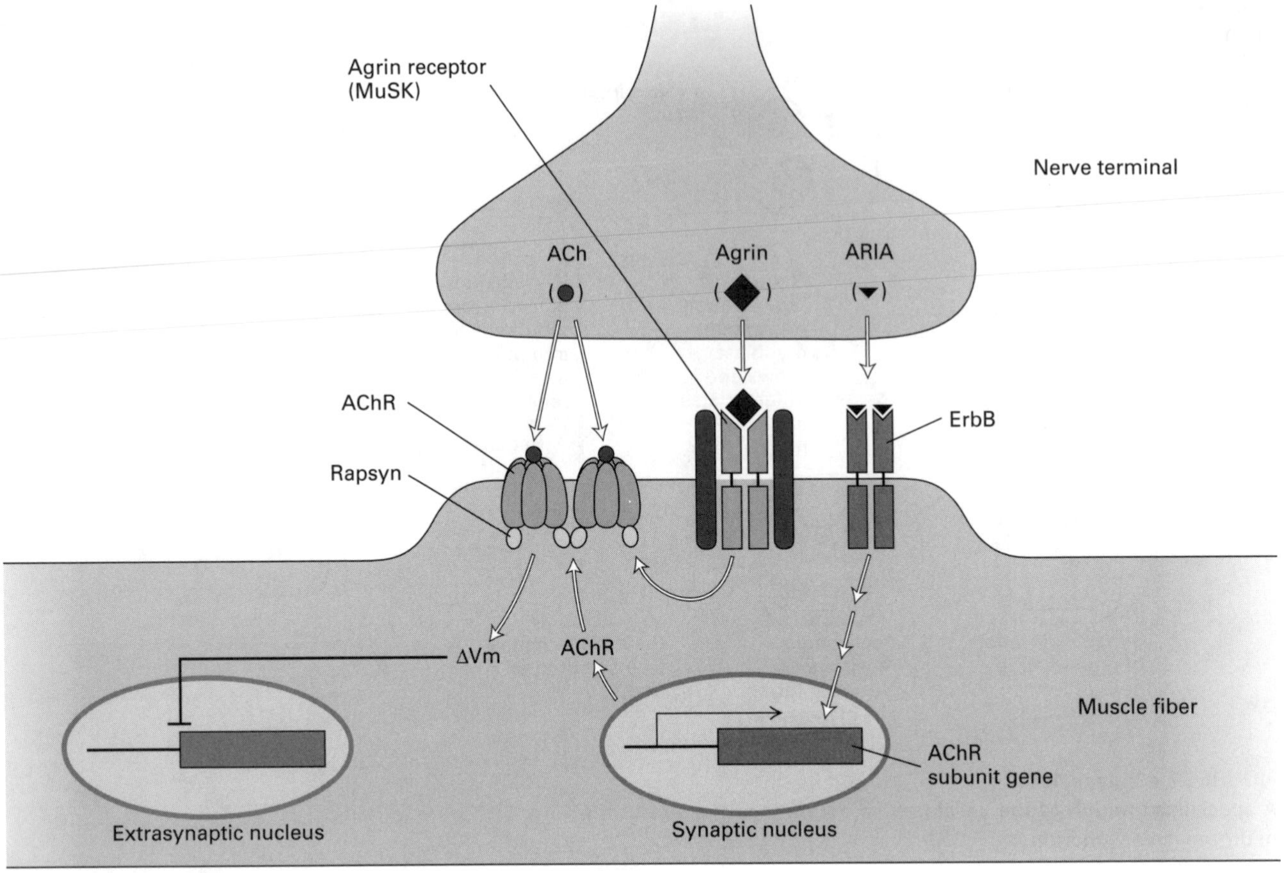

FIGURE 23-44, page 1044
Signals from the motor neuron regulate the expression and localization of the acetylcholine receptor (AChR).

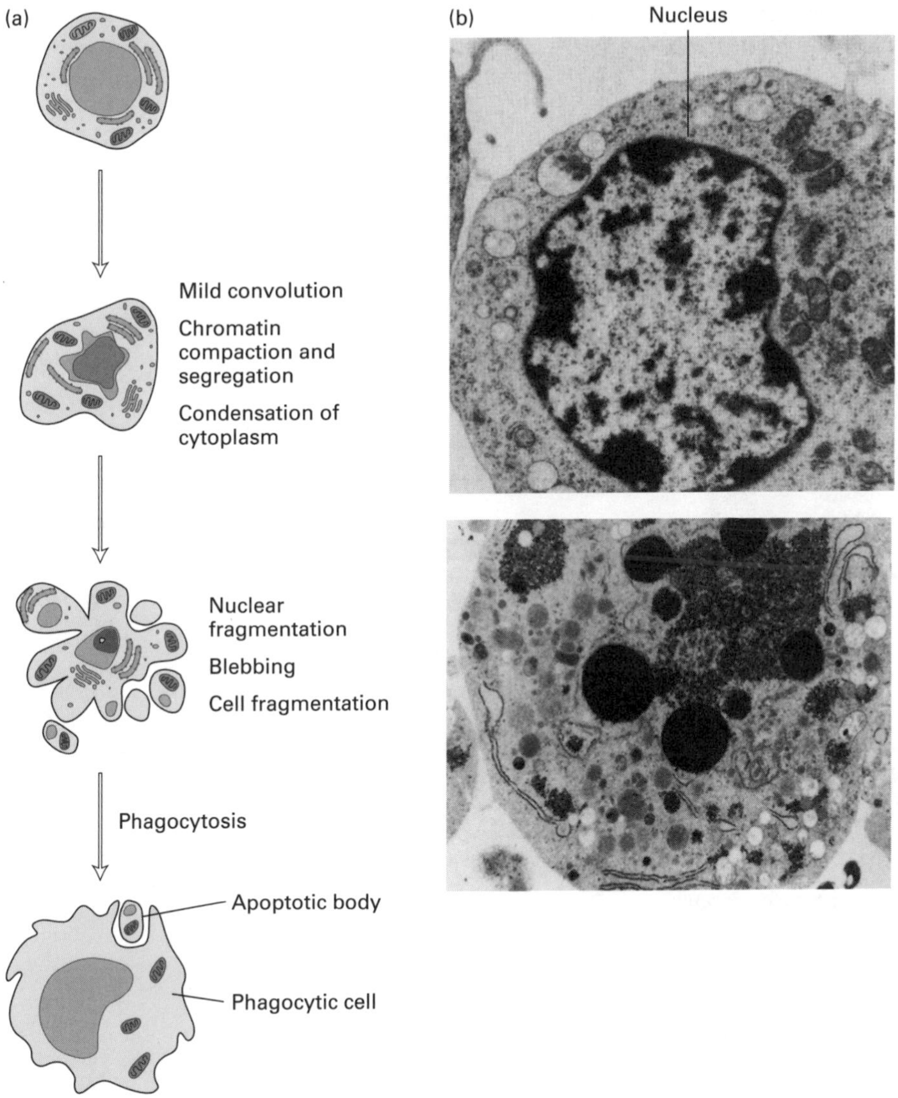

FIGURE 23-45, page 1045
Ultrastructural features of cell death by apoptosis.

MEDIA CONNECTIONS
Apoptosis

MEDIA CONNECTIONS Apoptosis

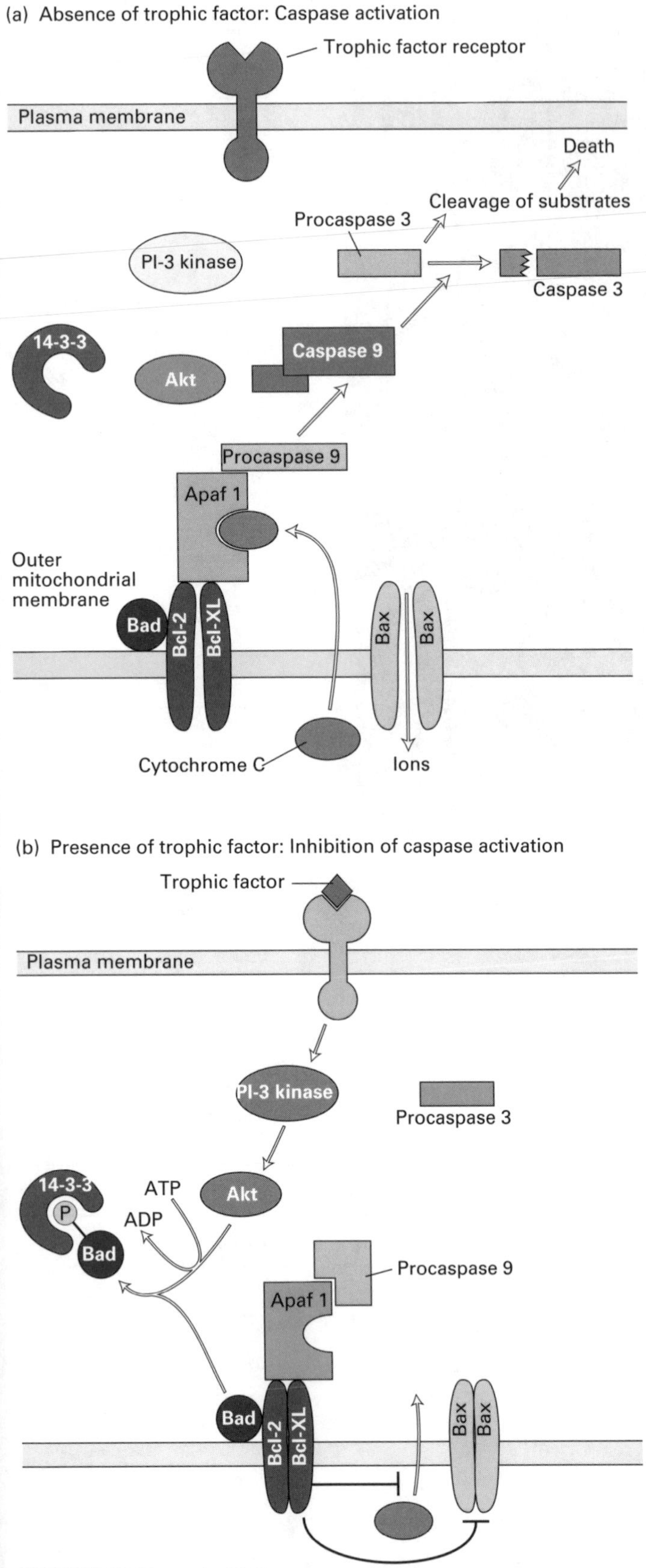

FIGURE 23-50, page 1049
Current models of the intracellular pathways leading to cell death by apoptosis or to trophic factor–mediated cell survival in mammalian cells.

Additional Notes

Cancer

24

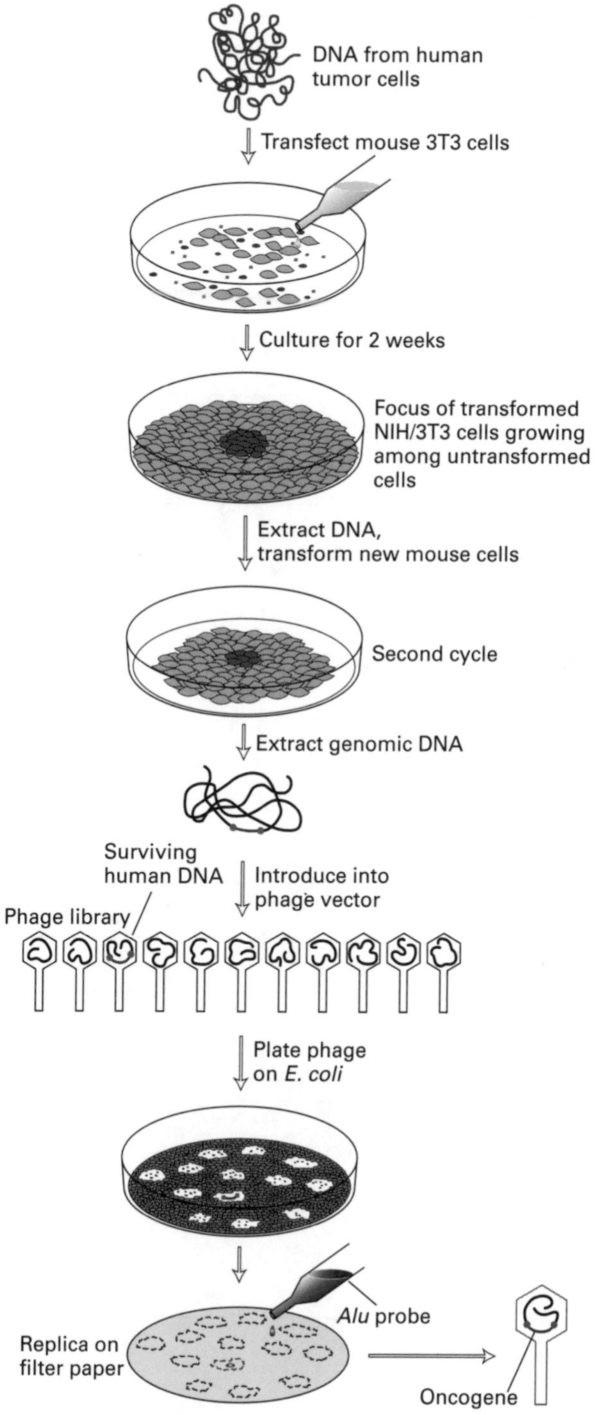

FIGURE 24-4, page 1058
The identification and molecular cloning of the *ras*[D] oncogene.

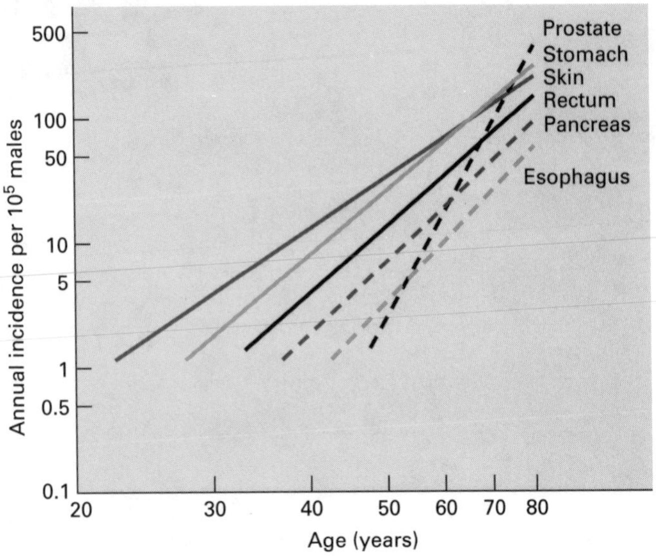

FIGURE 24-5, page 1059
The incidence of several human cancers increases markedly with age.

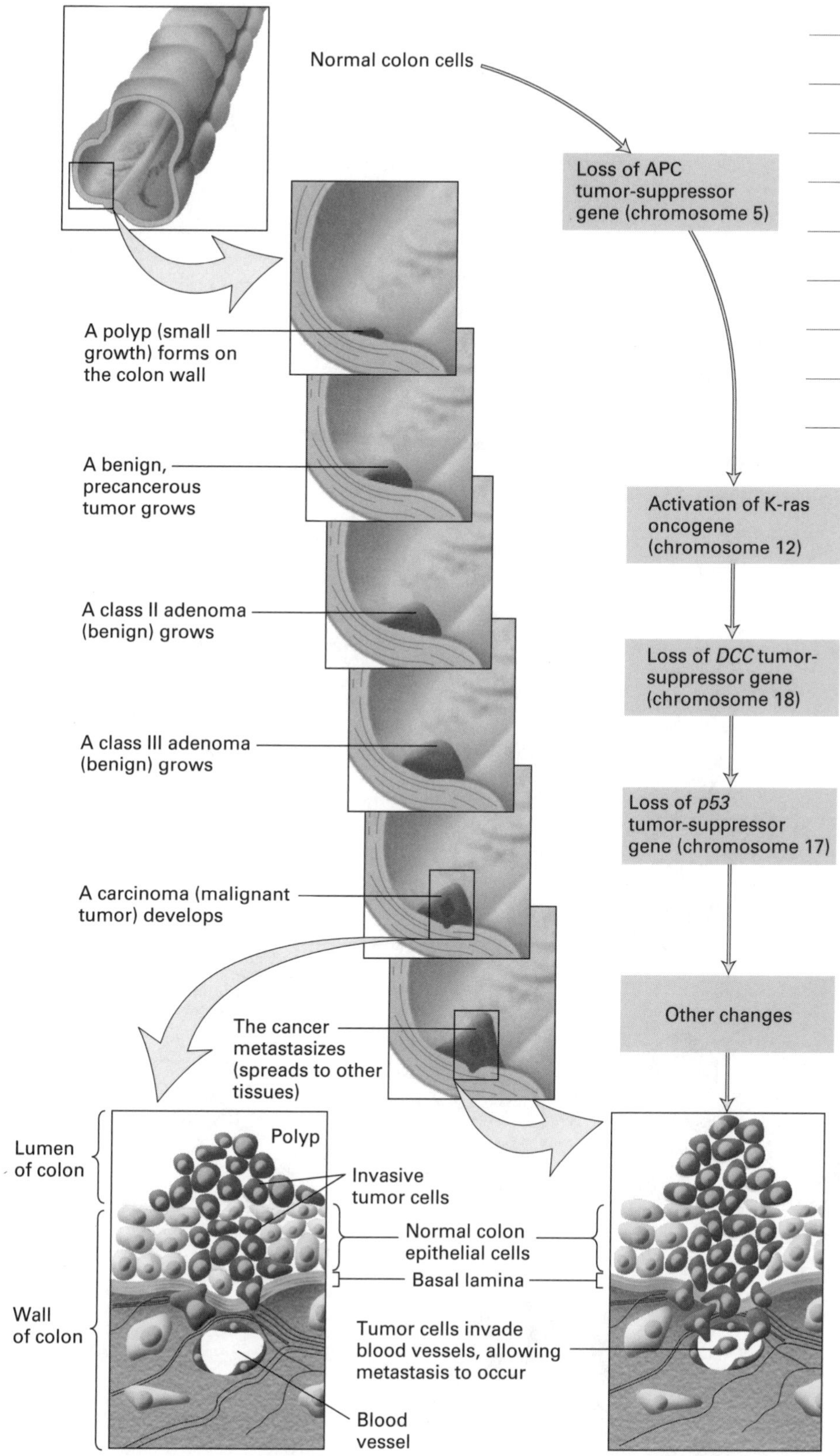

FIGURE 24-6, page 1060
The development and metastasis of human colorectal cancer and its genetic basis.

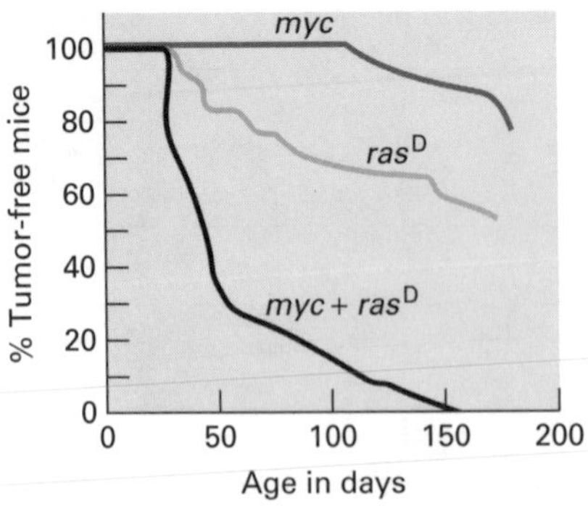

FIGURE 24-7, page 1061
Kinetics of tumor appearance in female transgenic mice carrying transgenes driven by the mouse mammary tumor virus (MMTV) breast-specific promoter.

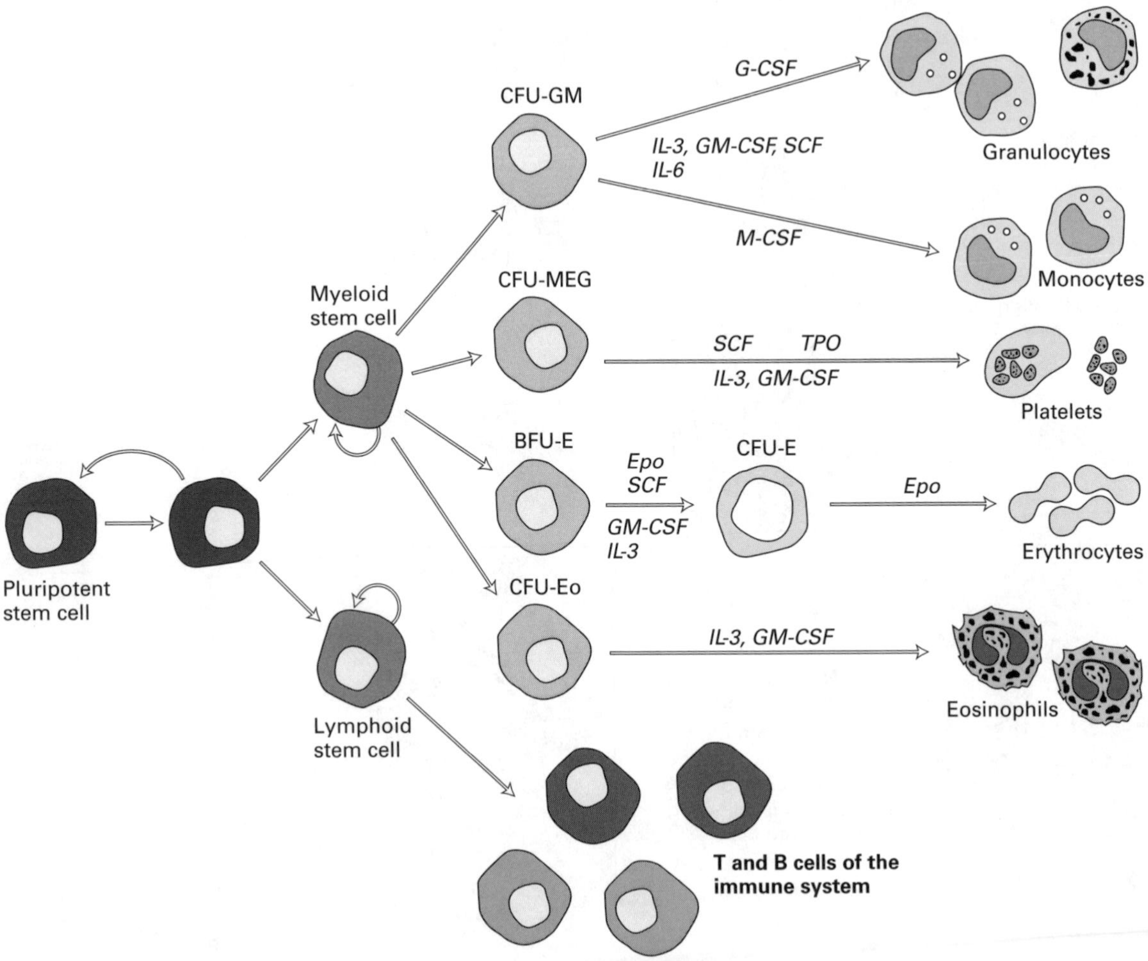

FIGURE 24-8, page 1063
Formation of differentiated blood cells from hematopoietic stem cells in the bone marrow.

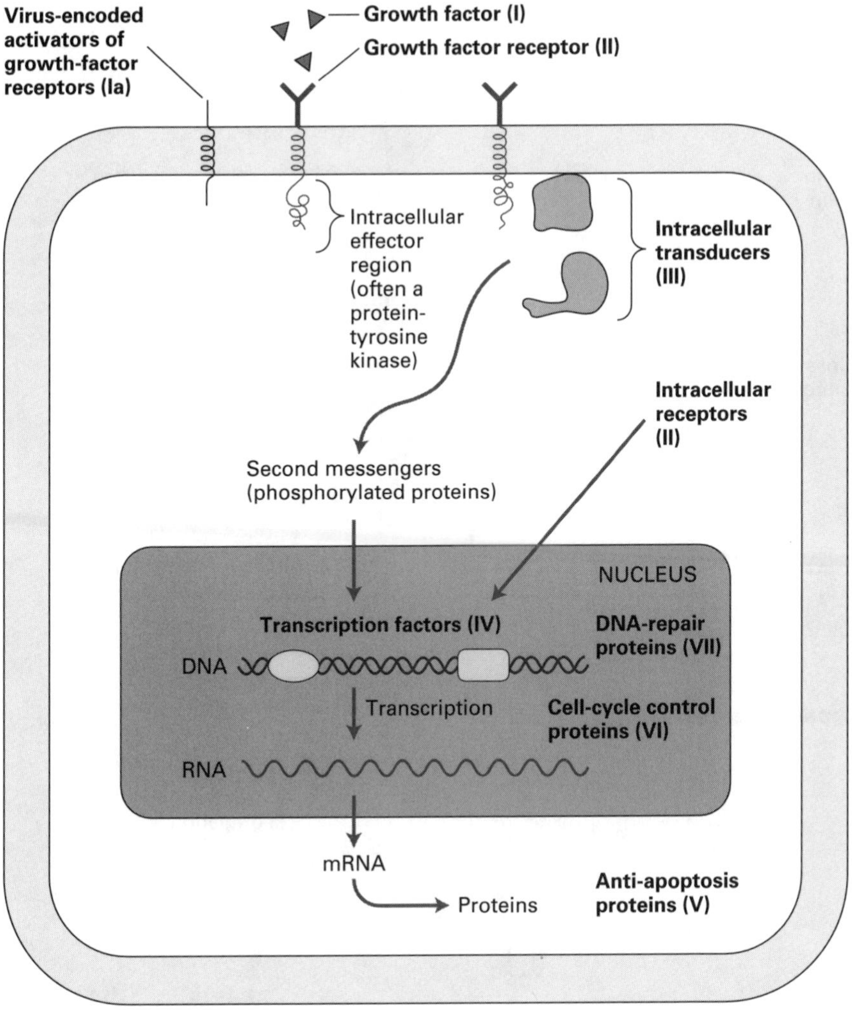

FIGURE 24-9, page 1064
The seven types of proteins that participate in controlling cell growth.

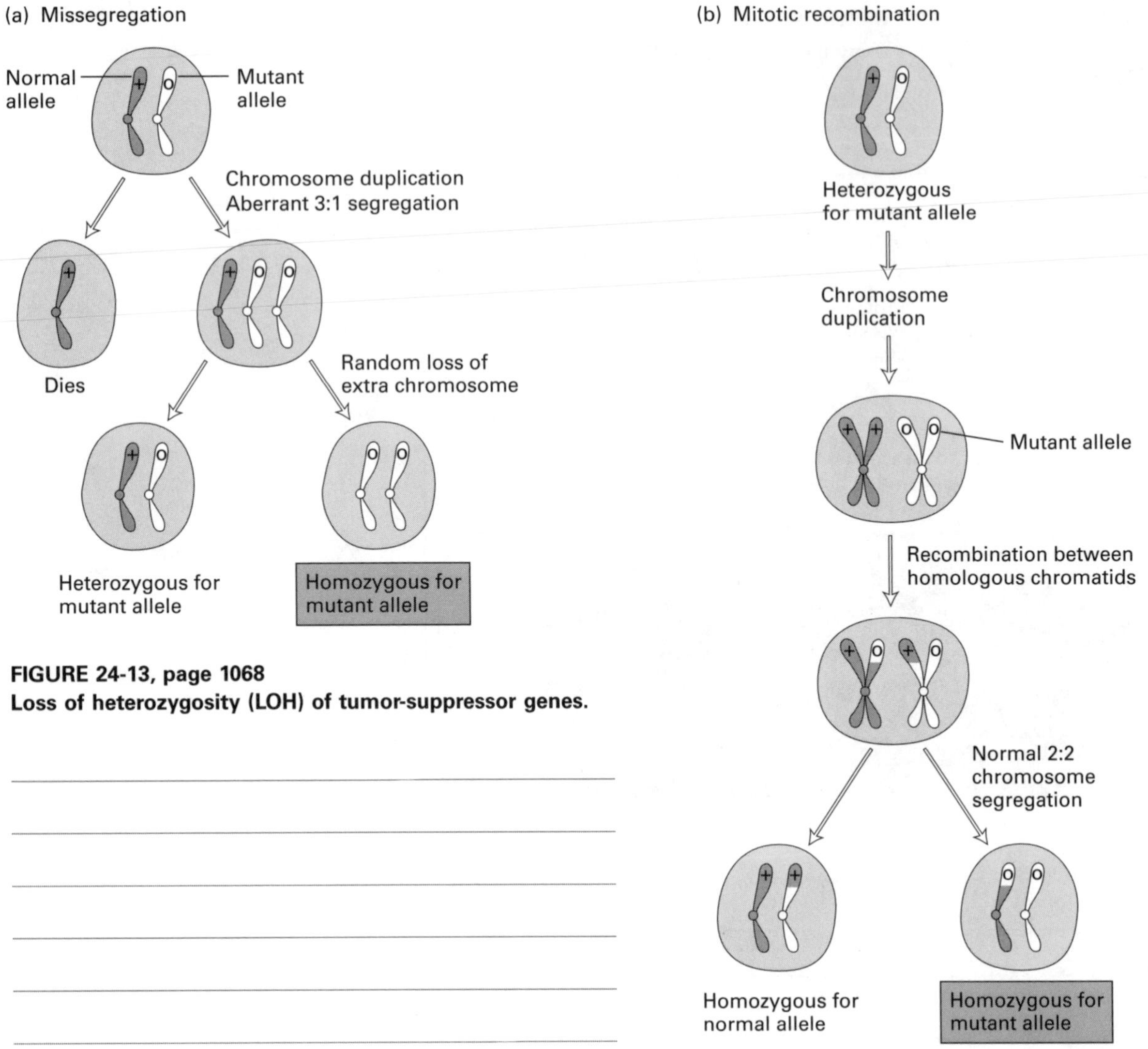

FIGURE 24-13, page 1068
Loss of heterozygosity (LOH) of tumor-suppressor genes.

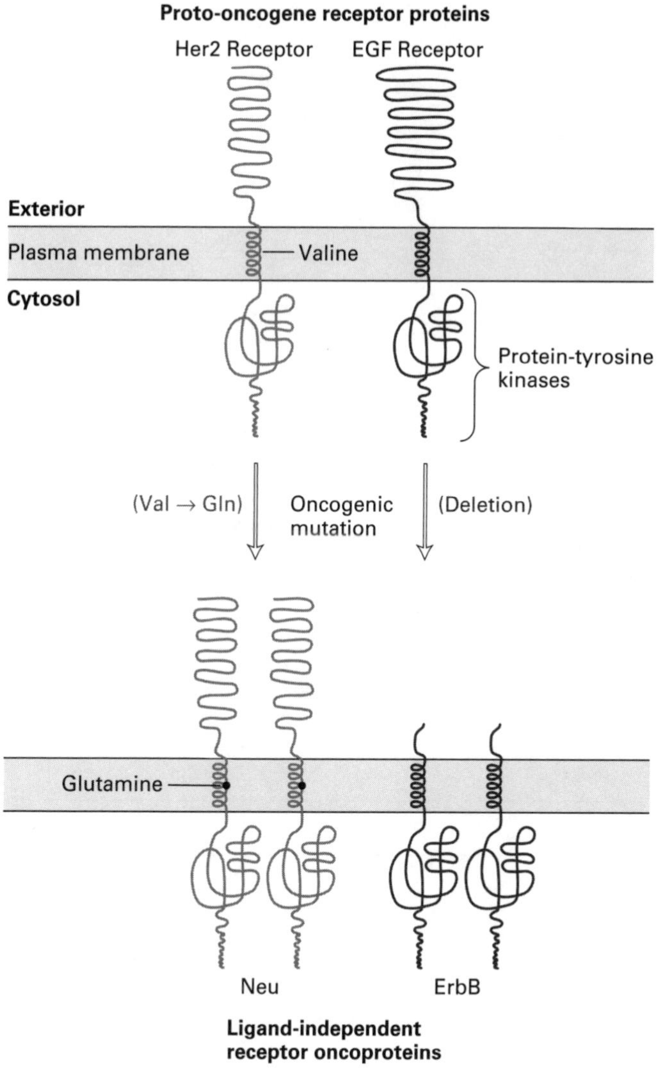

FIGURE 24-15, page 1071
Effects of oncogenic mutations in proto-oncogenes that encode cell-surface receptors.

MEDIA CONNECTIONS
Cell-Cycle Control

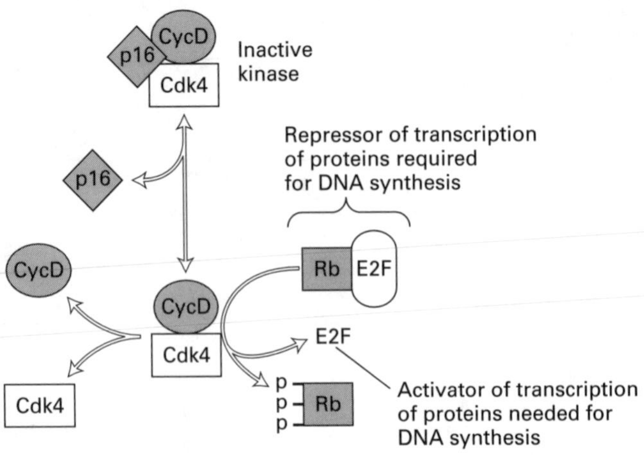

FIGURE 24-19, page 1075
Restriction point control.

MEDIA CONNECTIONS
TGF β Signaling Pathway

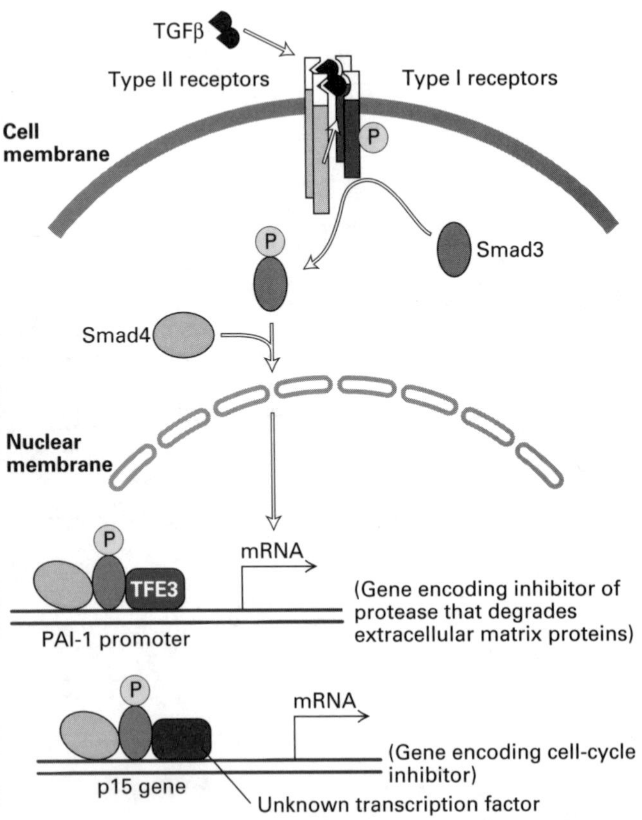

FIGURE 24-20, page 1076
TGF β signaling.

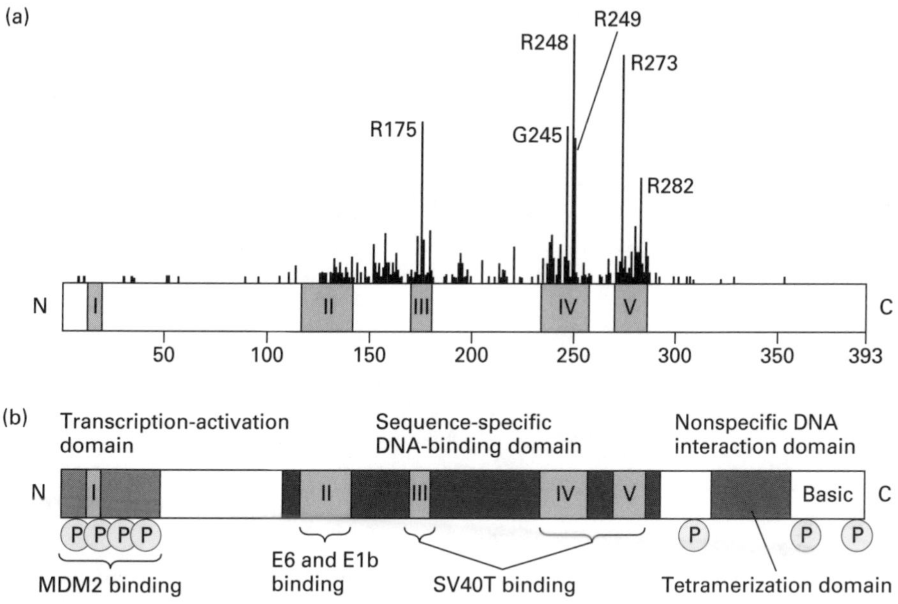
(a)
R249
R248
R273
R175
G245
R282
N
I
II
III
IV
V
C
50
100
150
200
250
300
350
393
(b)
Transcription-activation domain
Sequence-specific DNA-binding domain
Nonspecific DNA interaction domain
N
I
II
III
IV
V
Basic
C
P
P
P
P
P
P
P
MDM2 binding
E6 and E1b binding
SV40T binding
Tetramerization domain

FIGURE 24-21, page 1077
The human p53 protein.

Additional Notes